Toward a Science of Command, Control, and Communications

Edited by
Carl Jones
Naval Postgraduate School
Monterey, California

Volume 156
PROGRESS IN
ASTRONAUTICS AND AERONAUTICS

A. Richard Seebass, Editor-in-Chief
University of Colorado at Boulder
Boulder, Colorado

Published by the American Institute of Aeronautics and Astronautics, Inc.,
370 L'Enfant Promenade, SW, Washington, DC, 20024-2518

Copyright © 1993 by the American Institute of Aeronautics and Astronautics, Inc. Printed in the United States of America. All rights reserved. Reproduction or translation of any part of this work beyond that permitted by Sections 107 and 108 of the U.S. Copyright Law without the permission of the copyright owner is unlawful. The code following this statement indicates the copyright owner's consent that copies of articles in this volume may be made for personal or internal use, on condition that the copier pay the per-copy fee ($2.00) plus the per-page fee ($0.50) through the Copyright Clearance Center, Inc., 21 Congress Street, Salem, Massachusetts 01970. This consent does not extend to other kinds of copying, for which permission requests should be addressed to the publisher. Users should employ the following code when reporting copying from this volume to the Copyright Clearance Center:

1-56347-068-3/93 $2.00 + .50

Data and information appearing in this book are for informational purposes only. AIAA is not responsible for any injury or damage resulting from use or reliance, nor does AIAA warrant that use or reliance will be free from privately owned rights.

ISSN 0079-6050

Progress in Astronautics and Aeronautics

Editor-in-Chief
A. Richard Seebass
University of Colorado at Boulder

Editorial Board

Richard G. Bradley
General Dynamics

Allen E. Fuhs
Carmel, California

George J. Gleghorn
Rancho Palos Verdes, California

Dale B. Henderson
Los Alamos National Laboratory

Carolyn L. Huntoon
NASA Johnson Space Center

John L. Junkins
Texas A&M University

Daniel P. Raymer
Conceptual Research Corporation

Martin Summerfield
Princeton Combustion Research Laboratories, Inc.

Charles E. Treanor
Williamsville, New York

Jeanne Godette
Director
Book Publications
AIAA

Table of Contents

Introduction 1
    Carl R. Jones, *Naval Postgraduate School, Monterey, California*

Abstracts 13

**Task Identification With and Without Feedback in
    Hierarchical Teams** 21
    Gregory Burton, *Seton Hall University, South Orange, New Jersey,* and David
      L. Kleinman, *University of Connecticut, Storrs, Connecticut*

Headquarters Effectiveness Assessment Tool 39
    Robert Choisser, *Defense Information Systems Agency, Reston, Virginia,* and
      John Shaw, *APLHATECH, Inc., Burlington, Massachusetts*

No New Mathematics! No New C2 Theory! 63
    John T. Dockery, *Pentagon, Washington, DC,* and A. E. R. Woodcock,
      Synectics Corporation, Fairfax, Virginia

Formal Theory of C3 and Data Fusion 97
    I. R. Goodman, *Naval Ocean Systems Center, San Diego, California*

Statistical Mechanics of Combat and Extensions 117
    Lester Ingber, *Lester Ingber Research, McLean, Virginia*

**Impact of Organizational Structure on Team Performance:
    Experimental Findings** 151
    Victoria Y. Jin, *AT&T Bell Laboratories, Holmdel, New Jersey,* and Alexander
      H. Levis, *George Mason University, Fairfax, Virginia*

Problem Solving Systems: A New Concept for a Science of C2 167
    Carl R. Jones, *Naval Postgraduate School, Monterey, California*

Colored Petri Net Model of Command and Control Nodes 181
    Alexander H. Levis, *George Mason University, Fairfax, Virginia*

**Modeling and Evaluation of Variable Structure Command and Control
    Organizations** 193
    Jean-Marc Monguillet and Alexander H. Levis, *Massachusetts Institute of
      Technology, Cambridge, Massachusetts*

Command and Control Reference Model 221
 Israel Mayk, *U.S. Army Communications-Electronics Command, Fort Monmouth, New Jersey*

Stochastic Modeling, Analysis, and Calibration of C3 Systems 239
 Izhak Rubin, *IRI Corporation, Tarzana, California,* and Israel Mayk, *U.S. Army Communications-Electronics Command, Fort Monmouth, New Jersey*

Author Index for Volume 156 273

List of Series Volumes. 275

Introduction

Carl R. Jones*
Naval Postgraduate School, Monterey, California 93943

Depending on the author, command and control as a separate knowledge domain begins somewhere in the mists of time. Much more recently, there has been an ongoing effort to develop a science of command and control. For the purposes of this introduction, no distinction is made among interpretations of the instantiations of $C^n I^m$. In 1981, Donald Latham, then Assistant Secretary of Defense (Command, Control, Communications and Intelligence) [ASD(C3I)], commissioned Robert Herman to survey the field of C3I and report the deficiencies he found. Among his findings were that 1) there was no coherent theory of C3, 2) there was no joint planning and review of service C3 technology bases, 3) there was no technology consensus among C3 Laboratories, and 4) there were no multiservice experiments or demonstrations.

The response to this was the formation of the technical panel for C3 under the joint directors of laboratories. The technical panel consists of the technical directors of the service laboratories. Based on the Herman report, the chosen objectives of the panel are 1) to develop a coherent theory of C3, 2) to develop a process for the joint planning and review of service C3 technology bases, and 3) to form affinity groups to review and exchange information in a given discipline and formulate and execute multiservice experiments and demonstrations. The affinity groups are decision aids, radios and links, distributed processing, data fusion, networks (with a subgroup on network and simulation support), and the basic research group. The first program manager for these activities, Victor Monteleon of the recently reorganized Naval Command, Control and Surveillance Center's Research and Development Division (NRaD), briefed the program to Secretary Latham. Since then the program has been funded by the ASD(C3I).

With the creation of the reliance program and the reorganization of Department of Defense (DoD) research at the end of the cold war, the joint directors of laboratories have become the coordination and management

---

Copyright © 1993 by the American Institute of Aeronautics and Astronautics, Inc. All rights reserved.
*Professor, Department of Administrative Sciences and C3 Academic Group.

agency for joint service technology base research. In turn, the technical panel on C3 has become the coordination and management agency for joint service C3 technology base research.

Since its establishment, the basic research group has been addressing the issue of a coherent theory of C3. The basic research group consists of 1) the C3 laboratories [NRaD, the Army's Communications and Electronics Command (CECOM), and the Air Force's Rome Laboratory], 2) the Joint Staff, 3) the Naval Postgraduate School, 4) the National Defense University, 3) the Defense Information Systems Agency, and 6) George Mason University. The major thrusts of the basic research group's research have been 1) developing mathematical models of the C3 process, 2) the development and implementation of the headquarters effectiveness and analysis tool (HEAT), developing C3 analysis tools and methodologies, and 3) sponsoring symposia and workshops to facilitate the interchange of ideas among researchers and practitioners. Since the other affinity groups of the technical panel on C3 focus on physical technology, this has not been the focus of the basic research group.

What Is Command and Control?

For there to be a science of command and control (C2), there must be an answer to the fundamental question: What is the command and control phenomenon? A reasonable place to understand the phenomenon of command and control is with the official DoD definition.

> [Command and Control is] . . . The exercise of authority and direction by a properly designated commander over assigned forces in the accomplishment of the mission. Command and control functions are performed through an arrangement of personnel, equipment, communications, facilities, and procedures that are employed by a commander in planning, directing, coordinating and controlling forces and operations in the accomplishment of the mission.[1]

The focus of this definition is on six basic factors of command and control: the legal authority, the commander, the physical means for command and control, the procedures for command and control, the functions of command and control, and the mission. In addition to this source, there are also the principles of war; the result of combining these ideas is an understanding of command and control as the subsystem of a combat organization that provides the commander with the capability for unity of effort in the accomplishment of a mission. The science of command and control is the understanding of this subsystem of a combat organization. Here we will discuss each of the six basic factors and the principles of war.

INTRODUCTION

The legal authority factor is a very important element of command and control. However, to date the research reported here, as well as elsewhere, does not focus on this factor. As a result, the legal authority factor is not considered in the book.

The concept of a commander (by implication, a single commander) has a long history. Given the information processing and cognitive limitations of human decisionmakers, the single commander concept is, in execution, a hierarchy of commanders. This hierarchy is commonly called the chain of command. This same human limitation creates the need for staffs. As a result, the science of command and control must include an understanding of the decisionmaking of individual commanders in the chain of command as well as the decisionmaking of a commander's staff. The systemic behavior of hierarchically inter-related individual commanders and staffs must also be understood: associated with a command and control structure and its processes are a variety of decision support (aiding) and communication systems.

When command and control activities are observed, some physical means is observed as being employed to exercise authority and direction. Some communication and decision support aspects of this have already been noted when discussing the concept of a commander. The Joint Chiefs of Staff (JCS) definition notes several generic classes of physical means that are used. These classes of physicals means (and others) are the physical hosts for the capacity to perform command and control. The science of command and control must include the science of these physical means of accomplishing the exercise of authority and direction. Because these aspects of the science of command and control are well covered in many texts and articles and were not the focus of the basic research group's efforts, this book does not focus on them.

The procedures of command and control (the processes used within the command and control structure) are the heuristics, the algorithms, and the programs that are employed to collect data and information, diagnose problems, structure problems, and solve and implement solutions. The formal procedures for problem solving are sometimes called standard operating procedures.

The functions of command and control (planning, directing, coordinating, and controlling) are not all defined in the Department of Defense Joint Chiefs of Staff Dictionary of Military and Associated Terms[2]. The following definitions are drawn from Ref. 2 on the management and leadership of organizations to supplement the official joint publications:

1) Planning: Selecting the mission and objectives as well as the strategies, policies, programs, and procedures for achieving them; decision making that is future oriented; the selection of a course of action.
2) Directing (and Leading): Clarifying, guiding, teaching, and encouraging participants in the organization to perform effectively and with zeal and confidence.

3) Coordinating: Achieving harmony of individual effort with group effort toward the accomplishment of group purposes and objectives. (The words individual and group in this definition should be interpreted to include organizational units and supra units as well.)

The following definitions of controlling are drawn from Ref. 1. (The complete definitions include mapping, charting, and photogrammetry usage, as well as the use in the intelligence community; these aspects will not be included here.) Controlling: 1) Authority which may be less than full command over part of the activities of subordinate or other organizations. [DoD] 2) Physical or psychological pressures exerted with the intent to assure that an agent or group will respond as directed. [DoD] 3) That authority exercised by a commander over part of the activities of subordinate organizations or other organizations not normally under his command, which encompasses the responsibility for implementing orders or directives. All or part of the authority may be transferred or delegated. [NATO] These definitions relate to authority and the power over subordinates to control their activities and make them do what is intended. In some cases, it is useful to suppress the authority aspects of the definition and focus on the control aspect. Such a definition could be the following: measuring and correcting the performance of subordinates' activities to ensure that organizational objectives and plans are accomplished. These command and control functions are performed by the command and control structure and embodied in procedures. Understanding the nature and behavior of these functions is another part of the science of command and control. This collection of functions has recently begun to be considered a separate field of study known as Coordination theory. (Coordination when used with theory is capitalized to distinguish it from the function of coordination.) This theory suggests that these functions are used to integrate the diverse activities needed to accomplish a mission. By drawing concepts from the literatures of computer science (mostly operating systems), economics, operations research, and organization theory, a set of fundamental generic ideas has been proposed.[3]

With these ideas from the joint publication concept of command and control in mind, the discussion can turn to the command and control aspects of the principles of war. The principles of war, adapted from Ref. 4, are as follows:

1) Objective (clarity of purpose): direct every military operation toward a clearly defined, decisive, and attainable objective.
2) Offensive (initiative): seize, retain, and exploit the initiative.
3) Mass: concentrate combat power at the decisive place and time.
4) Economy of force (efficiency): allocate minimum essential combat power to secondary efforts.
5) Maneuver: place the enemy in a position of disadvantage through the flexible application of combat power.

6) Unity of command (cohesiveness): for every objective, ensure unity of effort under one responsible commander.
7) Security (do not be surprised): never permit the enemy to acquire an unexpected advantage.
8) Surprise (surprise your opponent): strike the enemy at a time or place, or in a manner, for which he is unprepared.
9) Simplicity: prepare clear, uncomplicated plans and clear, concise orders to ensure successful operations through understanding.

The principles of war that are of particular importance to understanding command and control are the principles of unity of command and simplicity. The principle of unity of command provides the wisdom of the ages on the nature of the chain of command and the need for integration within a combat organization. The simplicity principle tells us to be uncomplicated in structure and procedures to accomplish unity of command. The other principles of war relate to thinking how to influence your opponent, or how not to be influenced by your opponent.

The principle of unity of command has two aspects: the concept of unity of effort, and a best command and control structure being embodied in a single commander. The unity of effort aspect can be embodied in the single word cohesiveness. By historic wisdom, combat effectiveness increases as the cohesiveness of the combat organization increases. Sometimes this idea is stated in terms of C2 being a force multiplier to capture the idea of a cohesive synergism among the elements and weapons of the combat organization. The one responsible commander idea is a specific choice of the best command and control structure. That is, the codified wisdom of history states that the best command and control structure in a combat organization is a single commander. Because of the cognitive and information processing limitations of human decisionmakers, a multilevel hierarchy is used to compensate for the limitations of individuals. This compensation occurs by partitioning large problems exceeding such limits into small subproblems that are within the limits. Then the subproblems are integrated in a hierarchy that yields the desired systemic solution. The word commander must be interpreted as a hierarchically structured set of commanders called the chain of command. From the literatures of economics and organization theory, it is known that the "best" command and control structure, i.e., the best problem solving "system of systems," is contingent on 1) the command and control technologies, 2) the weapons technologies available to all sides in a conflict, and 3) the nature of the opponent. Codified wisdom must be judged in terms of the variability of these determinants in the modern world. Thus the science of command and control must include understanding the determinants of command structure and processes that result in the most effective combat organization.

The last basic factor embedded in the JCS definition of command and control is the mission of the commander and, by implication, the combat organization. Missions vary by detail, political-economic constraints, technology, the physical environment, and the opponent's capability and intentions. The science of command and control must include an understanding of the impact of these factors. In more general terms, the impact of the nature of a mission in a specific scenario on the command and control structure, processes, and effectiveness must be understood. The notion of a scenario includes the complete environment of the combat organization including the physical environment, the opponent(s) capabilities and intentions (the threat), the friendly forces and support available, and the higher echelons of command to the combat organizations commander.

Thus the science of command and control must encompass an understanding of the six basic factors of command and control and the associated C4I systems that implement the factors in physical systems and decision making. Since command and control is largely the "tie that binds" in a combat organization, it is affected by the combat organization's environment, its weapons technology, its mission, its combat support technology, and the inherent limitations of C4I technology and human decisionmaking.

What Is a Science of Command and Control?

In this section, the focus is on the nature of science as it relates to command and control. The discussion of science (the philosophy of science) is a fascinating topic in and of itself that is both wide and deep. In this discussion, there is a much more modest focus of outlining the criteria and process of science as it relates to a science of command and control. Thus we will define the notions of science, engineering, and design to illuminate the idea of science. Then we will consider the process of science and the criteria for judging a scientific result. The research path strategies to a science of command and control and the possible perspectives for understanding the command and control subsystem of a combat organization provide the basis for one schema for understanding a science of command and control.

The dictionary definition of science that is most applicable to our study of command and control is "knowledge covering general truths or the operation of general laws especially as obtained and tested through scientific method."[5] In turn, scientific method is defined as "principles and procedures for the systematic pursuit of knowledge involving the recognition and formulation of a problem, the collection of data through observation and experiment, and the formulation and testing of hypotheses."[4] These ideas of a process for obtaining knowledge that is repeatable by all is different from engineering. Engineering is the art or science of practically applying science. The scientific aspect of the definition of engineering permits the scientific study of engineering practice. In a science of command and control, the interest is in using the scientific method to obtain general truths about the phenomenon of command and control.

The dictionary definition of the scientific method is correct but lacks a sense of the dynamics of the process. The ancient Sufi tale of the six blind men describing an elephant-developing a science of elephants-is an excellent metaphor to gain an understanding of the dynamics of the process. The first man feels the tusk and proclaims elephants are spearlike. The second man, feeling the side, proclaims elephants are wall-like. The third, feeling the ear, proclaims a fanlike elephant. The fourth, feeling the tail, proclaims a ropelike elephant. The leg, when felt by the fifth man, leads to a tree trunklike elephant. Lastly the sixth man, feeling the trunk, proclaims a snakelike elephant. More abstractly, six observations yield six empirical generalizations. With an appropriate dialogue, the six scientists could add in the geolocation of their observations and begin the process of synthesizing one or more general hypotheses. These more general hypotheses then lead them, after more observation, hypothesis creation, and dialogue to a middle range theory of, say, the forward third of the elephant. After additional work, they could create a general unified theory of elephant exteriors. This is, metaphorically, the process by which a science of command and control can be developed. In actual practice, there is a need for dialogue mechanisms and, of course, human emotions are involved. But the process is defined and the resulting knowledge must satisfy certain criteria.

The traditional criterion that a theory must satisfy is a falsification standard. That is, the theory must be constructed such that it is possible to empirically refute it by observation. This observation must be a construct logically derivable from the theory. There may be some abstract concepts developed and used that are not directly testable but are testable by directly observable derived variables. One must also recognize that all theories are unrealistic in that they are all abstractions of reality. Theories are designed for economy of thought. What is important is that the theory captures the essence of the phenomenon's variables and these variable's interrelationships. That said, it is important to note that a theory is never proven, it is only not falsified. Thus in a science of command and control, there must be empirical generalizations that are falsifiable, middle range theories that are also falsifiable, and last, but not least, a general unified theory of command and control that is falsifiable by observation. Thus in categorizing a contribution to the science of command and control, one dimension of the categorization is the falsifiability of the research outcome.

Given this criterion, there are multiple research strategies for individual research efforts that, when embedded in the overall scientific method process, lead to a science of command and control. These individual strategies can be listed as follows:[6]

1) Develop empirical generalizations.
2) Refine empirical generalizations.
3) Synthesize empirical generalizations into a middle range theory.
4) Synthesize refined empirical generalizations into a middle range theory.

5) Develop a middle range theory without using strategies 3 or 4.
6) Refine middle range theories.
7) Synthesize middle range theories into a general theory.
8) Synthesize empirical generalizations into a general theory.
9) Synthesize refined middle range theories into a general theory.
10) Synthesize refined empirical generalizations into general theories.
11) Develop a general theory without using strategies 7, 8, 9, or 10.
12) Refine general theories.

Of course, combinations of the above are possible. In a new field such as C3, the middle range and general theories of other fields are adapted for use in understanding C3. The now classic C3 loop is an adaptation of the control cycle paradigm. In judging the development of a science of command and control, the research strategy used by an individual researcher is one type of indicator. The entire portfolio of researcher's strategies is also an indicator. It is also important to understand that a scientific process is analogous to a genetic algorithm. Thus a large diversity of individual research strategies provides the highest probability of obtaining a science of command and control.

Another perspective on developing a science of command and control is the categories developed by Kuhn[7] of normal and revolutionary science. Normal science is puzzle solving, is a highly cumulative enterprise, is eminently successful in its aim, and involves the steady extension of the scope and precision of scientific knowledge. Normal science does not aim at novelties of fact or theory. In terms of search theory, it is a cumulative local search procedure. Revolutionary science, on the other hand, is the result of a global search. It is noncumulative, episodic, and replaces all or a major portion of current thinking.

Since command and control is a subsystem of a combat organization, it is useful to consider the alternative perspectives that have been used to study organizations. For our purposes, the categorization scheme introduced by Morgan[8] will be used. He partitioned the possible perspectives on an organization in eight classes: mechanistic, organic, self-organizing, culture, political, change, domination, and psychic. These classes are most easily understood in terms of metaphors. The machine perspective relates organizations to machines and mechanical thinking. The biological nature of an adapting organism in an environment that ecologically selects out the unadapted serves as the metaphor for the organism perspective. Self-organization has the brain as its metaphor. Thus it focuses on information processing, decisionmaking, learning, and cybernetics. The culture perspective relates to the organizations as social systems with norms, subcultures, and context. The political perspective involves governance, political activity, conflict, and power. Whereas the change perspective relates to selfproduction (autopoiesis), mutual causality, dialectical change, and evolution. Domination relates to the nature of the employee's association with the organization with exploitation being a theme. Lastly, organizations can be viewed as psychic prisons for the participants. Overall, Morgan has provided us a comprehensive list of the possible perspectives on an organization. In some grand general unified theory of organizations, they all would appear. Fortunately, the goal of a science of command and control is more modest. In the editor's opinion, command and control thinking focuses on the mechanistic, organism, self organizing, and change perspectives.

There is one last perspective on contributions to a science of command and control. This perspective relates to the basic unit of analysis in command and control studies. An example is the basic unit of analysis, the individual, e.g., commander as decisionmaker, or the system, e.g., interrelated decisionmakers in a problem solving system. Much of the command and control research focuses on individual and team decisionmaking. There is some work focusing on systems and systemic behavior. For example, models of command and control systems where the comparative response time of your system compared with the opponents system is the focus with the goal of being within your opponents command and control response capability.

The dimensions discussed were falsifiability, research process, normal versus revolutionary science, perspective, and the basic unit of analysis. The reader has the opportunity to classify each of the papers in this volume, and the editor will comment about the field overall. Before these comments, it is important to note that a focused effort to develop a cohesive science of C3 is barely a decade old. Also, the science of the technology, which provides the capability of command and control, is well known, well developed, and rapidly evolving. It is a normal science. The rest of the science of C3 is not yet that well developed. There has been some experimentation both in the laboratory and in the field. There is some middle range theory adapted from other disciplines and there are postulated general theories that have as yet to gain widespread acceptance. The approach discussed is to use a normal science focus and to seek a revolutionary science. The field provides a reasonable amount of falsifiable propositions, but more emphasis could be placed on this. The main perspectives that are used are those based in information and decisionmaking. Thus, the mechanistic, organic, self-organizing, and change perspectives are all used. The basic unit of analysis tends to be the individual, with some work with teams; little research is available on large-scale system systemic behavior. These comments are intended to provide a benchmark for a dialogue on the state of the art in the science of command and control. Such a dialogue is helpful in focusing research on promising areas of importance to practitioners.

In the discussion so far of the science of command and control, no attention has been paid to the communication channels used by researchers and practitioners to discuss research. Such communication channels include the annual C2 Research Symposium with its published proceedings, as well as focused workshops, some of which publish proceedings for wider distribution of ideas. However, there is still not a scholarly journal that provides the necessary focus on C3; this needs to be corrected for a science of command and control to flourish.

We are not yet at the stage of having a general unified theory of C3, but good progress has been made and there is hope for the future.

References

[1] Joint Chiefs of Staff, *Department of Defense Dictionary of Military and Associated Terms*, June 1986.

[2] Koontz, H., O'Donnell, C., and Weihrich, H., *Management*, 7th ed., McGraw Hill, New York, 1980.

[3] Malone, T. W. and Crowston, K., "What is Coordinative Theory and How Can It Help Design Cooperative Work Systems?," *Groupware: Software for Computer-Supported Cooperative Work*, Marca, D. and Bock, G., ed., IEEE Computer Society Press, Los Alamitos, California, 1992.

[4] FMIOO-5, "Principles of War," *Operations*, Washington DC, GPO, Appendix A, pp. 173-177.

[5] *Webster's New Collegiate Dictionary*, 1981, p. 1026.

[6] Swamidass, P. M., "Empirical Science: New Frontier in Operations Management Research," *Academy of Management Review*, Vol. 16, No. 4, 1991, pp. 793-814.

[7] Kuhn, T. S., *The Structure of Scientific Revolutions*, University of Chicago Press, Chicago, Illinois, 2nd ed., 1970.

[8] Morgan, G., *Images of Organization*, Sage Publications Inc., Beverly Hills, California, 1986, pp. 11-17.

Acknowledgments

The creation of this volume has involved many individuals. The authors created their contributions over and above their normal duties and responsibilities. Their commitment to building a science of C2 is much appreciated. Such a volume could not have appeared without the staff of the C3 Academic Group at the Naval Postgraduate School who performed above and beyond. A special thanks to Milena Cochran for her support. Finally, thanks to the editors of the Progress Series and the Publications Staff of the AIAA for their purposive action in making this volume possible.

How the Book Is Organized

Organizing a set of papers for a book is always a challenge; it is doubly so in a new field. The hypertext model used in this book permits the reader to easily select the papers of interest and gain an appreciation for the others beyond the information in a title. Thus, this volume includes a Table of Contents, as a one-line information source; abstracts, which provide an extended abstract view of the papers; and the papers themselves.

This book is a report on the products of research sponsored by the basic research group. Some of this sponsorship was financial. The papers in the book were provided by the authors without specific funding for their preparation. The editor was also not funded.

Abstracts

Task Identification With and Without Feedback in Hierarchical Teams

Gregory Burton
Seton Hall University, South Orange, New Jersey 07079
and
David L. Kleinman
University of Connecticut, Storrs, Connecticut 06269

An experiment investigating the effects of feedback in teams engaged in a threat identification task is described. Hierarchically organized teams of three decision-makers were required to judge whether each of a series of contacts was a threat or a neutral. This identification was made on the basis of sensor data, which were of higher quality for the subordinates than for the team leader. The leader made the team's final decision based on the judgments submitted by his subordinates and their reported confidences that these judgments were accurate. As the sessions progressed, half of the teams received increasing amounts of performance feedback after each trial, whereas the remaining teams did not receive such information. In a final "crisis" condition, the confidences reported by the subordinates were withheld from the leader. The results indicate that those teams that had received feedback suffered less of a decrement in performance and confidence in their judgments than those that had received no feedback. The possibility that feedback to subordinates exerts a benefit to the entire team, through the fostering of better mutual mental models, is discussed.

Headquarters Effectiveness Assessment Tool

Robert Choisser
Defense Information Systems Agency, Reston, Virginia 22090
and
John Shaw
ALPHATECH, Inc., Burlington, Massachusetts 01803

This chapter describes the development and application of the headquarters effectiveness assessment tool (HEAT). HEAT's purpose is to enable a team of observers and analysts to objectively assess and quantify the performance and effectiveness of a military headquarters and to make recommendations for improving its design. The HEAT method consists of measures for quantifying headquarters performance and effectiveness, together with procedures for computing those measures through data collected from military exercises and laboratory experiments. The chapter includes discussions of the C3 theory that underlie the tool and specific findings from applying the tool to exercises and experiments, as well as a description of the tool itself.

No New Mathematics! No New C2 Theory!

John T. Dockery
Pentagon, Washington, DC 20318
and
A. E. R. Woodcock
Synectics Corporation, Fairfax, Virginia 22030

For this chapter the title is the abstract; it asserts our position in no uncertain terms that progress in building command and control theory is at an impasse absent the application of modern mathematical insights. The chapter itself catalogs our major efforts in the exploration of mathematical disciplines adequate for the job of letting some fresh air into the construction of command and control theory. Incidentally, each candidate mathematical technology described in the chapter is tied to a previously identified requirement for some aspect of an eventual comprehensive theory of command and control.

Formal Theory of C3 and Data Fusion

I. R. Goodman
Naval Ocean Systems Center, San Diego, California 92152

This chapter treats C3 processes from a formal theory viewpoint. The approach is microscopic in nature, using a time-slice model, as opposed, e.g., to the outcome path approach of Petri nets and their generalizations. The usual SHOR paradigm plays the central role in the structuring of nodes, although knowledge-based information also plays a role. These intranodal relations, as well as internodal relations in the form of signals and communications through medium noise, are combined into a single large-scale formal model. In addition, uncertainty in the form of nonstochastic information, such as through linguistic sources, is taken into account in the data fusion aspect. The basic model consists of axioms representing the various conditional relations among the C3 SHOR paradigm variables, such as input signals, detection states, manpower, supply levels, damage levels, hypotheses of situations, and decisions and reactions/responses. The choice of the actual functional distributional relations among these variables relative to the axiom constraints can be interpreted as a C3 design move within a zero-sum game theoretic context. The basic loss function here consists of some prechosen moe/mop of the state of "health" of the friendly and adversary C3 systems. In turn, the health of each side is determined from an averaging procedure over all of the node states of the individual node state distributions in conditional form following SHOR paradigm signal processing cycles. These node state distributions are obtainable as outputs of the basic model described.

Statistical Mechanics of Combat and Extensions

Lester Ingber
Lester Ingber Research, McLean, Virginia 22101

Previous papers have established the first theory of statistical mechanics of combat (SMC) baselined to empirical data gleaned from the National Training Center (NTC). Using new methods of mathematical physics developed in the past decade, a Lagrangian formulation has developed nonlinear stochastic generalizations of Lancaster theory to confront realistic combat scenarios. Using new numerical techniques of very fast simulated reannealing (VFSR) first developed in 1988, these theoretical constructs can be calculated and further graphically developed into decision aids faithful to realistic combat data. This methodology has been further tested successfully in other complex systems, ranging from nuclear physics, to neuroscience, to finance. This project also has established a Janus(T)-NTC computer simulation/wargame of NTC, providing a statistical "what-if" capability for NTC scenarios. This mathematical formulation is ripe for various extensions, e.g., to include human factors. This paper focuses on compelling arguments for the necessity of this approach in combat systems.

Impact of Organizational Structure on Team Performance: Experimental Findings

Victoria Y. Jin
AT&T Bell Laboratories, Holmdel, New Jersey 07733
and
Alexander H. Levis
George Mason University, Fairfax, Virginia 22030

A multiperson, model-driven experiment has been designed on the basis of a mathematical model of distributed tactical decisionmaking. Team performance, measured in terms of response time and accuracy, was compared for a parallel and a hierarchical organizational structure. The results show that interaction among decisionmakers compensates for differences in individual performance characteristics. Individual differences have more influence on performance in the organization in which decisionmakers have more autonomy in making decisions than in the organization in which individual decisions are coupled with the decisions of other organization members. When available time decreases, time pressure is introduced into the organization and decisionmakers have to adjust their processing rate. The experimental results confirm a hypothesis which predicts that with decreasing available time a significant degradation of performance occurs first in the organization which has the highest minimum feasible workload.

Problem Solving Systems: A New Concept for a Science of C2

Carl R. Jones
Naval Postgraduate School, Monterey, California 93943

Observably operational commands are organized to solve problems by having problem solving systems that operate with a subset of the command's environment. For example, in a Navy Battle Group problem solving is organized around such warfare areas as ASW, ASUW, AAW, and SEW. The purpose of this chapter is to develop the concepts needed to understand viable problem solving systems designed to operate in a subenvironment of the command's overall environment. The major determinants of the systemic behavior of problem solving systems are 1) the degree of uncertainty concerning the relationship of actions taken to outcomes obtained and 2) the degree of consensus about the problem being solved. The variables of interest in understanding problem solving systems are the problem's structure, the solution procedure, interdependence, decisionmaking, communication, perception of common objects, and coordination technique(s) (direct supervision, mutual adjustment, standardization, and doctrine) employed. Four stereotypical problem solving systems are described. They are labeled management science, incremental, carnegie, and garbage can in conformance with the literature.

Colored Petri Net Model of Command and Control Nodes

Alexander H. Levis
George Mason University, Fairfax, Virginia 22030

The organizational structure that implements the command and control process can be represented by a set of interconnected intelligent nodes that carry out tasks in coordination. The intelligence in the nodes may be due to humans, to machines, or to a combination of human and machine, such as a human with a workstation. An earlier model of an intelligent node was based on ordinary Petri nets with switches and led to several approaches for the design and evaluation of architectures with fixed structures. That model has been extended to accommodate the execution of multiple roles by the same intelligent node; this leads to architectures with variable structures. The Colored Petri Net formalism is used to develop a compact representation of this model.

Modeling and Evaluation of Variable Structure Command and Control Organizations

Jean-Marc Monguillet and Alexander H. Levis

Massachusetts Institute of Technology, Cambridge, Massachusetts 02139

Distributed decisionmaking organizations with variable structure are those in which the interactions between the members can change, or which can process the same task with different combinations of resources. Variable structure could be a possible design solution when no fixed structure organization can meet the requirements of the mission. A modeling methodology based on the theory of Predicate Transition Nets is introduced to represent variable structure organizations. Decisionmaking organizations are then viewed from a new perspective in which the types of interactions that can exist between the decisionmakers are first considered without taking into account the identity of the decisionmakers themselves. The latter are represented by individual tokens (instead of subnets of a Petri net) moving from one interaction to the other, and as such are treated in the same manner as any other resources needed for the processing of a task. Interactions, resources, and tasks are modeled independently, i.e., the representation of the interactions, resources, and tasks is done separately in separate modules, and modifications in one module can be made without affecting the others. The methodology is illustrated by an example of a three-member decisionmaking organization carrying out an air defense task.

Command and Control Reference Model

Israel Mayk

U.S. Army Communications-Electronics Command, Fort Monmouth, New Jersey 07703

The command and control reference model (C2RM) embraces, in an integrated fashion, analogous architectures for all key types of interactions subject to command and control. The C2RM provides a seven-layer structure for generic C2 application entities and a complementary seven-layer structure for generic implementations of each application entity. Applications provide the semantic shell and implementations provide the syntactic shell for C2. Applications are layered according to the level of conflict, presentation, operation, procedure, network, link, and asset services. Implementations are layered according to the level of experience, knowledge, information, object, tool, equipment, and supply services. As such, the C2RM provides a universal structure for C2 and underlying services which follow the open systems interconnection (OSI) RM to the maximum degree practicable. The C2RM may be used as a framework to define one's own or adversary force structure and viewpoints at multiple echelons. Clearly, services which embed human decisionmaking require extensive research and development and a high-level maturity of understanding before any agreement may be reached for the purpose of standardization.

Stochastic Modeling, Analysis, and Calibration of C3 Systems

Izhak Rubin

IRI Corporation, Tarzana, California 91356

and

Israel Mayk

U.S. Army Communications-Electronics Command, Fort Monmouth, New Jersey 07703

An architecture is presented to describe and model a C2 system and its dynamic evolution under a specified mission and scenario condition. This architecture has been presented as a stochastic C2 graph (SCG) model and implemented as an object oriented (OOSCG) structure and program. The graph consists of nodes, where each node describes the states and methods associated with a group of (friendly and/or adversary) resources within a time-space cell. Key events are used to characterize the nodal processes and the flows between nodes, in accordance with the underlying maneuvering policy and battle(s) conditions and results. A scheduling graph is used to describe the temporal (in terms of battle stages and combat phase) and spatial evolution of the battle process. Interaction graphs describe the interactions between the group objects during a combat phase. Actions, interactions, and resources are classified into four main categories, in accordance with the C2 reference model (C2RM): infliction (firing, weapons), identification (detection, sensing), transportation (maneuvering, vehicles), and communications (data distribution, coordination). Stochastic process-based analytical models have been developed, calibrated, and tested to statistically describe the outcomes of battle processes and the effectiveness of the underlying engagement strategies, decision rules, maneuvering policies, and resource allocations. These models are used as the basic methods employed by each group object and permit (faster than real-time) effective calculations of the outcomes of interactions between the SCG nodes. Three such stochastic C3 models are presented and analyzed: 1) a semi-Markov attrition model that involves the employment of a general maneuvering policy at the time of occurrence of each key event; 2) an outer-air/inner-air battle model which involves two battle stages and multiple combat phases per stage; and 3) a multimission area C2 (MMAC) resource allocation program, which evaluates the effectiveness of allocation of diverse resources among randomly occurring missions. It is demonstrated that following a calibration process, the OOSCG model can be used to provide extensive modeling, analysis, and tradeoff evaluation, serving as an efficient training, planning, and real-time decision aid tool.

Task Identification With and Without Feedback in Hierarchical Teams

Gregory Burton*
Seton Hall University, South Orange, New Jersey 07079
and
David L. Kleinman†
University of Connecticut, Storrs, Connecticut 06269

I. Introduction

IN large-scale systems (e.g., military command and control, electric power distribution, air traffic control, etc.) that involve humans, machines, and computers, the problem scope and complexity often require that the decisionmaking function be *distributed* over several humans. Quite often such systems have a *team* of human decisionmakers (DMs) who are geographically separated, but who must coordinate to share their limited information, resources, and actions in order to attain a common goal in what is generally a dynamic and uncertain environment. These teams are usually hierarchical in nature, with a leader and several subordinates. As a consequence of limited communications, each DM has access to only a portion of the information available to the team. Moreover, even the total information set is often incomplete and inaccurate due to lax updating, missed detection of events, and errors in measurements.

A critical issue regarding information dissemination in a hierarchical team is the amount and nature of the feedback that is provided to members of the team, i.e., the degree to which the leader should keep his subordinates informed of the team's and each individual's performance. Most of the psychology research on feedback has analyzed aspects of feedback to individuals or to members of *groups*. It has been found that feedback to members of organizations can have a variety of effects, not all of which are necessarily beneficial (e.g., Ilgen et al.[1]). Even when the effects of feedback on performance are essentially positive, the gains may not outweigh the time expended by the leader in providing the feedback and by the subordinates in assimilating it.[2]

Copyright © 1993 by the American Institute of Aeronautics and Astronautics, Inc. All rights reserved.
*Professor, Department of Psychology.
†Professor, Department of Electrical and Systems Engineering.

Generally, it is believed that feedback enables units to develop expectancies about the consequences of their decisions. If these units are members of *teams* with specific roles and responsibilities for their members that are not usually present in *groups* (where members are usually homogeneous[3]) these expectancies may be critical for appropriate coordination. Team members may require information not only about the ultimate results of individual decisions, but of more local results, namely the responses of other team members (particularly, but not exclusively, the leader) to individual actions. In some contexts, this framework of expectancies is referred to as "mutual mental models" (e.g., Athans[4]). To further elaborate the salient issues in team decisionmaking, Ho points out (see Ref. 5, p. 644) that, "Each person must take into account in his decision what the other [decisionmakers] may decide based on his information. More specifically ... the main ingredients of a theory of team decisions are: 1) the presence of *different but correlated* information for each decisionmaker about some underlying uncertainty; [and] 2) the need for *coordinated* actions on the part of all decisionmakers in order to realize the *payoff*."

The research reported in this paper is an attempt to investigate some effects of feedback on the performance of three-person hierarchical teams engaged in a target identification task. Teams were required to identify abstract targets either as threats or as nonthreats. One set of teams was provided with feedback reports of increasing detail on a target-by-target basis. Their response characteristics were compared to those of teams that received virtually no feedback on their overall performance. It was expected that teams in these two conditions would behave differently and that these differences could be attributable to the different amounts of detail in the mental models potentially developed by team members about likely consequences of individual decisions. However, the current design does not depend greatly on the concept of mental models, nor on their development in the specific context studied, but makes only the conservative prediction that mental models will be more effective if more detailed information is available for their development.

The following sections describe the target identification problem that we examine in this research, with emphasis on experimentation with human teams. The specific experiment that was designed to investigate the effects of feedback is presented. (For the results of a companion study that examined other aspects of the same hierarchical information processing experiment, but from an analytic/modeling perspective, see Ref. 6.) The experimental results follow, along with our discussion. Basically, the results indicate that feedback can have an insulating effect on a team by helping members to compensate for conditions in which the team suffers a loss of information; under normal conditions the effects of different levels of feedback, or of feedback vs no feedback, are not significant. Our conclusion is that teams exposed to feedback will be more robust than teams that have not been so exposed.

II. Problem Description and Approach

A. Threat Identification in Hierarchical Teams

In a hierarchical information processing/decision structure, two (or more) subordinates make lower-level decisions and report their results or assessments to a higher-level commander. The subordinates may disagree on their assessment of the situation, in which case the team's success hinges partly on the ability of the commander to fuse their separate opinions. Two of the factors the commander must consider are the previous reliability of each subordinate, and the "urgency" of the current opinions. For example, a commander might favor the opinion of a subordinate who is usually less reliable if that subordinate claims to be "100% sure" of his assessment. But on the other hand, if that subordinate is always reporting a confidence of 100%, this opinion will have less of an effect on the commander's decision.

One of the more critical situations in which hierarchical decision and information structures are salient is military threat identification, in which subordinates (who may have more detailed and more specialized data than their commanders) must report assessments on whether or not an approaching contact is a threat. Failures of such threat identification processes are not uncommon in military history. For example, during the Civil War the Confederate general Thomas "Stonewall" Jackson was killed when he was accidentally shot by his own troops. In the midst of some confusion, they incorrectly identified the group Jackson was riding with as a group of Northern soldiers.[7]

Even before the Gulf War, events in the Persian Gulf underscored the critical importance of correct threat identification by teams. On May 17, 1987, the crew of the U.S.S. Stark incorrectly identified a belligerent Iraqi plane as a friendly. Despite a warning from a nearby AWACS post, the Stark commenced an involved engine test, failed to lock the radar onto the incoming planes, failed to locate and inform the commanding officer of the vessel, and failed to warn away the Iraqi jets until they were already within firing range. The Iraqi jets sent two Exocet missiles into the ship, starting a fire that claimed 37 lives.[8,9] On July 3, 1988, the U.S.S. Vincennes incorrectly identified a neutral Iranian commercial plane as an F-14 and shot it down, killing all 290 passengers. A Navy report attributed the mistake to the human error of the ship's operations officer.[10]

In situations such as these, a team must process uncertain information and come to a definite decision. There is often prior information (i.e., the fact that Iraqi jets often approach U.S. ships quite closely before veering off harmlessly) and current information (the fact that an Iraqi jet was currently approaching, the warning from the AWACS station) which must be integrated, and a final decision made by the commander (whether the target should be treated as a threat or a neutral). Before the commander is required to make his/her final decision, the subordinates must make a number of decisions of their own (e.g., whether to report a suspicious blip on the radar screen, whether to attempt to sway the commander if at first he does not agree, etc.). This hierarchical decision structure is sometimes vulnerable to mistakes and, consequently, its analysis has proven to be an interesting and challenging research area for decision scientists. Developing an understanding of the process of human/team decisionmaking in such military contexts could ultimately lead to improvements in the underlying decision and/or information systems.

B. Empirical Approach and Hypotheses

Our main empirical tool is a unique paradigm for studying team distributed decisionmaking and coordination in a controlled laboratory setting.[11,12] In this distributed dynamic decisionmaking (DDD) paradigm, multiple targets with different deadlines, processing times, attributes, and values or priorities arrive at random. Coacting decisionmakers who are physically separated must process distributed information to 1) estimate various target attributes, 2) identify target class, and 3) determine and schedule the resources needed to process any specific target. The DMs can obtain target information from local or global databases, from communication exchanges with other team members, or by "probing" to actively seek additional information. The DDD paradigm is implemented on a network of SUN workstations, which provide real-time control and on-line data collection, interactive display/interface media, and a computerized intrahuman communications subsystem (using a preformatted message set for request and transfer of information, resources, and actions between team members) within which delay can be manipulated.

Bushnell[13] was the first to investigate a target identification task using a variation of the DDD paradigm. In her paradigm, each target displayed on the screens was either a threat (requiring attack) or a neutral. Each target was characterized by two attributes; different attribute profiles typified a threat and a neutral, but the data given on each trial were degraded by a varying amount. A team of two players, neither of whom had authority over the other, had to coordinate their individual pieces of information to come up with a judgment of whether the target was a threat. In the current experiment, we altered this paradigm so that a hierarchical three-player team could be tested. The information available included global information (given to all three subjects) and also separate or local data, of better quality, that were provided only to the two subordinates. Each subordinate could not send his actual data values to the commander; he could send only his individual assessment of the target (threat or neutral) and a report of his confidence in his judgment, defined nominally as the probability that his assessment is correct.

This confidence report is important because it provides subordinates with a means to influence the leader. One aspect of feedback is that it provides information on the necessity of corrective action. For example, Hogarth (Ref. 14, p. 200) observes that "[F]eedback plays a crucial role in [an] organism's capacity to make adaptive responses by reducing the commitment implied by any particular action." Thus, feedback may allow a team member to judge when "corrective action" must be taken to increase the team's success. In the experiment being considered, the corrective action that subordinates can take is rather limited, namely, varying the confidences they report to the leader on their assessments. For example, if a subordinate perceives that her reports are being undervalued and that the leader is consequently making wrong decisions, a response to this perception might be to increase the reported confidence. This sort of usage does not have to be planned ahead of time or even deliberate to be effective; it could be implicit as when a subordinate attempts to "sway" his leader. Ansari and Kapoor[15] (see also Yukl and Falbe[16]) found that the strategies used by workers to sway leaders were affected by the leadership style of the superior and the goals of the subordinate (for example, whether the subordinate wished to get more help in the office or wished to be promoted).

Thus, decisionmaking teams that received increasing levels of feedback were contrasted to control teams that received virtually no feedback. In addition, since feedback was expected to have an impact on the robustness of the team's performance, a "crisis" condition was added to investigate whether teams that had been strengthened by feedback could survive a loss of information better than other teams. In this final experimental stage, the confidences reported by subordinates were no longer passed on to the leaders. This manipulation drastically reduced the information available for the leader to make his identification and the means for the subordinates to influence the leader's opinion.

It is worth emphasizing that "confidence" in this experiment has somewhat of a dual nature. During most of the experiment, it is available to the subordinates as a "signaling" tool to inform and potentially influence the leader. However, when the confidences are no longer passed on to the leader, these reports no longer serve a signaling function (team members were, of course, informed of the nature of the final conditions) and, presumably, they more accurately reflect the level to which the subordinate feels secure that he has made the right decision. This is an interesting distinction, possibly mirroring the difference between the confidence that a subordinate might report to his superior and his true feeling of success. Thus, even though the same reporting procedure is employed by subordinates in each condition, the confidences they report in the final frame may be different from those that would be transmitted to the leader.

In this experiment, the leader also reported confidence judgments. However, in the experiment there was no superior or higher level to which to send these confidences. As such, these numbers presumably index the emotional confidence of the leader in his ability to make the final decision based on the information sent by the subordinates. Without the confidence reports from his subordinates, the leader has less information for his decision and must rely partly on his expectancies of what the subordinate assessments "mean." (For example, one subordinate may report "threat" more commonly than the data warrants; the leader, in the absence of other information, may have to preferentially trust the other subordinate if there is a conflict.) The degree to which the leader's confidence can "survive" this final condition serves as another indication of the robustness and/or fidelity of his mental model.

In general, it was hypothesized that the availability of feedback allows the team members to develop accurate mental models of the other players and results in stronger teams. This effect might be observable in several ways. First, we expected performance to improve as the level of feedback increased for teams that received it. Second, we also expected feedback teams to suffer less of a performance decrement when information was abruptly reduced in the final crisis condition. Because feedback could potentially provide data for subordinates to attempt to sway their leader (by increasing their confidences), we expected higher confidences reported by the subordinates as feedback increased; and since the feedback loses its signaling role in the final condition, the confidences reported might also decline in this condition. Of course, if feedback effects a general increase in team cohesiveness, we might also expect an increase in the confidences reported by the leader, even though in our experiment he/she is not transmitting confidence reports on to a higher level of the organization.

III. Experimental Design

A. Team and Information Structure

The team structure and its information flow are shown in Fig. 1. The team is hierarchical with a leader (DM0) and two subordinates (DM1 and DM2). Given corrupted attribute information on a target, the objective of the team is to correctly decide the target type: a neutral (N) or a threat (T). The team has access to conditionally independent attribute measurements y_i from sensors s_i, ($i = 0, 1, 2$). The measurement y_0 is global and is available to all three DMs, whereas the measurement y_i is accessible only to DMi (i = 1,2). Each of the measurements y_i is a noisy version of a two-dimensional vector of target attributes a. In addition to being noisy, the data obtained by the subordinates can sometimes be ambiguous, e.g., if the sensor breaks down the measurements provide absolutely no information on the target attributes.

The two subordinates have differential expertise: DM1 is an expert in detecting threats, whereas DM2 is an expert in detecting neutrals. The expertise is operationalized by adjusting the measurement noise covariances associated with y_1 and y_2. The subordinates (DM1 and DM2) report their opinions (u_1, u_2) and their associated confidences (c_1, c_2) in their decision. The leader (DM0) makes the final event decision u_0 (N or T) based on the team's common measurement data, y_0, the transmitted opinions (u_1, u_2), and (when provided) the confidences (c_1, c_2) of the subordinates.

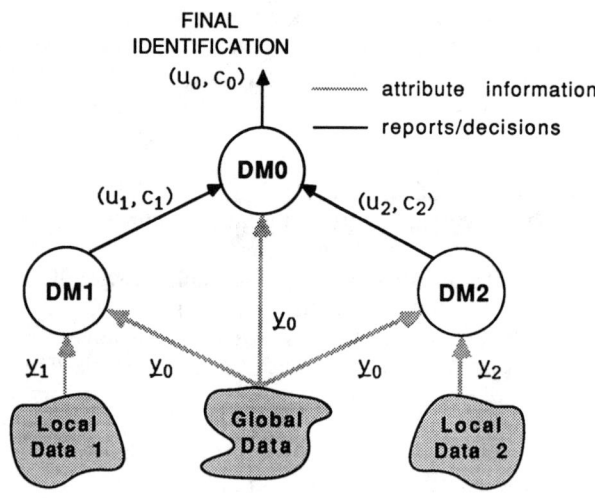

Fig. 1 Team and information structure.

B. Experimental Procedure and Method

Each player (DM) sat at a separate SUN Microsystems Inc. graphic workstation and interacted with his/her screen by means of a mouse. The three terminals were geographically separated but were synchronized and networked, so that communication lines within the team were maintained. Figure 2 shows how the screen looked for the team leader (designated as DM0) at one point in play; the screens for the subordinates DM1 and DM2 looked similar but differed in the amount of information that was presented and only allowed communication with DM0. For each trial, a questionmark icon (representing an approaching target) appeared at the top of the target window; in the attribute window, measurement data appeared on two arbitrary attributes of the target. (The two attributes of incoming contacts were completely arbitrary in this experiment. In the training manual, they were designated "speed" and "evasiveness" so that subjects could understand that they represented distinct aspects of the menace of the contact.) Subjects were to decide on the basis of these measurements whether the target was a threat or a neutral. In truth a threat had values of 5.6 for both attributes, whereas values of 4.4 defined a neutral. Subjects were informed, however, that the *measurements* of these attributes were corrupted by the addition of a certain amount of random noise such that the range of both attributes was effectively 0.0–9.9 for both threats and neutrals. This noise provided a Gaussian envelope around the actual attribute values; the standard deviation of the noise (σ) varied according to whether or not the subject was designated as an "expert" on

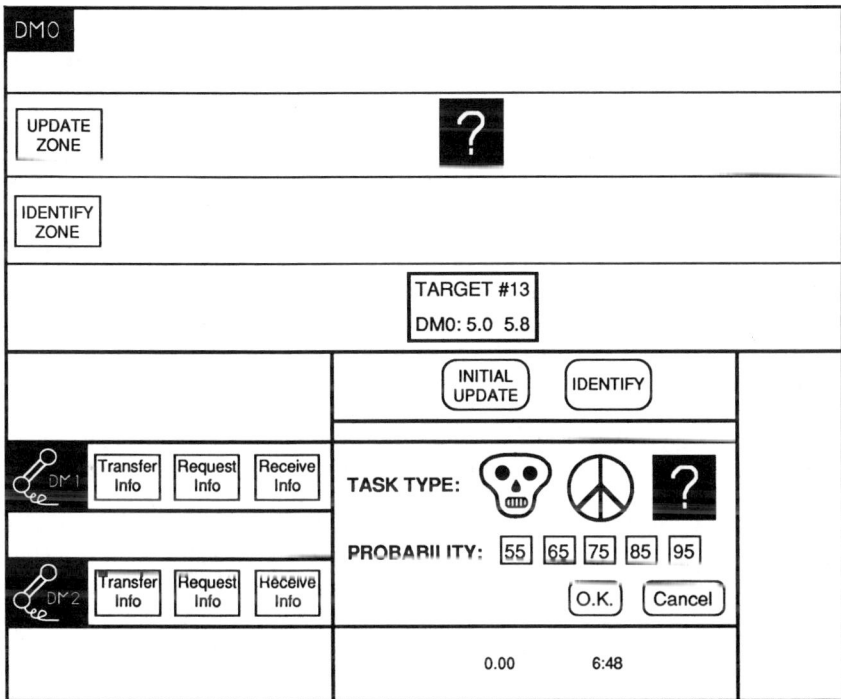

Fig. 2   Graphics display interface on the leader's terminal.

a particular sort of target. For the subordinate (DM1) who was designated an expert on threats, his private/local measurement data on incoming threats would usually vary between 4.0 and 7.2 ($\sigma = 0.8$), but his data on incoming neutrals would vary between 0.4 and 8.4 ($\sigma = 2.0$). Thus, this subject was more likely to correctly identify a threat than a neutral. The situation was reversed for the neutral expert, DM2. In addition to their local data, each subordinate also had access to global data (which were also available to the commander). The global measurement data of the target's two attributes varied more widely for both threats and neutrals, generally ranging from 0.4 to 9.6 ($\sigma = 2.0$).

The subjects were 18 engineering graduate students that participated to earn $4 an hour for the experiment, plus a bonus based on their team's final performance score. These 18 subjects were assigned to six teams. Mindful of the warnings of Hill[17] about using subjects who are already known to each other, we considered random assignment. However, it was deemed more practical to use teams consisting of people who already knew each other, because the experiment entailed a considerable time investment (roughly 12–13 h) during which all three team members had to be present. Also, previous experience with extended experimentation with multiperson teams suggested that members of randomly assigned teams would be much harder to schedule and more likely to not show up for a session.

During preliminary screening tests, subjects were presented, one by one, with a series of attribute values for a target and were required to identify it as a threat or a neutral. These presentations were alphanumeric, not graphic as in the main experiment. In addition, different versions of the screening test presented data of different expertise (i.e., favoring threats, favoring neutral, or inexpert on both), and the subjects who scored highest on the screening tests were able to discern the assigned expertise condition. To insure uniformity in the main experiment, subjects were directly informed as to whether they were experts on threats or neutrals, and so this knowledge did not have to be inferred.

Each of the experimental sessions included 15 trials, each trial being initiated by a questionmark icon that appeared in the "update" zone. As it appeared, the subordinates saw measurements of the target's attributes from both global and local information sources. On the basis of these two information sets, the subordinates identified the target as a threat or a neutral and also reported their confidence that their identification was accurate. (This reporting was effected by clicking on one of a series of five boxes, as in Fig. 2. The values ranged from 55 to 95 and each represented a 10-point range in probability of being correct.) Then the subordinates were required to transmit this information to the commander; if a subject clicked on his (or her) identification and his confidence but failed to transmit this information to the commander, the commander would not receive the information. Trials in which one or more players did not complete all of their steps were considered "failures," and the team would not get credit for correct identification, even if the commander's final decision was accurate. This requirement assured that all three players were involved in the identification of each target. While the icon was in the update zone, the commander (DM0) was required to make an initial assessment of the target's identity, including his (or her) initial confidence. The commander only saw his own data (i.e., the global data) at this time. This initial commander's decision was recorded but was not transmitted anywhere.

When the icon moved into the "identify" zone, the commander received the assessments transmitted by the other two players. Based on all of the information available (DM1's and DM2's assessments and the global data), DM0 was then required to make his final decision on the target's identity, including his confidence in this decision. The team received a score of 10 points for each target correctly identified, and 0 points for each incorrect identification. During the time that the icon was in the identify zone the subordinates were required to make a prediction of the commander's decision. Subjects were told that these predictions would be used when computing the bonus as multipliers on the scores for each target, if the final scores of the various competing teams were fairly close. (It turned out to be unnecessary to make these computations.)

Finally, the icon moved off the screen, and the commander of the team was informed, in the leftmost window, of the target's actual identity, and the team's updated/accumulated score based on his decision. The results seen by the subordinates after each trial depended on the specific experimental condition.

C. Dependent and Independent Variables

A full factorial design was conducted employing the three independent variables 1) level of feedback, or stage, 2) team type (with or without feedback), and 3) ambiguity. The different feedback levels, or scenarios, consisted of the following:

1) No feedback (NF): only the commander saw the final result after each trial. However, the entire team saw their final/total score after each full session of 15 trials.

2) Partial feedback (PF): on a trial-by-trial basis, all three players saw the results and their team's score. Also, the two experts were informed of the decision that was made by the commander.

3) Full feedback (FF): the same information was provided as in PF, and in addition, each expert was shown the assessment made by the other expert on that trial.

Three of the six teams were "experimental" teams and received increasing levels of feedback: there were eight sessions of NF, eight sessions of PF, and eight sessions of FF. The remaining three teams were "control" teams and received NF throughout the entire experiment. A fourth scenario of eight sessions was also conducted for all six teams, in which the confidences reported by the subordinates were not provided to the team leader (condition NC). This was included as a "crisis situation" in which the effects of feedback might be particularly critical. For the experimental teams, these scenarios were otherwise FF, i.e., the subordinates were informed after every trial of the team score, the results, the commander's decision, and the other subordinate's decision.

Since the experiment was conducted to learn if providing feedback yielded long-term benefits to the team, randomizing the order of the scenarios was inadvisable. A team that had received FF previously and currently received NF could not be equivalent to a team that received NF first. Thus, the scenarios were conducted in the above order, and any improvement in the scores of the experimental teams might have been influenced by increased practice as well as by the facilitating effects of feedback. It is this possible confounding that motivated the inclusion of control teams; the difference between the patterns of the control teams and the experimental teams across scenarios are indications of whether the increasing feedback had an effect over and above that of practice.

Each team was given a series of practice sessions that were slightly shorter than the main sessions (10 trials). Teams practiced until they understood and could perform the experimental procedure effectively; the number of failures per session served as an indicator of each team's readiness. Experimental teams were also given 1–3 practice sessions when the level of feedback was changed and all teams were given practice sessions before the final stage, in which the subordinate confidence reports were recorded but were no longer sent to the commander.

Thus, the primary independent variables were team (control vs experimental) and stage (control: NF1, NF2, NF3, NF-NC; experimental: NF, PF, FF, FF-NC). In addition, the level of ambiguity in the data was varied. Ambiguity refers to the percentage of the time that one of a team member's two pieces of data was randomly selected from a uniform distribution U [0,1], and thus had no correlation with the actual target values. Four of the eight sessions in each stage presented data with no ambiguity; in the other four sessions the teams were faced with data that had a 20% ambiguity level. Subjects were informed of the ambiguity condition of each session. However, this manipulation proved to have little effect on team performance and will not be discussed further.

The DDD software is designed to collect a large pool of dependent variables for each game. The following variables are of interest here: the team's final score (SCOR), which was proportional to the number of correct identifications made by the team, the confidence reported by the leader in his final decision (BPR1), and the combined confidences of the experts in their decisions (EXPR). In all cases we normalized these variables by their corresponding values in stage 1, where the (NF) treatment received by both experimental and control teams is identical. In this way the results more clearly show the effects produced on different teams as a function of stage.

IV. Results

Analyses of variance were conducted on the dependent variables with respect to the independent variables of team, stage, and ambiguity, and their interactions. The teams' final scores were significantly different depending on whether the teams were control or experimental teams ($p < 0.01$). The interaction of the team with the feedback stage was not significant overall; however, a Tukey test indicated that the conditions were significantly different in the final stage ($p < 0.05$). Thus, the first hypothesis was not confirmed, but the second hypothesis, that teams receiving feedback would be more robust in the face of a loss of information, was supported. Figure 3 shows that the lack of feedback did not entail a disadvantage for the control teams during the first three stages of the experiment, but in the fourth condition, when the confidence reports were no longer shown to the commander, the mean final score, normalized by the first condition (NF) score, was significantly lower for control teams than for their counterpart experimental teams. In this experiment the beneficial effects of feedback were not apparent during the "normal" conditions, but when the task facing the team became more difficult, the feedback received by the experimental teams seemed to have "insulated" them somewhat from the loss of effectiveness suffered by the control teams.

The leader's final confidence was also significantly different for the two conditions ($p < 0.001$); the interaction with the feedback stage was also significant ($p < 0.01$). Figure 4 shows the normalized BPR1 graphed against team and stage. For experimental teams, the confidences rendered at the four stages is fairly uniform. There is a slight dropoff of leader's confidence for the NC stage, but this difference was not significant, according to the Tukey test. On the other hand, the confidences rendered by control team leaders varied much more and declined much more drastically for the final stage. The overall BPR1s for all four stages were significantly different from each other ($p < 0.05$), and in particular, according to the Tukey test, the confidences for the two team conditions during the final stage were also significantly different. This effect provides another example of the insulating effect of feedback. The presence of feedback seemed to preserve the experimental teams' leaders from losing confidence in their decisions, which is interesting, considering that it was not the team leader, but the subordinates, who received additional feedback during the later stages. Thus, as predicted, feedback can have beneficial effects on the team as a whole, not simply on the particular members of the team who are receiving the feedback.

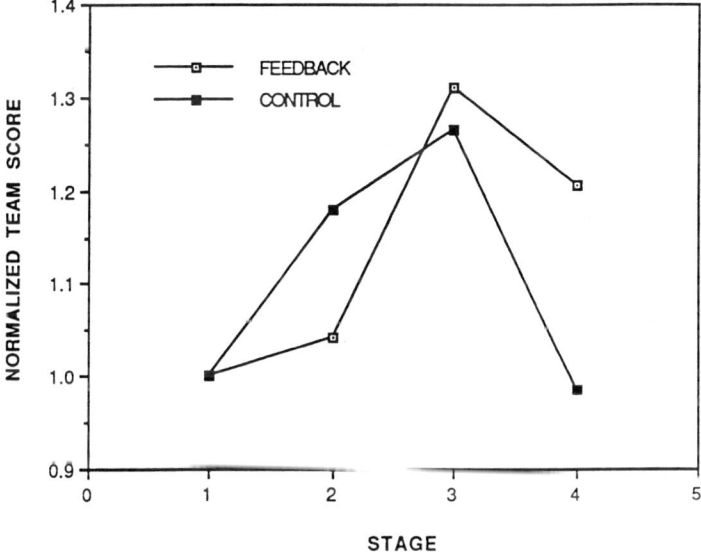

Fig. 3 Team score as a function of stage and the availability of feedback; scores are normalized by the scores of the first stage.

The difference between the pooled confidence of the subordinate experts in control vs experimental teams was quite small on an absolute basis. However, over all conditions, the pooled confidence was significantly higher in the control teams than in the experimental teams ($p < 0.01$), but the interaction was not significant. Figure 5 shows the normalized values for the two team conditions during the four stages. There was a pattern of steady increase in confidence for the control teams, with a mild decline during the final conditions; but there is no clear pattern for the experimental teams. None of the eight points were significantly different from each other, according to a Tukey test. In effect, there was a general increase in confidence for control teams, but the location of this difference cannot be pinned down with certainty. Contrary to our hypothesis, the greater confidences were found for the control teams; the benefit of feedback was manifest mainly in the less drastic reduction in confidence during the crisis stage that was exhibited by members of the experimental teams (a difference that, to reiterate, was not statistically significant).

V. Discussion

Various mechanisms have been postulated as the means by which feedback affects performance variables. For example, some researchers suggest that feedback serves to motivate workers (e.g., Hundal[18]). This has some relevance to the current research in that the control team subjects, during debriefing, sometimes complained that without knowledge of results, it was more difficult to stay interested in the task. In the field of group or team performance, another role for feedback seems to be the provision of data for the construction

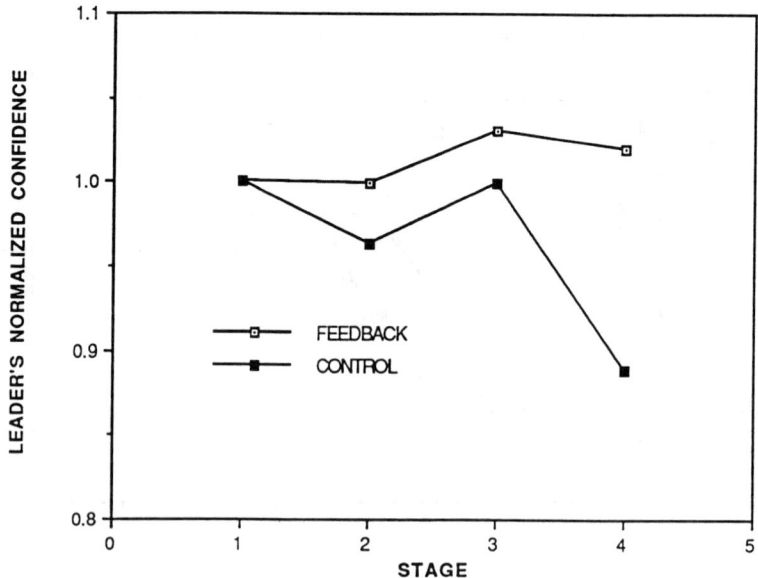

Fig. 4 Leader's final confidence as a function of stage and the availability of feedback: values are normalized by the values of the first stage.

of expectancy sets, or mental models of the team's probable reactions to task contingencies. The current experiment manipulated the availability of information with which team members—both leaders and subordinates—could build their possible models. The results of this experiment indicated that the feedback provided to teams had beneficial effects, but that these effects were subtle. In particular, the teams that received feedback did not seem to be at an advantage during the first three stages, under normal conditions, when team leaders received assessments and confidences from their subordinates. However, in a crisis situation when the amount of data given to the leader was drastically reduced, the effects of feedback were manifest; the availability of feedback seemed to have insulated the teams that received it, so that they could continue to operate at or near normal levels of performance. This result, in retrospect, is reasonable if teams are only forced to rely on mental models in the absence of sufficient current information.

The availability of feedback also gave experts some information to use for calibrating their own confidence reports when attempting to sway the leader. When the confidence reports were taken away, the teams may have known enough about their own behavior so that they could continue to operate efficiently. However, the dependency of subordinate confidence on the availability and detail of feedback was not clear in these results. The dual nature of the confidence reports, as discussed above, may have reduced the sensitivity of this variable for detecting effects of feedback. Moreover, the teams who had received no feedback had a (slightly) higher confidence level than did their counterparts. The difference in experts' confidences is understandable when the amount of information available to the different teams is considered. Members

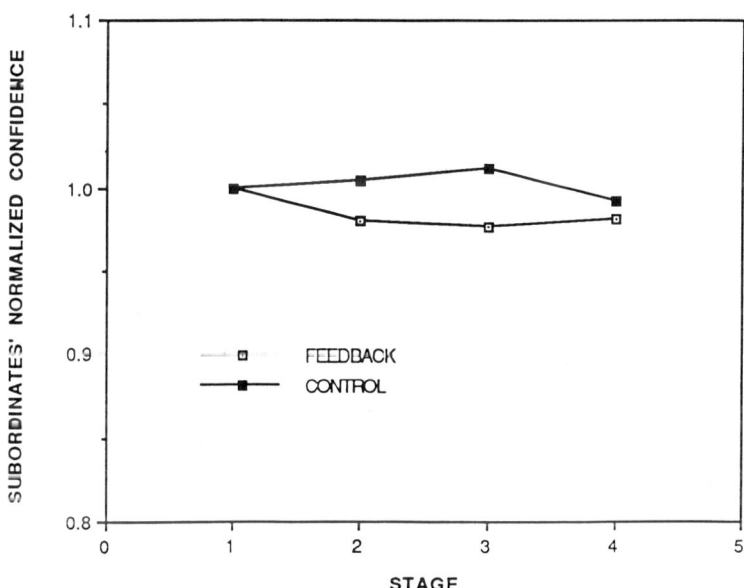

Fig. 5 Subordinates' pooled confidence as a function of stage and the availability of feedback: values are normalized by the values of the first stage.

of the control teams had access only to global results: how the team performed over 15 trials. Each expert could not, for example, assess how accurate his/her own decisions had been, or how often the team leader had acted on his/her advice. That this is an effect of inadequate feedback is also mentioned by Ilgen et al. (Ref. 1, pp. 361, 362): "In the absence of explicit information about performance, the individual often is left to infer what is the desirable behavior from the outcomes (positive or negative) that are administered.... More specific information about the behavior emitted should aid the learning of instrumentalities by providing individuals with a clearer perception of the behaviors with which the source of outcomes is concerned." In the experimental teams, experts had access to more local information, to an extent increasing with each stage. These experts could attribute their team's success (or lack of success) to their own decisions and may have had a more realistic picture of the accuracy of their personal decisions. The control team experts did not have the information to calibrate their own decisions in this manner. This explanation assumes that experimental team subordinates were realistically reducing their confidence reports in response to negative feedback. The direction of feedback is an important feedback variable; it might be profitable to control the favorability of feedback in some future experiments.

The confidence of the team leaders showed the same pattern as the overall team performance, being largely preserved in the final conditions for the teams that had received feedback relative to those that had not. The confidence reports submitted by the leader are unambiguous in nature because they were not passed on to any other persons in the hierarchy; thus, they should index the leader's subjective confidence as usually defined. This robustness due to feedback is interesting in that it was not the leaders, but the subordinates, who were receiving varying amounts of feedback. Apparently, the leaders of these teams were assured that the assessments of their subordinates had been sufficiently calibrated by feedback so that the leader could "know what to do about it."

This increase of leader confidence suggests that the provision of feedback affected the behavior of the entire team, not simply the persons who received it. In a sense, by providing feedback, the leader is serving to more evenly distribute the responsibility for decisions. Subordinates receiving only global information on team results have little ability to change specific aspects of their decisions. As discussed before, subordinates who gain more local feedback may alter their decisions in attempts to influence the leader, perhaps reducing the threshold for reporting a particular decision; the subordinate who acts in this fashion has thus taken a share of the responsibility of decisionmaking that accrues to the team. This process suggests some classical results in social psychology. In the middle decades of this century, social psychologists, following Marvin Shaw, investigated the benefits of various types of communication systems.[19] It was found that fully connected information networks were not maximally effective; it is inefficient for all knowledge to be available to all members of the net. It is also ineffective for all knowledge to go through a central person, suggesting that the leader of a team should not be an information clearinghouse. The leader does not have to know everything the team knows; in fact, if he did, there would be a natural limit on the knowledge capacity of a team equal to the capacity of its best member (even assuming that the leader is always the most knowledgeable member) and, if this were the case, the advantages of using a team may be dissipated.

It is generally suggested that an authoritarian approach to leadership, in which the leader makes all of the decisions, is less effective in most realistic contexts (e.g., Finch[20] and Hunt[21]). Moreover, leader consultation of subordinates (particularly in the setting of goals) is an important tactic of leadership (e.g., Yukl and Falbe[16]). This may reflect the same problems that apply to centralized communication networks; to reiterate, the optimal communication network is not the centralized one, in which the leader knows all of the team's information. Perhaps it is also true that a centralized decision network, in which the leader makes all of the decisions, is also suboptimal for the same reason: it limits the team's decisionmaking ability to that of its leading member. In realistic team contexts, a decentralized network, in which decisions are made at various levels, may be more effective.

Providing feedback to team subordinates is one way to support the decentralization of decisionmaking; when feedback is available, subordinates can, for example, alter their confidence reports in an attempt to sway the leader. Of course, there are more concrete ways of decentralizing the decisionmaking of a team, but even this mild step seemed to have some benefits to our experimental teams. The benefits were subtle, but were visible when the team faced a crisis. If our feedback manipulation did serve to increase the decentralization of decisionmaking in the teams, perhaps it is less surprising that the confidences of the leaders, as well as the subordinates, were preserved in the crisis condition for teams that received feedback. Though feedback is only presented to the subordinates, the benefits of decentralization can be global in nature, serving the team as a whole.

In the Bushnell experiment[13] with task identification, confidence judgments were significantly affected by an increase in the task load on the team members. This result anticipates the behavior of our control team leaders but not the leaders of our experimental teams. If the current results are comparable to Bushnell's, the leaders of teams that received feedback apparently perceived the crisis condition to be less of an increase in task difficulty than did the leaders whose teams received no feedback.

After all of the sessions were completed, team members were given debriefing questionnaires. Subjects in the no-feedback teams clearly felt they were at a disadvantage. In one control team, both subordinates wrote that they would have liked more feedback in the experiment and one felt the other teams had an advantage in that they were able to calibrate their performance with feedback. This commentary is interesting considering the result of significantly higher confidences for control team subordinates; they had little data to go on in assessing their performance on-line, other than the lump sum report at the end of each session. Also interesting is the fact that the subjects who made these comments belonged to the highest-scoring team overall. In another control team, subjects also wished for feedback and commented that the experiment became somewhat tedious when they did not know their results on a trial-by-trial basis.

In general, the results of the current experiment indicate a subtle but important role for detailed performance feedback in hierarchical teams. The paradigm in which our subjects were placed included task information (the data on the approaching contacts) and team information (feedback). When task information was adequate, team members showed little apparent sensitivity to team information in that the availability of feedback did not affect obvious performance variables. However, the performance of the two kinds of teams was distinct when task information was inadequate, a difficulty that is often a feature of real-world hierarchical team situations. Previous reviews on feedback (e.g., Ilgen et al.[1]) have underscored the importance of variables such as the direction (positive or negative), the timeliness, the detail, and the modality (quantitative or qualitative) of feedback. The specificity of task information available to an information-processing team seems to be another critical variable when it comes to the manifestation of beneficial effects of team feedback. This dependency seems intuitively reasonable in that mental models of team contingencies are likely to be unnecessary if task information is complete.

Acknowledgments

This research was made possible by Grants N00014-87-K-0707 and N00014-88-K-0545 from the Office of Naval Research. The authors would like to thank David Koenig for the bulk of the design and conduct of these experiments and Ranga Mallubhatla for assistance in that regard. The authors also acknowledge the advice and consultation of Elliott Entin, Robin Hogarth, Peter B. Luh, Xi-Yi Miao, Krishna Pattipati, John Payne, Daniel Sullivan, and Zhuang-bo Tang.

References

[1]Ilgen, D. R., Fisher, C. D., and Taylor, M. S., "Consequences of Individual Feedback on Behavior in Organizations," *Journal of Applied Psychology,* Vol. 64, No. 4, 1979, pp. 349–371.

[2]Ilgen, D. R., and Moore, C. F., "Types and Choices of Performance Feedback," *Journal of Applied Psychology,* Vol. 72, No. 3, 1987, pp. 401–406.

[3]Steiner, I., "Paradigms and Groups," *Advances in Experimental Social Psychology,* Vol. 19, 1986, pp. 251–289.

[4]Athans, M., "The Expert Team of Experts Approach to Command-and-Control (C2) Systems," *IEEE Control Systems Magazine,* Sept. 1982, pp. 30–38.

[5]Ho, Y.-C., "Team Decision Theory and Information Structures," *Proceedings of the IEEE,* Vol. 68, 1980, pp. 644–654.

[6]Mallubhatla, R., Pattipati, K. R., Tang, Z. B., and Kleinman, D. L., "A Model of Distributed Team Information Processing under Ambiguity," *EEE Transactions on Systems, Man, and Cybernetics,* Vol. 21, No. 4, 1991, pp. 713–725.

[7]McCain, D. R., "Stonewall Rode into History on Yankee Horse," *The Hartford Courant,* Aug. 24, 1988, p. B1.

[8]Moore, M., "Command 'collapse' left Stark vulnerable, Navy says," *Washington Post,* Oct. 16, 1987, p. A1.

[9]Wilson, G. C., "Courts-martial suggested for two Stark officers," *Washington Post,* June 27, 1987, p. A1.

[10]Wilson, G. C., and Moore, M., "One officer faulted in jet downing," *Washington Post,* Aug. 14, 1988, p. A1.

[11]Kleinman, D. L., Serfaty, D., and Luh, P. B., "A Research Paradigm for Multi-Human Decisionmaking," *Proceedings of the American Control Conference,* IEEE Press, New York, 1984, pp. 6–11.

[12]Kleinman, D. L., and Serfaty, D., "Team Performance Assessment in Distributed Decisionmaking," *Proceedings of the Symposium on Interactive Networked Simulation for Training,* Orlando, FL, 1989, pp. 22–27.

[13]Bushnell, L. G., "Team Information Processing," Master's Thesis, Department of Electrical and Systems Engineering, Univ. of Connecticut, Storrs, CT, 1987.

[14]Hogarth, R. M., "Beyond Discrete Biases: Functional and Dysfunctional Aspects of Judgmental Heuristics," *Psychological Bulletin,* Vol. 90, No. 2, 1981, pp. 197–217.

[15]Ansari, M., and Kapoor, A., "Organizational Context and Upward Influence Tactics," *Organizational Behavior and Human Decision Processes,* Vol. 40, No. 1, 1987, pp. 39–49.

[16]Yukl, G., and Falbe, C. M., "Influence Tactics and Objectives in Upward, Downward, and Lateral Influence Attempts," *Journal of Applied Psychology,* Vol. 2, 1990, pp. 132–140.

[17]Hill, G. W., "Group Versus Individual Performance: Are N+1 Heads Better Than One?" *Psychological Bulletin,* Vol. 91, No. 3, 1982, pp. 517–539.

[18]Hundal, P. S., "Knowledge of Performance as an Incentive in Repetitive Industrial Work," *Journal of Applied Psychology,* Vol. 53, No. 3, 1969, pp. 224–226.

[19]McGrath, J. E., *Groups: Interaction and Performance,* Prentice-Hall, Englewood Cliffs, NJ, 1984.

[20]Finch, F. E., "Collaborative Leadership in Work Settings," *Journal of Applied Behavioral Science,* Vol. 13, No. 3, 1977, pp. 292–302.

[21]Hunt, E., "A Cognitive Science and Psychometric Approach to Team Performance," Improving Team Performance, *Proceedings of the Rand Team Performance Workshop,* edited by S. E. Goldin and P. W. Thorndyke, Rand Corp., Santa Monica, CA, 1980, pp. 94–107.

Headquarters Effectiveness Assessment Tool

Robert Choisser*
Defense Information Systems Agency, Reston, Virginia 22090
and
John Shaw†
ALPHATECH, Inc., Burlington, Massachusetts 01803

Introduction

This chapter describes the Headquarters Effectiveness Assessment Tool (HEAT) and its underlying theory, along with its applications and associated research findings to date.

HEAT has been an ongoing program since 1982; it is the largest single project sponsored by the JDL/TPC3 Basic Research Group. The purpose of the HEAT program is to develop and apply a method to improve the effectiveness of C2 operations conducted by military headquarters.

We present in the following section of this chapter an overview of the HEAT program. In the subsequent three sections we discuss the theoretical underpinnings for the measurement tool, the measures embodied in the tool, and representative applications of the tool. We close this chapter with some thoughts on future research.

Overview of HEAT

HEAT was developed in response to a critical need; namely, the lack of objective methods for measuring the effectiveness of military headquarters. Traditionally, headquarters have been evaluated in the context of military exercises. Such evaluations, however, have tended to be subjective and nonreproducible, and have not focused on the major determinants of effectiveness. Since the actions of a military headquarters obviously exert strong influence on mission accomplishment, the value of an objective method for measuring headquarters effectiveness is self-evident. Accordingly, in 1982, the Defense Communications Agency initiated the HEAT program to develop such a method.

The problem addressed by the HEAT program is to answer the question, "What should be done to improve the effectiveness of a military headquarters?" The types of improvements that are candidates for consideration include changes directed toward the headquarters staff (staff size, staff roles and organization,

---

Copyright © 1993 by the American Institute of Aeronautics and Astronautics, Inc. All rights reserved.
*Group Leader, Large-Scale Systems Group.
†Special Assistant for Strategic Planning, Disjoint Interoperability and Engineering Organization.

training, procedures, etc.), as well as those directed toward the command center within which the commander and his staff operate (displays, automation, decision aids, communication links, etc.).

The approach to this problem taken by the HEAT program consists of three closely related parts: the theory, the tool, and the applications.

Theory plays an important dual role in the HEAT program. First, it aids the design of the measurement tool by identifying those attributes of a headquarters that are likely to be first-order determinants of effectiveness. Second, it suggests hypotheses for experiments and exercises. Since little C2 theory pertaining to headquarters effectiveness previously existed, the HEAT program developed a theory of headquarters effectiveness.

The HEAT measurement tool consists of a set of variables (or "measures") that collectively capture important determinants of effectiveness. HEAT also includes a rigorous set of methods and procedures for applying measures to experiments and exercises, and for analyzing the results. The HEAT measures fall into three broad categories:

1) *Process* measures that describe how command staffs seek and use information, arrive at decisions, and coordinate among themselves and with other commands;

2) *Performance* measures that describe how well internal headquarters processes are carried out in terms of accuracy, timeliness, consistency, and completeness; and

3) *Effectiveness* measures that gauge whether or not the headquarters accomplishes its mission.

HEAT applications result in the assignment of values to these measures. These applications consist of both experiments and exercises. Experiments involve the testing of one or more hypotheses using repeated trials of an automated wargame under controlled conditions. Hypotheses are often framed in terms of the operational benefits of some change in a headquarters, and are tested through statistical analysis of the experimental results. Exercises involve the resolution of issues critical to an operational command or defense agency. They cannot be replicated and their results are less generalizable. On the other hand, exercises provide richer operational contexts, and can thereby bring to light factors that need to be examined more carefully. HEAT applications also serve to validate both HEAT Theory and the HEAT measurement tool.

The desired end result of the HEAT program is a set of specific guidelines for the design of effective headquarters (in response to the basic question posed earlier). Over the long term, this means providing both the operational community and the command center design community with insights into the types of improvements that are likely to make the greatest contribution to effectiveness. Although preliminary findings along these lines are now available, full achievement of this objective requires the accumulation of data over a larger number of experiments and exercises. Toward this end, the results of all experiments and exercises have been archived in a relational database.[1] Over the short term, the results of exercises will continue to illustrate to commanders the relative strengths and weaknesses of their specific headquarters. The results of experiments will continue to help command center designers assess whether specific pieces of equipment are likely to offer significant operational benefits.

The HEAT program has undergone two distinct phases. Phase 1 extended from 1982 to 1986, with Defense Systems, Inc. as the contractor. This phase resulted in the initial version of the HEAT measurement tool, the basic theoretical foundation for the tool, and the application of the tool to seven exercises and three experiments. The present Phase 2 commenced in 1987, with ALPHATECH, Inc. as the prime contractor and Electrospace Systems, Inc. as the subcontractor. One objective of this phase is to improve the fidelity and accuracy of the HEAT measurement tool, and to extend it to both lower echelon headquarters and networks of headquarters. A complementary objective is to expand HEAT theory. Additional objectives are to apply HEAT to experiments and exercises, and provide specific guidelines for the design of more effective headquarters.

Two variants of HEAT, Phase 1 have been institutionalized within the armed services. The Army Research Institute has developed a version of HEAT called ACCES, which is tailored to the Army's needs and is now in use at Fort Leavenworth. In addition, the Navy's Second Fleet has developed a tailored and simplified version of HEAT and is routinely using it for the evaluation of its Battle Force In-Port Training (BFIT) exercises.

HEAT Theory

The premise behind HEAT is that effective headquarters differ from ineffective ones by the manner in which they approach planning, decisionmaking, and execution. HEAT Theory draws primarily on two schools of thought to identify the properties of effective planning. The first is *military theory*, which distills the lessons learned from military history into specific prescriptions for preparing and executing plans on the battlefield. These prescriptions appear in operational manuals and student texts used by military services. The second is *normative systems theory*, which illuminates those characteristics that distinguish systems that strive to optimize performance from those that do not. These two schools of thought agree on the properties that exemplify effective planning and decisionmaking.[2] This agreement is fortunate because it allows us to assume that military commands strive to be "optimal." One aim of headquarters assessment using HEAT is to assess whether obstacles prevent a headquarters from reaching the "ideal" suggested by normative theory. In this context, obstacles can include deficient communications and data processing support, inexperienced staff, and incoherent coordination within the headquarters.

Systems theory describes properties that distinguish optimal estimation processes and control strategies from suboptimal ones. We do not presume that command and control is simply an optimal control problem. We do assert that a rational C2 process will make the same provisions for coping with uncertainty found in optimal estimators. A rational C2 process will also make the same provisions for anticipating contingencies found in optimal control strategies. These concerns emerge from the general model of closed-loop optimal control illustrated in Fig. 1. The control system observes the system (within the "observer") and develops an understanding of the system and its dynamics (within the "estimator"). It translates that understanding into a course of action to move the system toward a desired end (within the "controller") and executes it (within the "actuator"). The reader will note that the "observer," "estimator," and "controller" are for us in the same headquarters, and often in the same mind. The "system" encloses those things that are not under the direct command and control of the control system. These include terrain, weather, and enemy actions.

Several relationships shown in Fig. 1 deserve further discussion. First, it is important to point out that the actions applied by the control system can directly affect observations in the future. Equally important is the implication that the observability of the system may depend on the state of the system itself.

The relationship between the observability of the system, the state of the system, and the controls applied to that system occupies a central place in

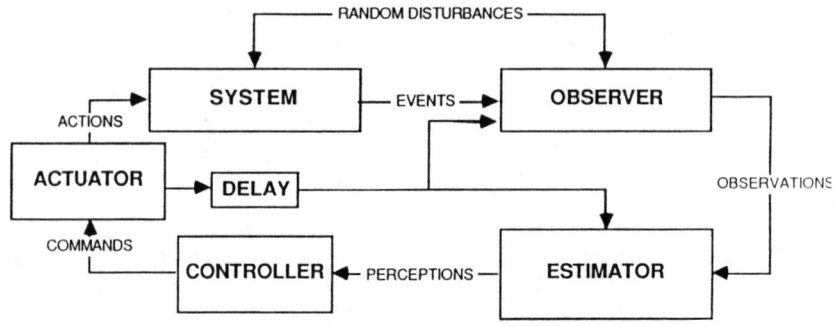

Fig. 1 Closed-loop feedback control system

contingency planning. Closed-loop control strategies *anticipate* that feedback will become available in the future. More important, control strategies of this sort anticipate that future information will allow the control system to *learn* more about the system than is understood at present. Learning does not merely report what information has been collected to date. Learning indicates whether or not events are unfolding according to expectations. Control strategies that anticipate learning recognize that they will continue to "calibrate" their models of the outside world. The aim of a good control action is not merely to place the system in a favorable state, but to also place it in a state that can be favorably observed. We will return to these ideas throughout this section.

HEAT Decisionmaking Paradigm

The most fundamental concept in HEAT theory is the HEAT decisionmaking paradigm, illustrated in Fig. 2. This paradigm asserts that the planning, decisionmaking, and execution activities conducted by a headquarters are analogous to a feedback control system that attempts to keep its external environment (e.g., a battlefield, a theater of operations, an airlift channel) within acceptable bounds despite the actions of the enemy. The control system analogy is widely accepted within the C2 community. Other paradigms exist that differ from the HEAT decisionmaking paradigm in appearance, but generally all share the analogy between a C2 system and a feedback control system. The HEAT decisionmaking paradigm expresses the control system in terms of six "process steps": 1) monitoring, 2) state estimation, 3) situation assessment, 4) option generation, 5) option selection, and 6) plan generation and direction. We describe these process steps in the following paragraphs.

Monitoring is responsible for obtaining data from the environment. Observations supporting this step may arise from *reactive sensing* (e.g., receiving status reports from subordinate headquarters), *proactive sensing*(e.g., executing a reconnaissance mission), or *probing* (e.g., observation obtained through the execution of a course of action that induces the enemy to reveal something about himself). Monitoring places no interpretation on what it observes but passes its observations to state estimation.

State estimation is responsible for estimating the values of physically realizable variables of interest to the headquarters. It combines current observations (from monitoring) with prior estimates to produce a current estimate of the variables of interest.

HEADQUARTERS EFFECTIVENESS ASSESSMENT TOOL

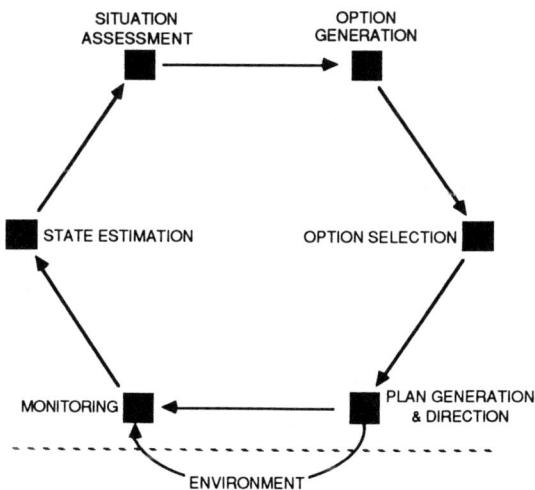

Fig. 2 HEAT C2 decisionmaking paradigm

Situation assessment is responsible for interpreting the meaning of the estimates prepared by the state estimation step. Alternatively, we can view this step as preparing estimates of parameters and variables that cannot be observed directly (e.g., enemy intent). It is at this step the headquarters compares present estimates with predictions made earlier to arrive at an overall assessment of the situation. These comparisons are made in the context of the current plan, and are used by the headquarters to establish whether or not current operations are proceeding according to plan. Situation assessment may also pass its hypotheses to state estimation for testing.

Option generation is responsible for generating feasible courses of action in response to the initiation of a new planning cycle.

Option selection is responsible for selecting the "best" course of action for implementation.

Plan generation and direction is responsible for generating verbal and/or written directions instructing component commands to execute the course of action selected in Option Selection.

The HEAT decisionmaking paradigm is simply a foundation for posing (and answering) questions about the factors that cause a headquarters to be effective or ineffective. No headquarters is organized in the manner implied by Fig. 2, and the six activities illustrated transcend traditional organization boundaries. Nevertheless, the model does permit us to distinguish the situation where a headquarters has an imperfect understanding of the situation from the situation where the headquarters does not provide for contingencies in its operational plans. Moreover, since an organization will "understand" a specific situation only as well as it monitors, estimates, and interprets the situation, the delineation of functions within the model permits us to distinguish the situation where a headquarters fails to monitor its environment from the situation where it lacks the experience to properly interpret the information conveyed on its maps and status boards.

The six blocks in Fig. 2 no more describe the workings of an effective headquarters than a collection of musical instruments describe an orchestra. To attempt an understanding of the workings of effective headquarters, we must peer behind the drawing and inspect the ways that effective headquarters orchestrate these elements of the HEAT decisionmaking paradigm. HEAT theory attempts precisely this goal, and its development continually strives to explain why some headquarters create symphonic movements while others create cacophony. Space does not permit us to give equal attention to all parts of the HEAT theory, and we choose here to present its major theme: the arrangement of military theory and normative systems theory into a "rationalist" view of effective C2. HEAT theory also deals with the cognitive limitations that prevent people from acting in a perfectly rational fashion;[3,4] the essential issue of time;[3] and the critical role of leaders in making military organizations resistant to stress.[2,3]

Developing an Understanding of the Situation and Its Dynamics

"If you 1) don't know the situation, and 2) understand its dynamics, anything you do will be right only by accident"
LTGen John H. Cushman

Monitoring the Situation

The fog of war presents every commander with imperfect data. Three critical sources of uncertainty or ambiguity arise in practice. First, observation errors introduce uncertainties in the observations of targets or variables of interest. These errors may arise due to environmental conditions (e.g., battlefield obscurants), faults in the collection equipment, or limitations in the observation procedures. Second, incomplete observations introduce uncertainty in the normal sense by denying information. A more insidious problem occurs if the observation process draws *inferences* from the data it does collect and substitutes those inferences for missing data. The double jeopardy is that recipients of the observation reports not only remain ignorant of omissions in the data, but inadvertently "double count" those data that have been collected. Finally, reporting delays introduce uncertainty by causing reported information to be out of date.

An optimal observation activity will recognize these limitations and account for them. The accounting can be done at the source in the preparation of observation reports. It can also be done at the destination in the interpretation of the observation reports.

Developing Estimates from Observations

Data normally do not reveal the "state" of the system directly. An estimate of the state is needed nevertheless before one can interpret what it means for the system to be in that state. An intermediate step between observation and interpretation, then, is *estimation*. Functionally, the estimation process (as we use the term here) seeks to answer "what is?", while situation assessment seeks to answer "what does it mean?" In practice, the products of the estimation process include map and status board updates and object tracks.

The objective of the estimator is to construct a conditional estimate of the state of the system; the estimate is conditional because it depends on the information available to the estimator. While it is true the system can be in only one state at a time, the best the estimator can do is postulate several hypotheses about the current state, and maintain for each possibility the likelihood it accurately describes the true state of the system.

In the absence of any new measurements the headquarters must use *prior* information to infer the current values of critical state variables. In the absence of any prior information on critical state variables the headquarters must rely on new measurements. When both prior information and new (but imperfect or unreliable) measurements are available, however, the headquarters can develop provably better estimates (say, in a least-squares error sense) by fusing both sources of data. An optimal estimator will consider not only the reliability of the new measurements, but the reliability of estimates that can be constructed using the prior information. The reliability of the latter depends not only on the time that has elapsed since that prior information was collected, but on the rate of change, or *dynamics*, of the battlefield. In addition, an optimal estimator will explicitly consider whether external observers are reporting events independently, or whether some observers are merely repeating observations reported by other sources.

Interpreting Data

No estimate derived from data obtained in the battlefield is complete without an interpretation about its meaning. Moreover, an interpretation is precisely that: assumptions and presumptions about the meaning of the information available. Whereas later events may indicate whether or not a particular interpretation is true, this is never known to the headquarters at the time the interpretation is made. A command can, however, anticipate that future events (including events brought about by its own courses of action) will ultimately confirm or refute an interpretation, and it can use that knowledge that learning will occur to develop a proactive strategy in response.

Although there is no single objective for trying to establish the meaning of data, U.S. military manuals that address this subject converge on three critical questions that a headquarters should try to answer:

1) What does this information tell me about the enemy's capability and intent?

2) Given it is true, what impact will it have on my current or future operations and plans?

3) Is it probable the enemy is aware that this information has been obtained by us and, if so, is he likely to make changes in his plans?

Operational manuals recommend using a dialectic method for correlating new information with currently held hypotheses about the situation. In this approach, new information serves not merely to support new hypotheses, but is the basis for confirming or refuting earlier hypotheses. These manuals also advise the reader that he will have ample opportunity to test whether he understands the dynamics of the situation as his hypotheses are accepted and rejected. A headquarters that discovers time and again that new data do not confirm its hypothesis about the enemy should recognize that it does not understand the dynamics of the battle. Maintaining a history of correct and incorrect predictions also helps a headquarters assign confidence levels to its hypotheses.

The U.S. Army's Intelligence Preparation of the Battlefield (IPB) process formally addresses these concepts using "templating." The development of the IPB involves the construction of three templates:

1) Doctrinal Template: a "model of how the enemy might look, according to doctrine and training, if he were not restrained by the weather, terrain, and combat losses."

2) Situation Template: a graphic portrayal that "shows how threat forces might deviate from doctrinal dispositions, frontages, depths, and echelon spacing to account for the effects of the terrain, weather, and combat losses."

3) Event Template: a graphic portrayal that "identifies and analyzes significant battlefield events and enemy activities which provide indicators of the enemy course of action."

Army doctrine also stresses the use of *named areas of interest* (NAIs) as part of the dialectic method underlying the IPB. To quote from page 4–1 of Ref. 5:

"The area of interest is a notional area developed jointly by the G3 and G2 and based on the commander's guidance. The area is dynamic in nature, measured in four dimensions: depth, width, height, and time. It is not depicted on a map. It includes all factors capable of affecting the unit's operations in the near future. It is at least as large as the area of operations and normally considerably larger. It is designed to focus information collection so the command can 'read the battlefield' and make proactive decisions."

This document goes on to say that a NAI is a "point or area where enemy activity or lack of activity will confirm or deny a particular enemy course of action."

Normative theory does not instruct us how to interpret data, but does instruct us how to maintain multiple hypotheses. The central idea is to generate a set of hypotheses, propagate those hypotheses forward to a time when new measurements are received, correlate the measurements to the hypotheses, incorporate the correlated measurements into each hypothesis, and then compute a conditional likelihood for each hypothesis based on the new measurements.

The computations used to compute conditional likelihoods reveal several important properties of multiple-hypothesis estimators, and are easily illustrated via Bayes method, the classic method for computing conditional likelihoods. Given I_k, the information that is known up to time k, the conditional likelihood the system is in state i, $P[i_k / I_k]$, is computed by:

$$P[I_k \mid i_k] = \frac{P[i_k] \, P[I_k \mid i_k]}{\sum_{j \in S_k} P[j_k] \, P[I_k \mid j_k]} \tag{1}$$

where S_k denotes the number of reachable states at time k. The term $P[i_k]$ denotes the prior likelihood the system could be in state i_k, and the term $P[I_k / i_k]$ denotes the conditional probability that the information received up to time k, I_k, would have been observed if the system were in state i_k. This last term plays a central role in the overall military situation assessment process, namely that a hypothesis (e.g., "we think the system is in state i_k") can be used to predict a range of observations the headquarters should be able to observe in the future. (More precisely, the hypothesis can be used to construct a *conditional* probability distribution for future observations.) This is the fundamental idea behind the concept of event templating described earlier. Bayes rule effectively instructs the estimator to update posterior likelihoods for each hypothesis by correlating the actual observations with the observations that would be predicted under each hypothesis, and moderating those correlations by the prior likelihoods.

Developing Courses of Action

"True wisdom is shown by those who make careful use of their advantages in the knowledge that things will change"
Thucydides

U.S. military doctrine stresses mission analysis as the critical starting point for developing courses of action. Time is a foremost consideration in developing orders and plans. Experience and doctrine alike stress the obligation that higher headquarters have to subordinates to provide the commander's intent as early as possible to permit the latter to begin their mission analysis. The fulfillment of this obligation actually serves two important purposes. First, the commander's intent may be based on an understanding of the situation that is no longer correct (or not as correct as the understanding held by its subordinates), and early communication of his intent provides one way for the commander to check his assumptions. Indeed, a statement of intent that is vague and incomplete is very likely to be an indication that the commander has a flawed understanding of the situation.[6] Second, it permits the subordinate headquarters to step up their own staff coordination and information gathering activities (e.g., reconnaissance, spot reports from units at echelons below them) in anticipation of the start of their own mission analysis effort. To cite page 13 of Ref. 6, the National Training Center has driven these points home in the Army's quest to hone its Airland Battle doctrine:

"One very strong principle that has emerged in our doctrine, as part of the Airland Battle, has been the emphasis on understanding intent... The brigade concept of operations has been fuzzy on many occasions. They have good sets of graphics; they have a written order coming down from brigade. However, by the time somebody enunciates intent, the task force commander is well into his planning. We cannot have that. Another thing that has emerged is the importance of a warning order."

Time is an important consideration in the mission analysis phase of developing courses of action. As the controlling agent of a command and control system, a headquarters is in a race against time, and must ponder not only the time required by its forces to execute a course of action, but the time required by the enemy to respond in kind. One fundamental precept in mission analysis is being able to act within the "control cycle" of the enemy; the control cycle being the time it takes a Command and Control system to understand the situation and act on that understanding. This concept is stressed by US military doctrine, and is exemplified by the counsel offered on page 3-3 of Ref. 7:

"The [command] estimate process is continuous. The staff must be aggressive in anticipating the actions of the enemy. Only in this way can U.S. Forces seize the initiative by acting inside the decision cycle of the enemy. If the friendly decision cycle does not begin until triggered by action of the enemy, the enemy decision cycle is complete and friendly forces are reactive."

and by the principle articulated on page 2-3 of Ref. 5:

"An essential portion of Mission Analysis is the analysis of time, which identifies the amount of time available and begins the process of allocating that time to the various portions of the (command) estimate process."

A second fundamental precept in mission analysis is "reverse planning" wherein the command staff mentally works backward from the desired goal to the present state in order to comprehend the amount of time and the nature of force synchronization that will be required to fulfill the commander's intent.

U.S. Army doctrine stresses the completion of five steps in developing courses of action: 1. analyze relative combat power, 2. array initial forces (e.g., determine force ratios), 3. develop schemes of maneuver, 4. determine command and control means and maneuver control measures, and 5. develop course of action statements and course of action sketches. The objective in developing the array of initial forces is not to assign missions to units, but to comprehend what forces can be allocated to accomplish the mission. The objective in developing the scheme of maneuver is to establish *how* a course of action will be executed. It provides a means for employing forces and is the basis for analysis; it is here that the planner establishes how maneuver forces and fires will be incorporated into the scheme of maneuver. The purpose of the fourth step is to assure that control can be maintained during an operation, and that the span of control (especially the ability to coordinate the actions of subordinate forces) is not exceeded. Finally, the purpose of the last step is to visualize (war game) the battle and assess the results. The key here is to be able to identify how the immediate subordinate commander can employ his forces and how he can determine the composition of those forces. The war game should also identify critical decision points, named areas of interest (NAIs) and targeted areas of interest (TAIs), and make it clear to the command how it will synchronize forces.

Operational doctrine also considers the need both to determine when and how enemy responses to a course of action might be observed, and to anticipate future decision events in support of the execution of a plan. The U.S. Army stresses the use of event templates, described above, for the former, and the use of decision templates for the latter. The decision template is a graphic portrayal that "relates the details of event templates to decision points that are of significance to the commander. It does not dictate decisions to the commander, but rather identifies critical events and threat activities relative to time and location which may require tactical decisions" (page A–35 of Ref. 5).

In optimal control theory, the objective of the controller is to determine control activities that will induce the evolution of the system towards an acceptable goal. We use the term 'induce' here since there is no assurance the control can be *absolute* when the system cannot be controlled with precision. Stochastic optimal control theory yields a diverse collection of strategies for planning in the face of uncertainty. The diversity is achieved through the characteristics of the system that are incorporated into the control strategy, most notably 1) the treatment of uncertainty now and in the future; and 2) the treatment of feedback now and in the future.

Control theory distinguishes three broad types of control strategies[8]: 1) open-loop control; 2) open-loop feedback control; 3) and closed-loop feedback optimal control. Open-loop control (OLC) establishes a sequence of control actions, but does not use feedback to trigger compensatory actions should random events drive the system toward undesirable states. Open-loop feedback control (OLFC) uses feedback to develop compensatory actions, but simplifies the effort needed to select an action by assuming that action will be executed in an open-loop fashion. This type of strategy therefore takes advantage of the learning provided by past feedback, but cannot take advantage of the fact that learning will continue to occur. Finally, closed-loop feedback optimal control (CLFOC) anticipates that feedback will be available in the future. Here, the controller anticipates the range of possible future measurements, and develops a strategy wherein the specific control action taken in each future period will be determined by the measurements known at that time.

The distinction between OLC, OLFC, and CLFOC strategies is best illustrated by a simple example. Let us imagine a situation where a headquarters has reason to believe that an enemy company is making its way to a pick-up point on the river, and is deciding on courses of action to ambush that company along its expected route. Now imagine three different exchanges occur between the commander and his staff:

Exchange 1: Commander: (pointing to a map) "Let's set up an ambush team at this location."

Exchange 2: Commander (pointing to a map): "Let's set up an ambush team at this location. G3: "What if the team reports the company hasn't appeared by 1600 hours?" Commander: "Let's cross that bridge when we get to it."

Exchange 3: Commander (pointing to a map): "When do we anticipate the company will arrive at this location?" G2: "Approximately 1600 hours." Commander: "What can we conclude if our team doesn't report siting them by, say, 1630 hours?" G2 (pointing to another location on the map): "It could mean the company has taken this route, and will be picked up on the river at this location." Commander (pointing to the alternate route): "We can hit them with a JAAT (Joint Air Attack Team) along this route, right?" (Heads nod) "Okay, I want us to send an ambush team here along the expected route, have them report their status no later than 1630, and I want us to be prepared to send out a JAAT to the alternate route if we receive a negative confirmation."

The first exchange illustrates an OLC strategy in operation: a course of action is set with no alternatives contingent on future measurements. The second exchange illustrates an OLFC strategy in operation, but in a subtle form: even though the G3 has anticipated that new measurements will be available in the future, the ambush plan is made without contingent plans developed in anticipation of those measurements. The third exchange illustrates a CLFOC strategy in action: the Commander has anticipated that measurements will be made in the future, with the help of his staff has concluded the types of inferences that can be made when those measurements become available, and has incorporated a contingency action into his primary ambush plan.

The value of being able to anticipate that learning will occur is revealed in an important relationship between the three classes of control strategies discussed. Let us assume (with no loss of generality) that we seek a control strategy to minimize some cost. Let us denote by J_{CLFOC}, J_{OLFC}, and J_{OLC} the expected cost of executing a strategy prepared using CLFOC, OLFC, and OLC, respectively. The following relation holds:

$$J_{CLFOC} \leq J_{OLFC} \leq J_{OLC} \qquad (2)$$

The difference $J_{OLC} - J_{OLFC}$ represents the value of *receiving* feedback from the system. However, the difference $J_{OLFC} - J_{CLFOC}$ represents the value of *anticipating* that feedback (i.e., learning) will be provided in the future.

Notice at this point how the military and systems theories underlying HEAT blend together. They both converge to the *closed-loop* feedback control strategy as the standard for planning and decisionmaking.

Additional Elements of HEAT Theory

The analogy between effective command and control and optimal feedback control can only go so far. It works best when it prescribes how a headquarters should *manage* its operations. It is least satisfactory when used to explain how effective headquarters align their people - and subordinate commands - toward a common vision of the future and inspire them to work toward that future. This shortcoming becomes striking when we try to explain how headquarters continue to stay effective in the presence of the frictions and extreme stresses that descend upon them.

This has led us to open up a new front in the development of HEAT theory: the behavior of organizations under stress and the role of leadership in keeping the headquarters effective in the presence of stress. Our findings to date are conceptual for the most part, but are rooted in observations we have made of organizations operating under stress. The key postulate that has emerged from our work is that *leadership*, not merely *management*, allows an organization to cope with stress. We direct the reader to Shaw et al.,[3] and to Shaw and Athans[2] for a complete treatment of our developments to date on this subject.

HEAT theory also embraces queuing theory to describe the origins of time delays and resource contention, and the effects of process/task synchronization points, e.g., the need to coordinate course of action development among subordinate and lateral commands, within a headquarters. HEAT expresses queuing models using the Petri net paradigm, and supports the representation of Petri nets directly within the relational database that archives experiment and exercise results.[1] We direct the reader to Shaw and Powell[9] for a complete treatment of the Petri net paradigm within HEAT theory.

In addition, HEAT theory draws on cognitive science for an understanding of the cognitive biases that prevent people from performing optimally in the control theoretic sense. It also borrows from organizational theory for an understanding of the behavior of organizations.

HEAT Measures

HEAT measures provide answers to the following questions:

1) How well did the headquarters understand the situation and its dynamics, and how did it develop that understanding?

2) How well did the headquarters comprehend its mission, objectives, and the consequences of its plans and orders to accomplish its objectives, and how did it arrive at that comprehension?

3) How well did headquarters control and supervise execution of its orders?

4) How did the headquarters determine when to issue new plans and orders?

5) How well was the headquarters staffed and equipped to perform its functions?

6) Were operations hindered by problems in the availability of support equipment?

A multiplicity of measures is required to address each of these questions in order to pinpoint the specific strengths and weaknesses of a headquarters.

The HEAT measures were designed to be, insofar as possible, exhaustive and mutually exclusive; i.e., to collectively reflect all significant facets of performance and effectiveness, without overlap between them. The complete set of HEAT measures is necessarily large, due to the dimensionality of the problem. However, we view this set as a "menu" from which to select those individual measures that are appropriate to the application at hand; we have not yet experienced an application that required more than a small fraction of the totality. The value of having a common set of measures from which to choose is that each facet of performance and effectiveness is represented by a unique variable and, hence, comparability of results across different applications is enhanced.

During HEAT Phase 1, attempts were made to aggregate the various performance measurements into an overall figure of merit by scoring all measures on a utility scale, where "0" represents the minimum acceptable level of performance and "100" represents the maximum useful level. This approach was abandoned because of difficulties in establishing both the end points for the utility scale and the weighting factors to express the relative importance of the measures. Instead, headquarters performance is now assessed on a measure by measure basis.

The remainder of this section discusses, first, the HEAT process and performance measures, and then the effectiveness measures.

Process and Performance Measures

HEAT uses four types of measures in the process and performance categories[9]: 1)*Characteristic* measures, 2)*Coordination* measures, 3)*Queuing* measures, and 4)*Quality* measures.

The first two types can be characterized as process measures, while the latter two are performance measures.

HEAT presumes that effective headquarters differ from ineffective ones in the manner they approach planning, decisionmaking, and execution. Specifically, the agreement between the military theory and systems theory leads us to conclude that effective commands strive to be "optimal". The characteristic measures indicate whether a headquarters endeavors to be optimal, and permit the analyst to identify specifically how a headquarters differs from an "ideal" planning and decisionmaking operation. Characteristic measures indicate how a headquarters develops an understanding of the situation and its dynamics. They also indicate how a headquarters anticipates the future, including contingency actions and information needs.

HEAT defines unique characteristic measures for the six activities in the decisionmaking cycle illustrated in Fig. 2. Characteristic measures associated with one activity tend to complement characteristic measures associated with other activities. Estimation, for example, has a characteristic measure to record whether a headquarters recognizes it has an imperfect understanding of the system and its dynamics when it prepares estimates. An *optimal* control process can respond to this predicament by anticipating that future measurements of the system will permit it both to test how well it understands the dynamics, and to "tune" its model of the system to better match observed behavior. Thus, situation assessment complements this estimation measure with a measure of its own to record whether the headquarters anticipates future times or events for reassessing the situation. Option generation completes the cycle with two characteristic measures of its own. The first records whether the headquarters integrates intelligence collection with plan execution. The second records whether the headquarters evaluates its options by anticipating whether, how, and when it can expect to observe the response of the opposing forces to those options. Considered individually, each measure indicates how a headquarters copes with uncertainty. Used in concert, the four measures reveal how a headquarters reduces its uncertainty by anticipating its information needs and collecting that information as an integral part of its operations.

Queuing measures indicate the origins of time delays, critical processing paths and resource contention for processing that occurs within a headquarters. Queuing measures also shed light on the nature of task synchronization within a headquarters and across echelons of command. Whereas the characteristic measures are defined separately for each HEAT activity, the queuing measures apply to all activities. The queuing measures include the "obvious" measures one expects for a queuing system, including the elapsed time that a request for processing waits for processing to begin ("time in queue") and the elapsed time that a request spends in processing ("time in service"). The queuing measures also include resource utilization measures (including "resource availability" and "resource contention") to help indicate why time delays occur.

Coordination measures capture the degree of coherence in information, goals, and perceptions between cells within a headquarters and between multiple headquarters. We use the term information in the most general sense to include individual data items, assessments of the situation, mental and/or mathematical models used to describe the evolution of the system, or decisions. We use the term "coherence" to cover two separate considerations: 1) the relative amount of information that is common to two or more decisionmaking bodies (multiple headquarters, or cells and sections within a single headquarters), and 2) the consistency of the information between those bodies. Many of the coordination measures are simply counts on the number of requests for information that are made because some decisionmaking body lacks information that is available to another. Other coordination measures account for the proportion of times that information either gets to the bodies who are supposed to receive it, or coincides between two or more bodies.

Quality measures indicate whether performance approaches or deviates from a norm. We apply some caution here since we reject the proposition that a HEAT analyst can play "Monday morning quarterback" and judge whether a headquarters made the "right" decisions. Instead, we support the proposition that a HEAT analyst should try to determine whether a headquarters' understanding of the situation matched the truth, and whether its decisions led to the desired results.

The quality measures for the monitoring, estimation, and situation assessment activities account mostly for *accuracy*, be it in the information the headquarters receives (monitoring), the estimates it prepares (estimation), or the interpretations that are developed (situation assessment). Accuracy can be established unambiguously for monitoring and estimation by comparing what the headquarters perceives to the actual situation ("ground truth"). However, it is often much harder to quantify the accuracy behind interpretations that a headquarters develops as a part of situation assessment. Interpretations are inferences drawn from observations and estimations, and are not facts in and of themselves. Indeed, they often have no observable counterparts for measuring accuracy. Ponder, for example, how one might gauge the accuracy of the statement "current operations are in jeopardy because the enemy has the capability to use chemical weapons." The proposition cannot be tested for its veracity if one subsequently does not observe the enemy using chemical weapons. Interpretations must have their basis in fact, and HEAT can indeed illustrate to the commander whether his facts were accurate. The astute HEAT analyst does not try to extend the analysis of interpretations beyond that point.

The quality measures for option generation account largely for how well the headquarters satisfied constraints and foresaw the consequences of its decisions. The quality measures for planning and direction account mostly for the elapsed time that is required before a plan is prepared and distributed to subordinates.

Effectiveness Measures

Effectiveness is the "bottom line" payoff for good headquarters performance. HEAT equates the effectiveness of a headquarters to its *contribution* toward mission accomplishment. Accomplishment of a specific mission may not be possible if, for instance, the friendly forces are badly outnumbered. The effectiveness of a headquarters in this situation might still be assessed as "high" if it is responsible for seizing an orderly retreat from what otherwise might have been a rout.

We cannot, of course, measure directly the *contribution* of a headquarters to mission accomplishment. We can determine whether missions are accomplished, however, and can use controlled experiments, with a sufficient number of trials, to isolate the impact of a particular headquarters variable on mission accomplishment.

Mission accomplishment measures are obviously mission-dependent. For example, "FEBA movement" may be a good measure for a maneuver mission, whereas "number of targets destroyed" may be suitable for an air strike mission. For this reason, it would be impractical to try to develop a general list of all potential mission accomplishment measures. Instead, HEAT derives from the mission of the headquarters under study those measures that appear to best capture convergence to or divergence from mission accomplishment.

In addition to mission accomplishment measures, HEAT also uses two measures of overall plan quality as *indicants* of headquarters effectiveness. The first of these is termed "plan lifetime." This is the ratio of the length of time that a plan remains viable to the length of time that it was intended to remain in effect. The presumption here is that an effective headquarters will consider the many frictions of war that will work against its plan (including "first encounter with the enemy"), and will anticipate the decisive points when it will need to be changed. This measure does not imply that headquarters should strive for long lifetimes in their plans, but does imply that commanders and their staffs should plan their operations on a timescale that it consistent with their knowledge of the situation and its dynamics. When a plan is no longer viable (perhaps because of a faulty initial perception of enemy intent), this is known as "major incongruence" (between the actual battlefield situation and that on which the plan was based). By contrast, the second indicant of headquarters effectiveness measures the number of occurrences of "minor incongruence" during the interval in which the plan remains in effect. Minor incongruence is defined as an incongruence that can be quickly rectified without discarding the plan, by means of substituting new plan elements that were developed in advance in anticipation of the contingency that has occurred on the battlefield.

HEAT Applications

Applications are the mainstay of the HEAT program. They make possible the assignment of numbers to measures and, hence, the quantification of the research findings.

HEAT applications consist of both experiments and exercises. Experiments offer several advantages, including the ability to control all variables other than the independent variables of interest and the ability to perform statistical analyses through multiple trials. On the other hand, experiments necessarily lack realism and richness of context, and, thus, we cannot be sure that experiment findings will be fully applicable to real operations. Herein lies the critical value of an exercise. An exercise normally contains a high degree of realism, but, because of the multiplicity of uncontrolled variables, one cannot isolate cause-effect relationships. Thus, experiments and exercises are complementary, and each plays an important role in the HEAT program. Exercises can validate experimental findings and suggest independent variables for future experiments that otherwise might not have been considered. In all cases, the immediately useful product of a HEAT exercise is the insight that can be provided to the sponsoring command.

One of the most important aspects of any HEAT application, whether it is an experiment or an exercise, is observation. Every HEAT observer is given at least two days of specialized training, which includes fundamental HEAT concepts as well as the routine procedures of data collection. It is important that a HEAT observer keep the basic HEAT control theoretic paradigm firmly in mind in order to understand when a comment or action that does not appear on the data sheet (e.g., the time at which a commander voices the realization that what he thought was the enemy's main thrust was really a feint) is worth capturing. Questions such as where the observers should be positioned and what, specifically, they should measure are resolved in advance of an application.

The results of all HEAT phase 2 experiments and exercises have been recorded in a relational data base in order to foster comparability of results and the integration of research findings across the various HEAT applications.

The following two subsections provide more detailed discussions of experiments and exercises, respectively. Each subsection begins with a general description of the major elements and considerations common to all applications in that category. This is followed by a table summarizing each application that had been conducted at the time of this writing. The subsection *Experiments* provides a narrative description of a particular experiment to illustrate how a typical experiment is conducted. This subsection also provides a list of key findings, referenced to the experiment(s) from which each finding was derived. The security classification of exercises precludes publication of their findings herein. Instead, the subsection *Exercises* offers general "lessons learned" in exercises to date.

Experiments

A HEAT experiment consists of replications of a computerized wargame under controlled conditions. Each experiment involves one or more independent variables, such as the presence or absence of a decision aid, that are manipulated systematically among the experimental trials. Hypotheses are formulated concerning the effects of these variations on headquarters effectiveness, and are tested through statistical analysis of the experimental results.

Multiple teams of players are used in order to minimize learning effects and balance the variability in competence and style among the individual players. A team is normally subdivided into several command "nodes," which represent either different decisionmaking elements within a single headquarters or different headquarters that interact with each other.

Scenarios prescribe the operational setting, but the play is free and unscripted. The play is two-sided in the sense that the RED player (a member of the control team) is allowed to react intelligently to BLUE's actions.

An essential element of every HEAT experiment is the "simulator." This is a computer program that drives the map displays by performing function such as advancing moving objects at the proper speed and heading, implementing commands from the players, computing kills and accounting for terrain masking. A wide variety of simulators is available throughout the wargaming community. The differences between them are mainly in the type of warfare addressed, the echelons considered, and the level of detail incorporated. The HEAT program has used four different simulators to date.

A written test plan is developed for every experiment prior to the first experimental trial. It documents all aspects of the experimental conditions, including the placement of observers. The final report on the experiment covers additional topics such as the data analysis methodology, results and conclusions.

Table 1 summarizes the settings for the eight experiments conducted to date.

Illustrative Experiment: Investigating the Value of Shared Battlefield Graphics for Crisis Management (Experiment 8)

The objective of this experiment was to critically assess the proposition that shared battle graphics will make possible better crisis resolution. Experiences in

Table 1 Completed HEAT Experiments

No.	Site	Date	Simulator	Experiment objective
1	Naval Postgraduate School	1983	IBGTT	Determine effect of HQ connectivity on effectiveness
2	Naval Postgraduate School	1984	IBGTT	Determine impact of centrality on force effectiveness
3	Naval Postgraduate School	1985	IBGTT	Determine effect of role specialization on force effectiveness
4	Naval Postgraduate School	1987	JTLS	Determine whether contingency planning leads to better combat force effectiveness
5	Naval Postgraduate School	1988	JANUS	Determine whether contingency planning leads to better combat force effectiveness under stress
6	Naval Postgraduate School	1988	None	Determine whether cognitive biases can make tactical situation assessments sensitive to the order of presentation of information
7	Naval Ocean Systems Center	1989	RESA	Determine whether a prototype air strike planning aid helps experienced strike planners prepare more effective air strike plans
8	Naval Ocean Systems Center	1990	RESA	Determine whether shared battlefield graphics promotes better resolution of a crisis

the Persian Gulf with the Joint Operations Tactical System (JOTS), for example, had demonstrated that critical information concerning a crisis could be made available to the NMCC in near-real-time, thereby making crisis resolution across several echelons of command an achievable aim. While these specific experiences were successful, they did not offer proof per se that shared battle graphics (or any equivalent technology) made the difference, since there were no *comparable* crises that had been handled without shared battle graphics that could be used as a basis of comparison.

Nine retired and active-duty Navy commanders participated as subjects in the experiment. These subjects were organized into three teams of three decisionmakers each. In each team, two of the subjects executed the responsibilities of ship captains at the scene of a developing crisis, and the third member executed the responsibilities of a commander at the National Military Command Center.

Each team attempted to resolve two crisis situations, each with and without shared battle graphics, making a total of four scenarios presented to each team. Two of these scenarios were relatively difficult to solve, as they involved aggressors from countries not currently hostile to the United States, while the remaining two were relatively easy for the opposing reason. Each scenario opened with the team conducting an ongoing mission, followed shortly thereafter with a series of critical events serving to trigger a crisis. The crisis had to be resolved within the context of the on-going mission, and each team was obliged to decide how best to respond to the situation.

The teams were marginally better able to cope with the ongoing mission and with the crisis that emerged in that mission when they used shared graphics. The ship captains were able to undertake a significantly larger number of actions to attend to the ongoing mission and to resolve the crisis when shared graphics was available. Equally important, the effectiveness of the ship captains in this respect was identical in the easy and difficult scenarios when shared graphics was used, but their effectiveness declined significantly when going from the easy to the difficult condition when shared graphics was not used. We interpret this finding to indicate that team effectiveness in this regard was *robust* to the nature of difficulty of the scenario when shared graphics was used.

The teams were marginally better able to comprehend the situation when shared graphics was available. The teams were aware of a significantly larger number of the hostile and friendly platforms that would have bearing on their immediate and *future* mission when shared graphics was used. Curiously, the teams did not identify the aggressor(s) responsible for setting off the crisis better or faster when shared graphics was used. This finding suggested that shared battle graphics permitted the teams to develop a more comprehensive (but not necessarily more timely or more accurate) understanding of the situation and its dynamics.

On a final note, subjective evaluations provided by the subjects in a series of questionnaires administered immediately following each session and following all sessions were strongly in favor of the shared graphics. Subjects felt that their workload and their perceived understanding of the situation were improved when they used shared graphics.

Key Findings from Experiments

The key findings from the HEAT experimental program to date are outlined below. The numbers shown in parentheses refer to the numbers of the experiments that confirmed the underlying hypotheses.

1) In a network of headquarters, a "geographic organization" (specialized geographical sectors of responsibility for each headquarters) is superior to a "functional organization" (specialized functions for each headquarters) for discrete problems. (1,2)

2) The geographic organization is less vulnerable than the functional to communications disturbance. (1,2)

3) The geographic organization is more vulnerable than the functional to increased problem complexity. (1,2)

4) The functional organization is superior to the geographic for complex problems if no disturbance occurs, but the geographic organization's resistance to communications disturbance offsets the initial functional superiority. (1,2)

5) Changes in the functions performed by a headquarters staff have greater impact on effectiveness than changes in either structure or capacity. (3)

6) An operations plan with multiple options will be more effective under conditions of uncertainty than will a plan with a single option. (4,5)

7) A strong recency bias (a tendency to place too little emphasis on prior information and too much on new information) was exhibited by subjects performing situation assessment in a tactical setting. (6)

8) Naval air strike plans made with the N-KRS decision aid are more accurate and can be prepared more quickly than similar plans made without it; however, this advantage is offset by a tendency for planners using the aid to neglect option

exploration and to predict strike outcomes with less accuracy (a possible consequence of their being less involved with the details of the planning process). (7)

9) In a crisis situation, sharing battlefield graphics among echelons of command, i.e., providing the "big picture" to lower echelons and the detailed battlefield picture to higher echelons, increases crisis management effectiveness. (8)

Exercises

HEAT is most relevant to the operational community when it is used in support of exercises. The products of a HEAT application in an exercise are usually analysis reports and briefings prepared for the sponsoring command that identify what needs fixing, what does not, and what needs fixing now. This analysis is normally prepared from systematic reconstructions of the "command estimation" process within the headquarters that led up to and followed critical decisions made during the exercise. The objective of these reconstructions is to distinguish factors common to command decisions that were ultimately successful from factors common to decisions that were not. A reconstruction will normally concentrate on the "command estimation process," and can cover activities spanning several hours or several days. A good reconstruction draws on HEAT theory to identify those factors in the command estimation process that most strongly determined mission effectiveness during the exercise.

HEAT phase 1 did not stress diagnosis through reconstruction of the command estimation process, but instead stressed evaluation using performance and effectiveness "scores." We found the scoring approach to be an abstract nicety that did not translate well into the real world. At best the scores merely indicated how close a command approached a measure of "goodness," but otherwise did not reveal *what* the command did (or did not do) to achieve that score. The alignment of HEAT phase 2 toward reconstruction of the command estimation process made the tool better suited to help commands analyze exercises to uncover deficiencies affecting mission accomplishment. Particular emphasis is placed on reconstructing the steps the command adopts to cope with uncertainty, anticipate contingencies, synchronize operations, and monitor plan execution.

The use of HEAT in support of an exercise involves four major activities: 1) Preparation, 2) Data Collection, 3) Data Analysis, and 4) Reporting.

The two most important objectives in *preparation* are to establish specific issues for analysis, and to identify the types of observations the HEAT analyst will require to support those analyses. Here, the HEAT analyst must *anticipate* the types of observations he will want the HEAT observers to record during the exercise, and *translate* those collection requirements into "observation cues" the observers will be trained to recognize during observer training.

Figure 3 illustrates the principal steps that are covered in a good HEAT preparation.

The two most important objectives during *data collection* are to assure that the data is accurate and factual, and to assure that the data continues to be relevant to the anticipated analysis effort that will follow the exercise. The astute analyst will promote the first objective by recording observations alongside his HEAT observers from time to time and then comparing journals. He will also correlate the entries in observation journals from several observers throughout the course of the exercise to identify discrepancies and possible biases in reporting. He will also promote the second objective by establishing observation goals with his observers at the start of every day or every shift. We refer to these goals as the "analyst's essential elements of information" (AEEIs).

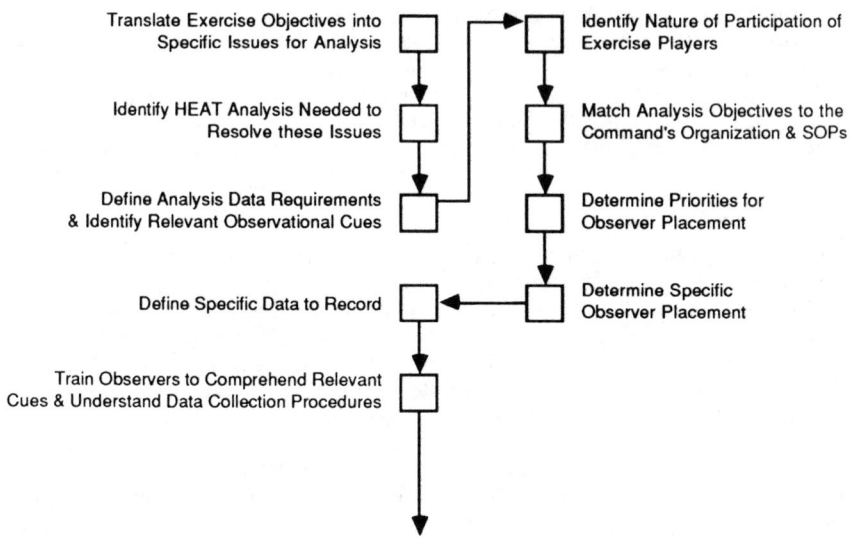

Fig. 3 Major steps in preparing a HEAT application for an exercise

We express our AEEIs in written checklists at the start of every shift. Appearing in the checklist is a description of the specific information the analyst wants recorded or the specific activity he wants the observers to track, relevant observational cues, and relevant antecedent events.

Data collection works best when the HEAT observers enjoy a good, objective rapport with the command and with the staff sections they are observing. Command staff members often work 12 plus hour shifts during an exercise, and we find that a good rapport develops when HEAT observers work alongside the staffs they are observing for their entire shifts. This practice leads to observations that tend to be more relevant both because observers observe the continuity of the command estimation process, and because staffs tend to share more information with observers who they accept as "colleagues in misery." We also find that observers who become deeply involved in an exercise tend to understand better the difficult problems staffs must grapple with during the command estimation process. As a result, they are less likely to inject subjective judgment in their observations.

The objective of *data analysis* is to develop an accurate chronology of the activities that preceded and followed major command decisions. The AEEIs we mentioned above often provide an initial basis for sorting out the major decisions and for matching observations to the event reconstruction. Indeed, the AEEIs are often sufficient to assemble a first-cut analysis for presentation at the "hot wash" immediately following the exercise. The analysis begins by establishing the major command decisions that were made during the exercise. These are arranged in chronological sequence and causal relationships between decisions are noted. The analysis then extracts observations appearing in the observer journals relevant to activities preceding or following each decision. Here, the HEAT analyst draws on HEAT theory to try to interpret the full meaning of the observations so extracted.

Several reports normally emerge from a HEAT application. The first is a "first look" report or briefing presented to the commander at the exercise hot wash. The purpose of this briefing is not only to inform the command, but to subject the analyst's findings to critical review. The second report is usually a draft version of the exercise analysis report that is presented to and reviewed by the command. The third report is actually a briefing presented to the commander summarizing the findings that will appear in the final report. As a courtesy to the staff sections that were observed, the HEAT analyst should also try to arrange a separate briefing to the command staff to permit a full airing of points of agreement and points of contention concerning the analysis findings. The final report is the exercise analysis report itself.

HEAT has been used in support of eight exercises to date; Table 2 summarizes the settings for these exercises. The exercises selected for HEAT application are normally command-post exercises (CPXs) that emphasize C2 and involve reactive play (limited or no scripted action). HEAT-trained analysts and observers have also participated in several Army brigade- and division-level exercises.

General Lessons Learned Through Exercises

Several characteristics of effective command and control emerge from our analyses of exercises. We summarize seven characteristics that appear foremost in our findings: 1) the command understands the current situation and its dynamics, 2) the command knows what information it wants to collect, and coordinates intelligence collection with plan execution, 3) the command strives to minimize ignorance of information and assumptions held by elements within it and at the next echelon below, 4) the commander's intent is understood by his staff and by at least the next echelon below, 5) the command understands the amount of time and the nature of force synchronization that will be required to carry out its plan, 6) the command actively monitors plan execution, and 7) the commander's concept of operations preserves flexibility to respond to contingencies.

Every one of these seven characteristics appears self-evident and might appear to need no further amplification. Nevertheless, they deserve consideration precisely because they are so frequently overlooked during actual and mock combat situations. A striking finding from our analyses of exercises attended under this program is that command staffs are often deficient in the second and fifth characteristics. We find it noteworthy that deficiencies in these two characteristics are cited as common causes of C2 failures by other studies. The

Table 2 Completed HEAT Exercises

No.	Exercise	Sponsoring command	Date
1	BOLD EAGLE	REDCOM	1983
2	WINTEX	AFCENT	1984
3	BORDER STAR	REDCOM	1985
4	Battle Force In-Port Training (BFIT) 2-85	Navy Second Fleet	1985
5	BFIT 1-86	Navy Second Fleet	1986
6	FLEETEX	Navy Second Fleet	1986
7	POM-87 WARGAME	Navy OP 95	1987
8	WINTEX	USTRANSCOM	1989

reasons for this are not mysterious if we consider that a growing number of the officers serving in command staffs have not engaged in actual combat, and therefore lack the experience needed to make these two characteristics an integral part of their decisionmaking repertoire.

At the risk of oversimplification, we believe the empirical evidence supports the following claims:

1) the optimal control perspective toward C2 presented earlier does indeed describe the traits of effective commands

2) experienced officers are more likely than inexperienced officers to include these traits in their decisionmaking repertoire

3) a cause of C2 failure repeated time and again is the failure of a command to understand what information it needs, and to coordinate the activities subsumed on left hand side of the HEAT cycle with planning and execution

4) a second cause of C2 failure is the lack of understanding of the amount of time and the nature of force synchronization that will be required to execute a plan.

An intriguing question in regard to the second bullet is whether inexperienced officers recognize these traits in principle, and simply lack the experience to apply them in practice, or whether they tend not to recognize these traits at all. We do not believe the available evidence favors one interpretation over another. Our observations from the exercises attended under this program, coupled with the lessons offered in the third and fourth bullets, tend to favor the former interpretation.

Future Research

In the future, the HEAT program will emphasize applications, rather than further improvements to the measurement tool or its underlying theory. HEAT applications may suggest needed improvements in the latter areas, however, and such improvements will be incorporated on an ad hoc basis.

An important exception to this will be an analytical effort to better understand the relationship between the various HEAT performance measures and overall headquarters effectiveness. Understanding this relationship will permit us to predict headquarters effectiveness from the performance measures and will identify the areas where improvements will yield the greatest payoffs. The approach to this problem will be an analysis of data on previous experiments and exercises that resides in our electronic database to determine which performance measures are shown to correlate most strongly with the effectiveness measures under a given set of conditions.

The HEAT experimental program will continue to test the key tenets of HEAT theory. However, the potential headquarters improvements that are introduced as independent variables will emphasize hardware and software, such as decision aids and improved displays, since the effects of organizational and functional changes were stressed in the earlier HEAT experiments.

Three topics that rank high on the research agenda of the JDL/TPC3/BRG will be investigated experimentally in the near future. These are

1) The investigation of the failure characteristics of a headquarters under extreme stress, and the exploration of ways to extend the stress level at which a headquarters fails.

2) The impact of improved man-machine interfaces on headquarters effectiveness.

3) The behavior of networks of headquarters and the identification of means for improving the overall effectiveness of headquarters networks.

The third topic will be investigated through distributed wargaming via JDLNET (a new data network sponsored by the JDL/TPC3 that will interconnect laboratories located within its member organizations).

A stronger coupling is planned between the HEAT program and the Defense Communications Agency's Command Center Improvement Program (CCIP) in order to promote operational implementation of the more promising HEAT findings. This coupling will be strengthened by conducting HEAT experiments in the new CCIP laboratory (which is interconnected with several other wargaming laboratories through JDLNET).

As a final note, a continuing effort has already begun toward reducing the number of observers required in a HEAT application by means of automatic audio and video recording.

Acknowledgments

The authors would like to extend their appreciation to several individuals for their many critical contributions to the development and application of HEAT. These individuals include Michael Athans, Lt. Gen. John Cushman (U.S., ret.), James Deckert, Elliot Entin, Richard Hartman, Richard Hayes, Paul Hiniker, William Powell, Mark Scher, Daniel Serfaty, David Signori, Conrad Strack, and Robert Tenney.

References

[1] Blitz, A. L., Entin, E. E., Shaw, J. J. and Vail, P.A. "Framework for Generalizing Test Results," TR-405, ALPHATECH, Inc., Burlington, MA, December 1989.

[2] Shaw, J. J., and Athans, M. "Modification of HEAT Theory from Test Results," TR-491, ALPHATECH, Inc., Burlington, MA, July, 1990.

[3] Athans, M., Madsen, K. R., Serfaty, D., Shaw, J. J., Tenney, R. R., and Wohl, J. G. "Critique, Modification, and Expansion of Headquarters Effectiveness Theory: Volumes I - IV," TR-408, ALPHATECH, Inc., Burlington, MA, November, 1989.

[4] Shaw, J. J., Entin, E. E., Deckert, J. C., and Athans, M. "Additional Hypotheses on Determinants of Headquarters Effectiveness," TR-449, ALPHATECH, Inc., Burlington, MA, November, 1989.

[5] U.S. Army Command and General Staff College, "The Command Estimate," Student Text 100-9, Fort Leavenworth, KA, 1986.

[6] Banks, J. H., and Meliza, L. L. "Observations from Three Years at the National Training Center," Research Product 87-02, U.S. Army Research Institute for the Behavioral and Social Sciences, Monterey, CA, Jan. 1987.

[7] U.S. Army Command and General Staff College, "Corps and Division Command and Control," Field Circular FC 101-55, Fort Leavenworth, KA, 1985.

[8] Bertsekas, D. P., *Dynamic Programming and Stochastic Control*, Academic Press, New York, 1976.

[9] Shaw, J. J., and Powell, W. S. "Critique and Modification of HEAT Measures," TR-432, ALPHATECH, Inc., Burlington, MA, July 1989.

No New Mathematics! No New C2 Theory!

John T. Dockery*
Pentagon, Washington, DC 20318
and
A. E. R. Woodcock†
Synectics Corporation, Fairfax, Virginia 22030

I. Introduction

This chapter outlines several years of research that has sought a mathematical idiom with which to express the fundamental concepts of command and control (C2). As the title suggests, we are committed to the proposition that the way forward in developing a theory of C2 lies with the development and application of new mathematical tools and methods. Take, for instance, the classical Lanchester equations. There is obviously a diminishing return to be achieved from cataloging yet again their admitted difficulties! It appears necessary and appropriate to move on. The choices are stark. Either discard differential attrition process models of combat, or reformulate them to meet the challenge of the modern combat and command and control environment.

Many of the approaches which we have chosen to present in this chapter borrow generously from advances in other research fields. Because we are convinced that combat is an example of a highly nonlinear system, we

---

Copyright © 1993 by J. T. Dockery and A. E. R. Woodcock. Published by the American Institute of Aeronautics and Astronautics, Inc., with permission. The views expressed in this paper are those of the authors and do not necessarily reflect those of the United States Government or of Synectics Corporation. Part of the materials in this chapter were presented by Woodcock in 1989 at the Third International C3MIS Conference, Bournemouth, England; parts of this chapter are also excerpted and edited from *The Military Landscape: Mathematical Models of Combat*, copyright © 1993 by J. T. Dockery and A. E. R. Woodcock, Abington Hall, Cambridge, Woodhead Publishing Limited.

*Currently, Defense Information Systems Agency, Reston, Virginia; also, Visiting Research Professor, Center for Excellence in C3I, George Mason University, Fairfax, Virginia

†Also, Visiting Professor, Cranfield Institute of Technology, The Royal Military College of Science, Shrivenham, England.

acknowledge a special debt to those research workers involved in developing an understanding of the behavior of nonlinear systems. We have derived considerable inspiration from studies of self-organization, particularly as this work has been applied to an understanding of ecological and evolutionary systems.

In our work, which is briefly reviewed in this chapter, the goal has been to embed any theory of C2 in a theory of combat. Much of the work appears in a series of papers which we refer to as the *Models* Series (Woodcock and Dockery[1], Dockery and Woodcock[2], Woodcock and Dockery[3], Woodcock, Cobb, and Dockery[4,5], Dockery, Anthony, and Woodcock[6], and Carvalho-Rodrigues, Dockery, and Woodcock[7]). We have perceived that the existing theories of combat are inadequate. For this reason we have been forced to develop an adequate theoretical understanding of some of the common processes and principles of combat before considering the role of C2 in these processes. As part of this effort, we have described tools and methods that are primarily concerned with combat, others that are concerned with command and control, and yet others concerned with the actions of the military commander.

II. Picking the Tools and Methods

Selection Schemes

The General Approach

Command and control is about structure. Since it concerns itself with providing structure to combat, its description should be in terms of structure. Our selection of candidate mathematical tools was not accidental. Rather it was the result of a deliberate review of the literature, which has attempted to match the claims of various mathematical theories to perceived need to express statements about combat with embedded C2 in precise terms (Table 1).

We have studied the anecdotal and historical record with a view to identifying fairly universal statements about conflict. Some statements eventually assume the status of assumptions, and perhaps even axioms of combat. The following are examples of such statements.

1. Combat is locally chaotic but possesses long range order.

2. The structure of combat is a product of the dynamics of the combat environment. Hence C2 paradoxically both shapes combat and is, in turn, shaped by combat.

3. Command is characterized by feedforward, control by feedback.

4. Without C2, combat is simply mob action and this implies that combat with embedded C2 is a self-organizing system.

5. C2 functions to enhance or diminish nonlinear effects in combat.

6. Many aspects of combat normally attributed to active C2 are actually the natural products of good discipline and training.

7. Commanders may operate as much on perceptions as on facts.

Mathematical tools included for discussion in this chapter are thus survivors of a long and complex selection process that began with a statement of requirements for a theory of command and control. Some elements of the selection process are described in Dockery[8-10]. We have been asked how we keep track of everything. The answer is through the formation and use of a collection of relational paradigms. As the selection process has matured, a number of taxonomic structures have emerged.

Four taxonomies which have proven particularly useful are presented in summary form in Table 1. These include a problem-relational taxonomy which specifies a list of mathematical tools which we have used to model combat and command and control; an organizational taxonomy relating to operational levels within the combat environment; a functional hierarchy taxonomy which considers levels of operational detail; and a state space taxonomy which considers increasingly complex levels of dynamic and structural detail.

Taxonomic Details

The first detailed taxonomy, which we introduce in Table 2, is simply our list of the mathematical tools and methods that we have found useful for modeling combat and command and control and the reasons for such usefulness.

Table 1 Combat and command and control modeling and analysis taxonomies

Organizational taxonomies	Explanation
Problem-relational (Presented in Table 2: mathematical tools)	For displaying the connection between selected mathematical tools and combat and related C2 processes in terms of the problem aspects being addressed
Combat-relational (Presented in Fig. 1: Mathematical overview as it relates to combat)	For inter-relating aspects of combat and related C2 processes in terms of the degree to which they explain combat itself
Functional hierarchy (due to Wohl[11]) (Presented in Table 3: C2 analysis hierarchy: micro to macro)	For identifying the key factor(s) with which a theory, and the related mathematical tools, must deal at various levels in a four tier hierarchy which leads from C2 component to complete system.
State space organization (Presented in Fig. 2: The catastrophe theory-based modeling and analysis paradigm)	For placing tools within a catastrophe theoretic framework of increasing complexity which moves from a consideration of stationary processes through to full representation of time and space dependence

Table 2 Listing of mathematical tools and methods used to construct a theoretical understanding of command and control

Mathematical tool	Description of modeling requirement
Catastrophe theory	To capture in both a time independent and time dependent manner the global nonlinear responses of combat in terms of a few control variables, therefore essentially presenting the commander's C2 perspective on the battle
Category theory	To define measures of effectiveness and to embed the whole C2 modeling process in a larger context as well as the C2 requirements statements
Cellular automata	To capture the self-structuring of combat based on minimal nearest neighbor rule sets, essentially to present the small unit C2 perspective of combat
Chaos theory	To capture the observed chaotic dynamical nature of combat and to identify attractors and to present the combat evolution perspective
(Fuzzy) differential equations	To describe attrition based combat processes when parameter and/or variable information is fuzzy
(Partial) differential equations	To describe attrition based combat processes with maneuver added, in essence to extend the Lanchester equations
(Stochastic) differential equations	To describe attrition based combat processes under uncertainty
Entropy computation	To capture the effects of casualties upon the structure of the fighting and to introduce variables appropriate system variables
Fractals	To capture deployment hierarchies and also the power law nature of combat data by describing both self-similar and self-affine military structure
Fuzzy sets	To capture the imprecision in all phases of the military command process from data to orders
Relativistic information theory	To capture the idea that organizations "move" relative to each other in a metaphorical sense of internal organizational efficiency
Iterated function systems	To examine the consequence of generating the details from random numbers and global considerations where the latter correspond to attractors expressed as transformation matrices
Lagrangian Formulation	To define classical, albeit heuristic, expressions such as "combat momentum" within a rigorous framework
Petri nets	To express the transactions that occur in a C2 system between its elements
Q-analysis methodology	To capture the fundamental idea of a C2 relationship and express it as a polyhedron
System dynamics	To capture feedback and feedforward aspects of C2 interacting with combat processes

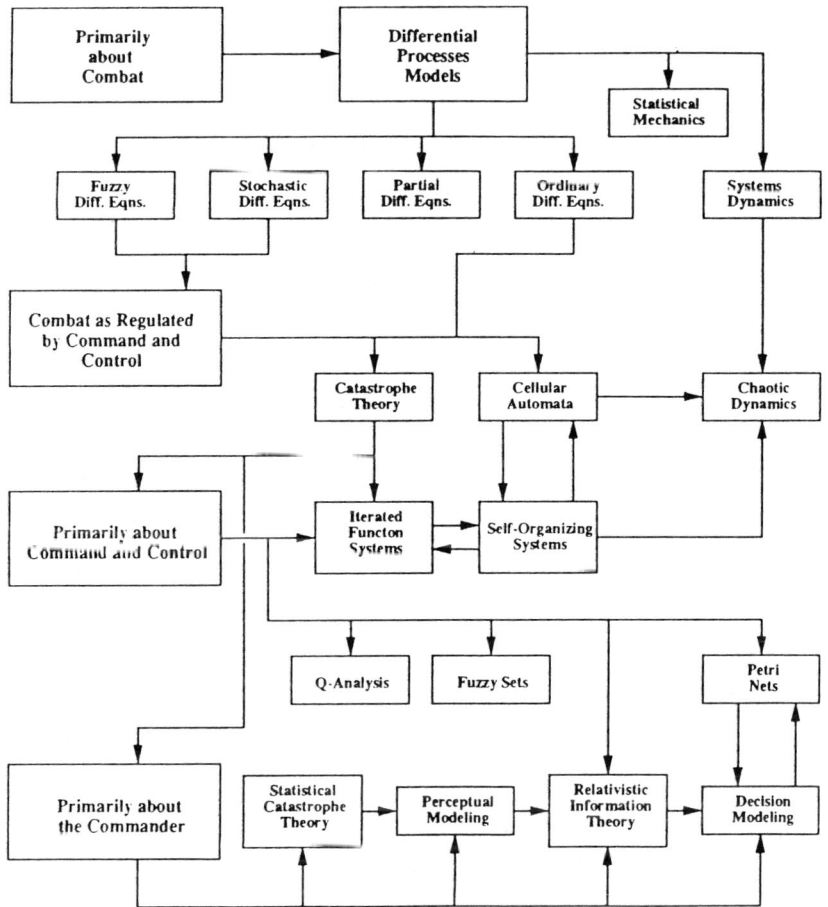

Fig. 1 Mathematical overview as it relates to combat.

Selected groups of these tools and methods have been used for specific tasks within the remaining three paradigms, as described below.

The second detailed taxonomy is organizational in nature (Fig. 1). This taxonomy identifies four levels of combat and command and control modeling and analysis activities which are involved primarily 1. with combat itself, 2. with combat as regulated by command and control, 3. with command and control, and 4. with the activities of the commander.

Only the links that we know to be established are shown in this figure. Other links are in process of being investigated. For instance, partial differential equations clearly are being coupled to C2 by the continuing work at Oak Ridge, particularly in the mapping of chaotic dynamics revealed in the solutions of the appropriate PDEs but the work is still in progress (Protopopescu et al.[12]).

We have used the functional hierarchy taxonomy originally developed by Wohl [11] to build a general structural definition of C2 systems (Dockery et al.,

Table 3 Analysis hierarchy for tools primarily related to a description of command and control itself

Level	C2 functional focus	C2 operational focus	Examples of appropriate tools
Micro	Nodes	Components and data	Classical information theory
Meso	Links between nodes	Equipment and systems	Q-analysis
Meta	Processes within Linked Nodes	Commander's staff	Fuzzy sets; Petri nets
Macro	Functions consisting of related processes within linked nodes	Commander	Catastrophe theory; Relativistic information theory

[13]). This third detailed taxonomy applies primarily to C2 specific tools, and is displayed in Table 3.

The fourth detailed taxonomy moves to increasingly complex descriptions of combat with embedded C2 beginning with stationary state descriptions and ending with descriptions which exhibit both time and space dependent behavior (Fig. 2). Catastrophe theory provides the overall organizational framework and serves as the point departure for use of additional mathematical tools and modeling techniques identified in Table 2.

Other Modeling Taxonomies

There are several other areas of active command and control research that are worthy of mention, but for which lack of space prevents any significant discussion which would do them justice. The work being performed by Goodman[14] and Girard[15], which is highly algebraic in nature, is notable in this regard.

III. A Catastrophe Theory Framework

We have studied historical aspects of combat, searched for partial analogs of combat, and specified requirements for selecting specific mathematical tools with which to develop a robust framework for performing mathematical analyses (Fig. 2). This has led us to develop an appropriate mathematical language for combat modeling which we have used to define models and translate them into computer software. Selected combat data have been analyzed in great detail and models and

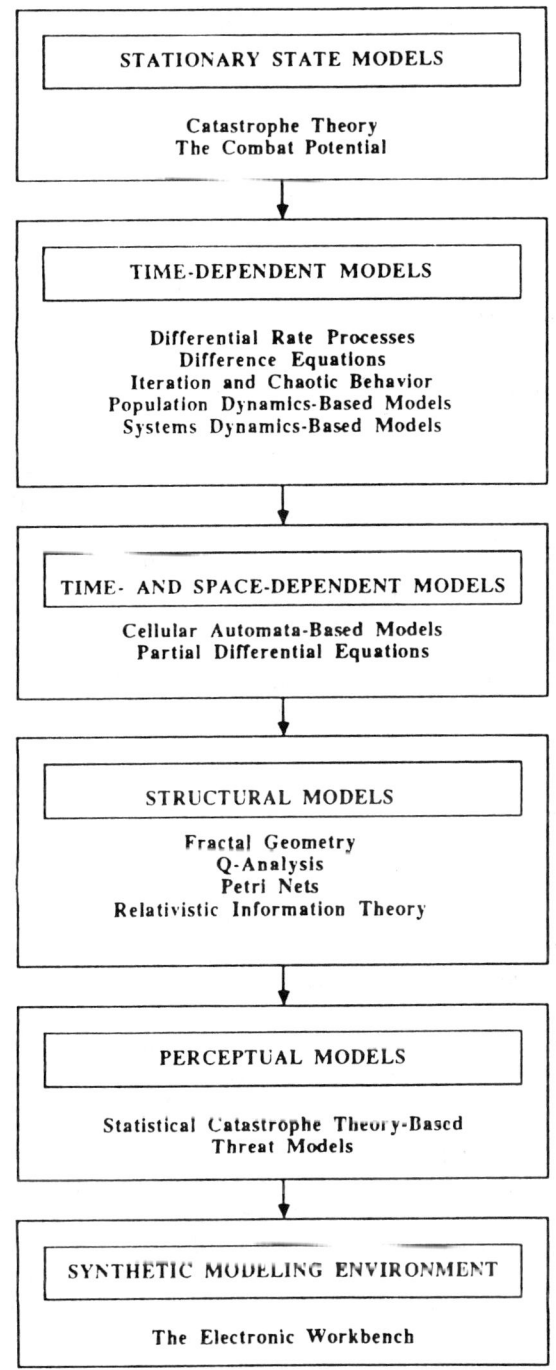

Fig. 2 Catastrophe theory-based modeling taxonomy.

theories of combat have been developed, which have made it possible to make predictions and reproduce examples of actual combat behavior.

A first or simplified group of models that we have developed describes stationary state aspects of combat behavior. Adding back layers of complexity to these models has produced other models that account for variations in combat properties as functions of time and space and describe aspects of the structure of the combat environment. The role of perception in the analysis of combat behavior has been considered, and a combat modeling and analysis facility called the electronic workbench has been developed. The framework provided by catastrophe theory, which has emerged from our research, appears to include the wider classes of dissipative and chaotic dynamical systems.

Stationary State Models

We have begun the construction of a theory of combat with embedded command and control by modeling the behavior of the stationary states of military systems. In searching for an appropriate mathematical methodology for this purpose, we were drawn to catastrophe theory. We were interested in this theory because it provides a proven framework for modeling the stationary state behavior of those systems whose behavior exhibits hysteresis, bimodality or ambiguity, and/or divergence, which seem to be important aspects of combat behavior. The theory is particularly useful when gradually changing forces, represented by the actions of small numbers of key influences, and can give rise to either gradual or sudden changes in behavior in the same system under different conditions. An excellent introduction to the topic is provided by Poston and Stewart[16].

Combat Potentials and the Military Landscape

Use of catastrophe theory provides an excellent example of the way in which a solid theoretical framework can translate military aphorisms and heuristics into concrete entities expressed in mathematical terms (Ref. 1, for example). Catastrophe theory provides a classification of systems in terms of the numbers of key system variables associated with functions known as catastrophe potential functions.

A key feature of the theory is the organizational framework that it provides for systems with limited numbers of key system control parameters. *Once the number of such parameters at work in a particular system has been identified then the theory provides a series of the simplest possible models that can describe the system behavior. All other models are topologically equivalent to one of the catastrophe models.* Under these circumstances, identification of the number of operational key control parameters can lead directly to the construction of models based on entities called catastrophe manifolds that can capture the qualitative aspects of system behavior. Additional activities, using statistical catastrophe theory and other techniques, can provide quantitative and statistically-verified models of system behavior (Woodcock, Cobb, and Langendorf[17], for example).

When used to model the behavior of combat systems, the catastrophe potentials could be referred to as *combat potentials*. The stationary states of such functions, which define the structure of the catastrophe manifolds associated with the combat system, represent stationary states of the combat process. An example of the application of the cusp catastrophe manifold to the modeling of military behavior is presented in Fig. 3. When used for this purpose, we refer to the catastrophe manifold as the *military landscape*.

Technically speaking the application of the cusp is as follows. The cusp catastrophe combat potential function [$V_c(x)$, Eq. (1)] consists of germ ($x^4/4$) which serves to organize system behavior and unfolding components ($ax^2/2 + bx$) which describe the impact of key system influences on system behavior. In modeling combat behavior, we have identified the control factors (a and b) with such properties as force strength, fire power, and command and control capabilities and the behavior variable (x) with a property that we have referred to as force survivability.

$$V_c(x) = x^4/4 + ax^2/2 + bx \qquad (1)$$

Stationary states of the cusp catastrophe function occur when the differential of this function with respect to the behavior variable x, that is, when $dV_c(x)/dx = 0$. This relationship describes a three-dimensional (x, a, b) curved surface (technically called the catastrophe manifold) that we have called the catastrophe landscape, which can serve as the basis of descriptive models of the combat process.

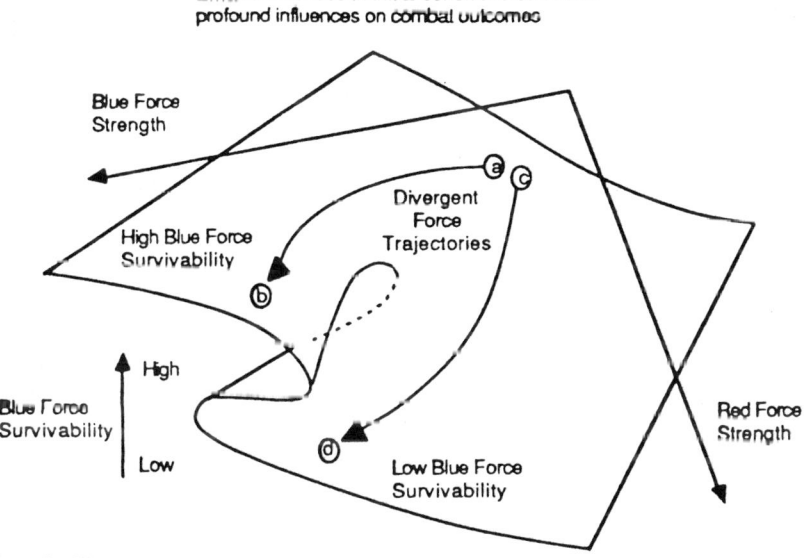

Fig. 3 The military landscape, a "militarized" cusp catastrophe manifold surface.

Higher-Dimensional Models

This kind of descriptive modeling is not limited to three dimensions. Powerful theorems tell us that the cusp can be embedded in a five-dimensional surface called the butterfly landscape. Three-dimensional projections of the butterfly landscape can sometimes look like the three-dimensional cusp manifold, which then deforms as values along the other two axes are varied. These descriptive combat models have a small number of control factors that resemble the properties or "axes" around which a military commander appears to organize his perceptions of the combat environment. One example of a two-control factor model (which is based on the cusp catastrophe) shows the impact of small changes in the relative strength of two adversarial ("red" and "blue") forces on the survivability of the blue force has been shown already in Fig. 3.

A more elaborate model with four input variable and one output variable can be based on the butterfly catastrophe (Fig. 4). Such a model shows the explicit interplay of force strength, fire power, and command and control capability parameters on force survivability. The model illustrates several well-known military phenomena, including the fact that relatively weak forces can be sustained by high levels of firepower and/or command and control capabilities.

Statistical Catastrophe Combat Models

Despite the fact that an infinite number of models with a like number of dimensions can be mapped onto one of the catastrophe manifolds, qualitative modeling still seems arbitrary to many. Fortunately it has proven possible to fit cusp surfaces to data. Such a more quantitative model of combat based on catastrophe theory was advanced by Dockery and Chiatti[18]. It describes an air defense front breakthrough. That model was generated through the use of a fitting program based on statistical catastrophe theory due to Cobb[19] to fit data generated by COMO III (which consists of an elaborate computer simulation of an air attack involving thousands of entities) to a catastrophe manifold surface (Fig. 5).

Bimodality of combat behavior as a function of time was discovered during these investigations. Such behavior is directly related to the problem of ambiguity in the data handled by a commander's staff and has important implications for the command and control of military forces during combat.

Time-Dependent Models

Lanchester Revisited

Considering the time-dependent behavior of combat systems is the first step in the process of putting back elements of system complexity that we had removed in order to create simple types of combat model which describe stationary state aspects of combat. We begin by reviewing Lanchester-type models, and then consider other types of time-dependent process models.

Lanchester-type combat models are differential rate process models. They have played a central role in the development of a theory of combat, have served as starting points for many subsequent investigations, and are extensively reviewed in Taylor[20]. The Lanchester equations (2) and (3) provide a mathematical justification for the combat principle of "force concentration,"

NO NEW MATHEMATICS! NO NEW C2 THEORY!

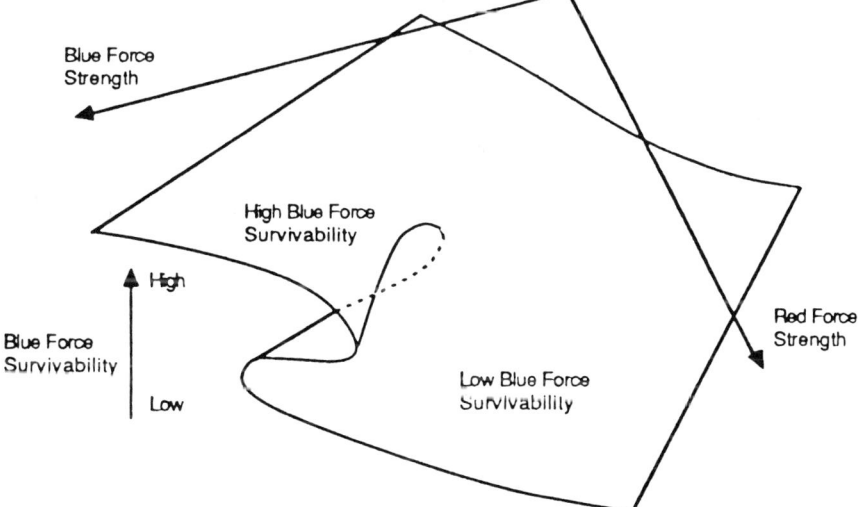

a) Low Blue Force Command and Control Capabilities.

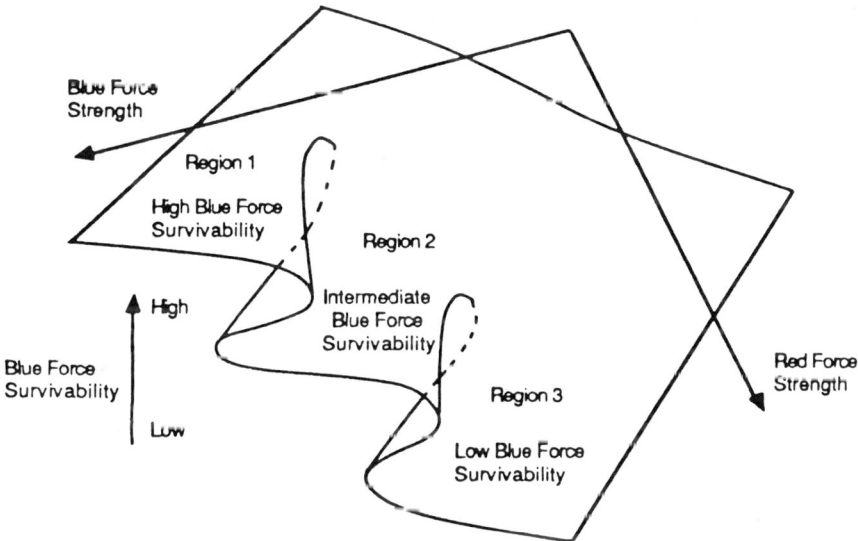

b) High Blue Force Command and Control Capabilities.

Fig. 4 The butterfly catastrophe-based military landscape.

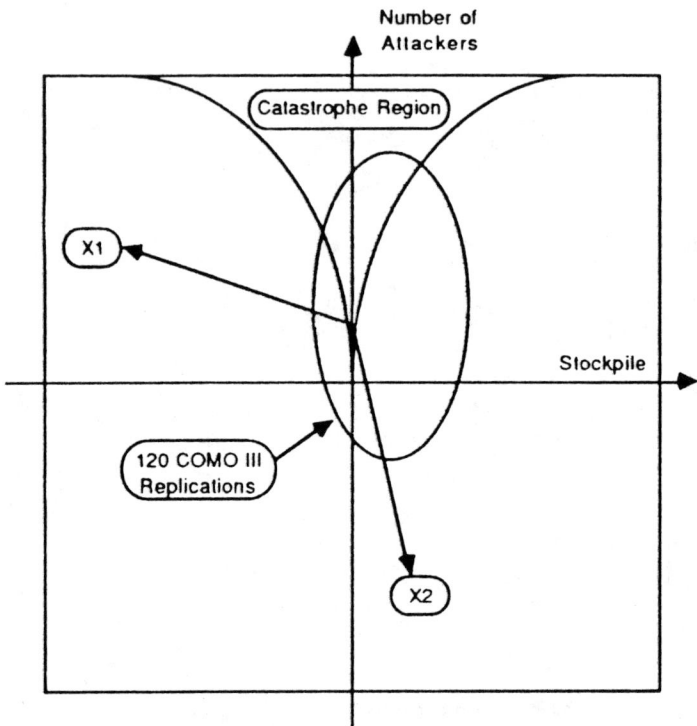

Fig. 5 Representative statistical cusp catastrophe fit of data generated by the COMO III system.

which is sometimes a goal of C2, but may have little else to do with C2 or other aspects of combat.

$$dx/dt = - a_1 y - a_2 xy \qquad (2)$$

$$dy/dt = - b_1 x - b_2 xy \qquad (3)$$

where the coefficients (a_i) and (b_i) are attrition coefficients. What is required for more accurate depiction of the interaction of C2 and combat are more inclusive representations. These must embody both stochastic effects (the short range chaos of combat) and the effects due to spatial coherence (the long range order). But first, what can we learn from embedding the Lanchester equations in our evolving framework.

Generating Lanchester Equations from Catastrophe Potentials

Guided by the catastrophe theory framework, we have demonstrated that the hyperbolic umbilic, double cusp, and double butterfly catastrophe functions can serve as generators of multiple element combat equations that resemble many of the typical Lanchester-type of combat attrition relations (Refs. 1, 2) (Woodcock and Dockery, 1988; Dockery and Woodcock, 1988). Such analysis has made it possible to establish relationships between actual command and control capabilities on the one hand and the parameters of the combat models on the other.

The double cusp catastrophe function [$V_{dc}(x,y)$] has the following form.

$$V_{dc}(x,y) = [x^4 + dx^2 + gx] + [y^4 + ey^2 + hy] \\ + ax^2y^2 + bx^2y + cxy^2 + fxy \qquad (4)$$

The function in Eq. (4) consists of two separate cusp catastrophes.

One cusp for (x): $(x^4 + dx^2 + gx)$

Another cusp for (y): $(y^4 + ey^2 + hy)$

And a series of shared terms: $ax^2y^2, bx^2y, cxy^2,$ and fxy

These separate cusps can represent those components of the combat environment controlled by the x and y force commanders, respectively, whereas the shared components represent factors that can be influenced by either force. The characterization of such behavior can provide guidance for the design of command and control systems since this can identify where the expenditure of C2 resources could provide benefits, for example.

Stationary states of the double cusp function occur when $\partial V_{dc}/\partial x = \partial V_{dc}/\partial y = 0$. The relationships $\partial V_{dc}/\partial x = - dx/dt$ and $\partial V_{dc}/\partial y = - dy/dt$ have been used to derive Lanchester-like multiple element combat equations (5) and (6).

$$dx/dt = - 2axy^2 - 2bxy - cy^2 - 2dx - fy - g \qquad (5)$$

$$dy/dt = - 2ax^2y - bx^2 - 2cxy - 2ey - fx - h \qquad (6)$$

These equations include contributions to an overall picture of combat which can involve ancient combat, modern warfare, area fire, and other types of attrition. We have referred to the latter as "smart weapons" and "combined forces" fire.

Time-Dependent Cusp Equations

Although not as well known as the stationary state potentials, there exist time dependent formulations of catastrophe manifolds. For example, the cusp has no less than 10 forms. The totality of time-dependent catastrophe surfaces has been explored by Wasserman[21]. Dockery and Woodcock (Ref. 2) have applied Wasserman's work to a description of combat and interpreted the result in terms of combat with embedded C2. In particular, the specific time-dependence of the control variables was examined. The time-dependent cusps are divided into two grouping. One is called the time-equivalent; the other, space-equivalent. These correspond, respectively, to all state (coordinate) solutions at a fixed point in time, and at a fixed coordinate for all time! The first can be identified with the offense and planning whereas the latter can be identified with defense. More startling was the possible existence of chaos in the coefficients and the observation that one of the coefficients in one of the equations was itself a cusp in the time variable. Some details are given in the next paragraph.

Observation of chaos occurs when the control variables were recast as time-dependent normal and splitting factors. Time must also be treated iteratively by which we interpret to mean cyclically. Thus, for example, let us take one of one of the space-equivalent catastrophe functions. Stationarity occurs when

$$\partial V(x,t)/\partial x = 4x^3 + 2x(u + t^2) + (v + t^2 + wt) = 0 \quad (7)$$

$$\partial V(x,t)/\partial t = x(2t^2 + 2t + w) = 0 \quad (8)$$

Setting $u^* = -u$ and $w^* = -w$ permits us to write the following functional relationships, which have the form of the well known quadratic and logistic maps.

$$f_1(t) = t_{n+1} = (u^* - t_n^2) \quad (9)$$

$$f_2(t) = t_{n+1} = w^* t_n(1 - t_n/w^*) - v \quad (10)$$

Iteration of these functions has led to the identification of chaotic properties associated with the catastrophe control coefficients. This type of behavior is illustrated in Figs. 6a and 6b which present results obtained from 20 iterations of Eq. (10) with selected values of the w^* and v parameters. It is quite clear that this iterative process has generated chaotic behavior for a negative value (-1.0) of the v control parameter and a positive value (3.2) of the w^* parameter.

We now consider another series of time-dependent combat models. In particular, we have developed models of low intensity combat which are based on the principles of population dynamics.

Command and Control for Low Intensity Conflict

Having established a theoretical basis for differential forms based on catastrophe theory, we turn now to related equations of population dynamics. We use techniques of system dynamics to develop time-dependent models of some of the interactions between security/military forces and insurgent forces. In this synopsis we draw heavily on the work reported in Ref. 3. The low intensity treatment draws upon ecological analogies. We construct a military ecology in which military and security forces can interact with insurgent forces in much the same way that biological species can interact as predators and prey in a biological ecology.

In one model (Fig. 7), the growth of a military or security force is restricted to a finite level by limiting the level of available resources or the size of the pool of individuals available for recruitment into these forces. Such growth (which is often referred to as density-dependent growth) can be described by equations of the form

$$dF_1/dt = m_1[1 - (F_1/S)] F_1 - (m_2 F_1) F_2 \quad (11)$$

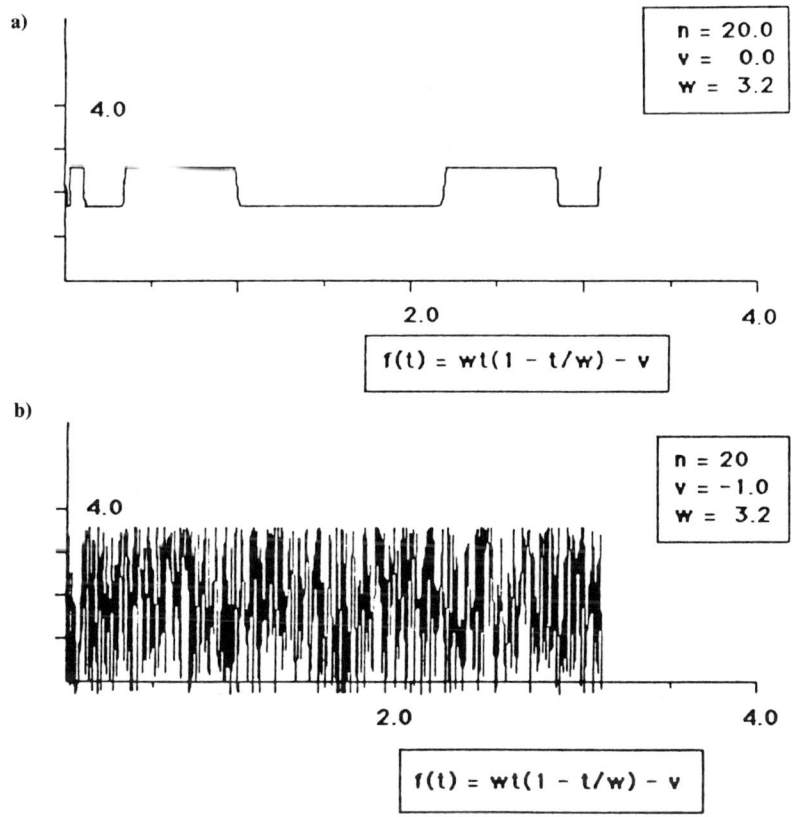

Figs. 6 Chaotic behavior generated by iterating the control coefficients of a time-dependent, space-equivalent, catastrophe.

$$dF_2/dt = m_3 (m_2 F_1) F_2 - m_4 F_2 \qquad (12)$$

where S is the size of the recruitment pool for the military or security force (F_1); m_1 the intrinsic rate of recruitment for the military or security force; m_2 the intrinsic rate of disaffection for the individual members of this force; m_3 the intrinsic rate of recruitment of disaffected individuals by an insurgent force (F_2); and m_4 the rate of loss of members of the insurgent force.

Numerical integration of the differential force strength equations was performed with a software system called STELLA™[22] whose functions are derived from systems dynamics (Coyle[23]). Under some circumstances, oscillations in the strength of both military and insurgent forces can occur which are rapidly damped and reach equilibrium at a relative force ratio of about 10 to 1. The insurgents are not eliminated but they are controlled (Fig. 8). An increase

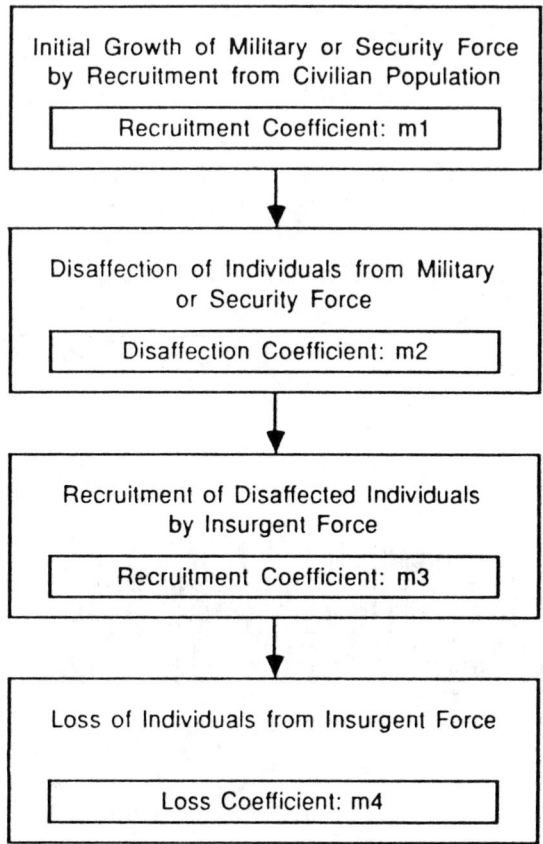

Fig. 7 The growth and disaffection of a military or security force and the recruitment of an insurgent force.

in the rate of loss of the insurgents, however, would tend to eliminate the insurgent force. The phase portrait in Fig. 9 shows a force trajectory that spirals into an equilibrium state known as a point attractor which is characterized by a fixed value for the ratio of the insurgent and military forces. The damped oscillatory changes in force strength values probably indicate the existence of a certain amount of confusion about relative force strengths during the initial phases of an insurgent operation.

Iterated Maps and More Chaos

Having been alerted to the unusual behavior of even simple equations when iterated, we choose to experiment with Eqs. (11) and (12). The first step was to rewrite them as difference equations as follows.

$$F_{1,t+1} = m_1(1 - F_{1,t}/S) F_{1,t} - m_2 F_{1,t} F_{2,t} \qquad (13)$$

NO NEW MATHEMATICS! NO NEW C2 THEORY!

$$F_{2,t+1} = m_3 \, m_2 \, F_{1,t} F_{2,t} - m_4 \, F_{2,t} \tag{14}$$

where $F_{1,t+1}, F_{2,t+1}, F_{1,t}$, and $F_{2,t}$ are the strengths of force 1 and force 2 at times $(t+1)$ and (t), respectively and the coefficients S, m_1, m_2, m_3, and m_4 are as specified above.

Iteration of these equations with a range of different values for mobilization, disaffection, recruitment, and loss coefficients reveals a surprisingly elaborate pattern of behavior. We begin by setting $m_1 = 3.9$, $m_2 = 0.265$, $m_3 = 10.0$, and $m_4 = 0.01$. Changing the value of the disaffection coefficient m_2 from 0.265 to 0.330 (that is, by only 0.065 units) shows what is possible (Figs. 10a-10d). Changing m_2 to 0.275 results in the transformation of a point attractor where the relative force sizes eventually reach fixed values (Fig. 10a) into a limit cycle where sustained force strength oscillations occur (Fig. 10b). Additional changes in the value of the m_2 coefficient to 0.290 transforms the limit cycle into a period-five attractor where five different sets of force strength values are repeated in sequence (Fig. 10c) Further m_2 coefficient changes to a value of 0.330 leads to the creation of a strange attractor where the strengths at a particular time step would give no indication of the strengths after the next time step (Fig. 10d).

These changes in force strength and system behavior that we have generated suggests that the military planning process breaks down as the levels of recruitment and disaffection reach abnormal levels. This form of analysis can perhaps be used to suggest strategies for exerting or regaining control of a chaotic low intensity conflict environment, for example.

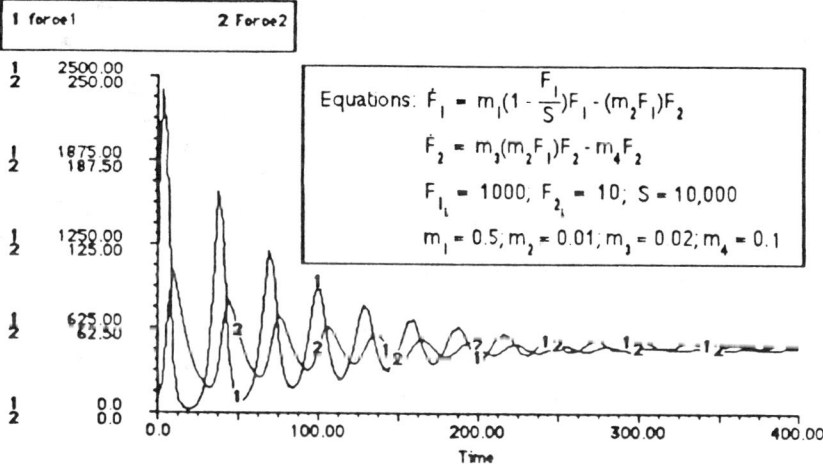

Fig. 8 Time-dependent behavior of the strengths of a military or security force and an insurgent force.

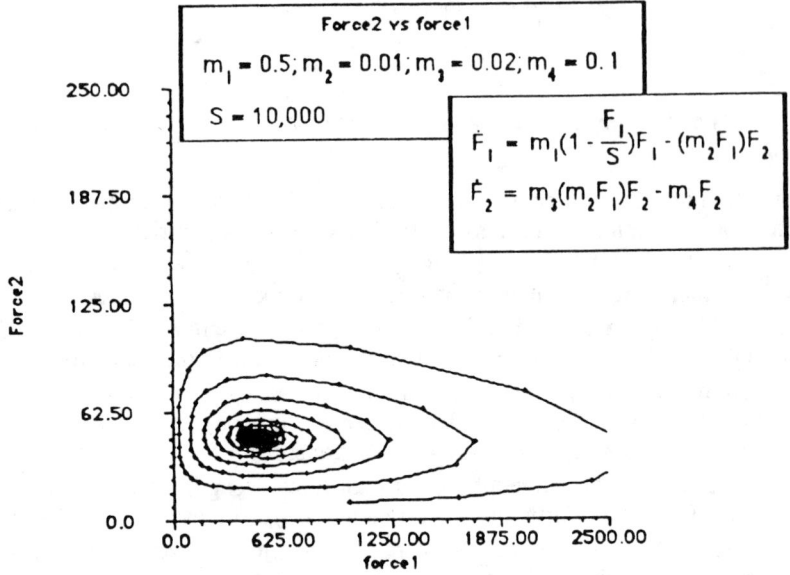

Fig. 9 Force strength phase portrait diagram of the simultaneous strengths of a military or security force and an insurgent force.

Time- and Space-Dependent Models

We now turn to the problem of modeling the spatial as well as the temporal aspects of combat by describing combat models based on cellular automata and partial differential equations. Cellular automata are attractive for this purpose because many aspects of combat behavior involve nearest-neighbor interactions which are also key feature of cellular automata rules (Ref. 5). Partial differential equations also provide a framework for studying the impact of spatial variation on system behavior (Refs. 12, 24).

Cellular Automata as Models of Combat

A cellular automata is a form of "machine" with rules for getting from an initial to a final state What is of particular interest is that the state of the cell under examination is found by examining the states of nearby cells. This can be linked to the situation governing small unit interactions in combat, or so we postulated. The question then becomes what is the maximum structure to be observed from a minimum number of rules. And if significant structure is observed, what indeed does that say about the trade-off between explicit C2 and the value of discipline and training which instills the minimal rule set. A software implementation was built by Woodcock, Cobb, and Dockery, and is described in Ref. 5.

Our cellular automata-based system involves an hierarchical control structure involving "colonels," "captains," and "enlisted personnel." The system supports

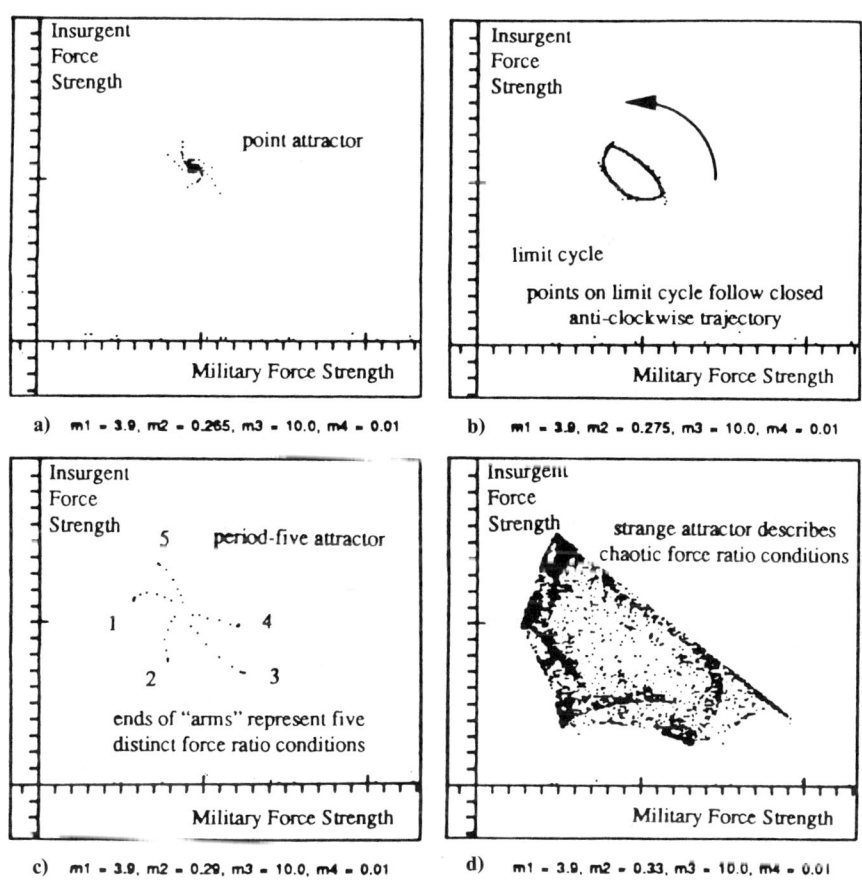

Figs. 10 Iterative solutions of difference-equation models of the strengths of a military or security force and an insurgent force.

the coordinated movement of opposing forces through the use of selected rules of engagement over notional terrain. In the cellular automata-based simulations, the "battlefield" is represented as a rectangle. The members of the two opposing armies are represented by round and square symbols, respectively, with the colonels represented by open (round or square) symbols, the captains by open symbols with a central spot, and the enlisted personnel by filled symbols (Fig. 11).

We have extracted key combat principles of actual land combat engagements (including frontal attack, penetration, oblique or echelon order, flanking movements, and envelopment) and have produced cellular automata-based simulations which appear to reflect the operation of these principles. Because they are generally less complicated than modern battles, early-modern and ancient combat examples were used as benchmarks against which to perform an initial series of tests of the cellular automata-based combat simulation software.

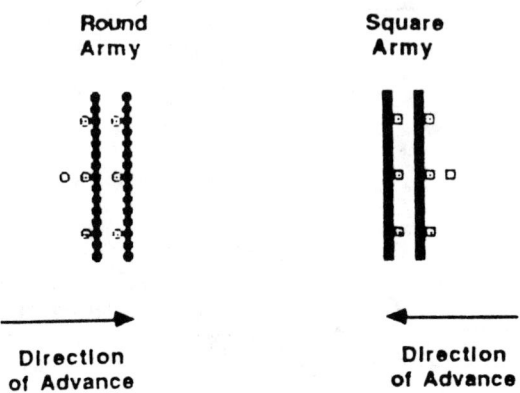

Fig. 11 The cellular automata battlefield.

Subsequent tests could involve examples of modern combat activities with the explicit involvement of C2.

The following example involves a cellular automata simulation of flanking and enveloping maneuvers where "round" and "square" armies each with a single colonel, six captains, and 30 other combatants are initially arranged in two parallel lines (Fig. 11) The round and square armies advance towards each other in frontal attack formation, and by time 29 (Fig. 12a) the armies have come within range of each other. The round army (strength 37) has deployed in an encircling attack (located along the front a-b) against the square army (with a strength of 29 as the result of simulated attrition, and located at position c).

Attrition of the square army reaches a level where a retreat is ordered by the square army colonel. By time 39 (Fig. 12b), the round army (strength 28) is arranged along the front (d-e) and is in pursuit of the square army (strength 18, located at position f). At time 60 (Fig. 12c), the round army (strength 28, located at positions g and h) is exerting a flanking type of attack on the square army (strength 11, located at position i) which has retreated into the corner of the battlefield into a structure that can be considered to represent some form of natural barrier (such as a river or box canyon) to continued movement.

These and other analyses have demonstrated to us that the behavior exhibited by interacting automata is seldom predictable solely from knowledge of the individual automata rules. Such phenomena may constitute the central core of combat behavior and provide the basis for future theories of complex battle interactions. The results of some of these simulations have been fitted by us statistically to a Lanchester-type differential rate process combat model and this type of analysis may provide a basis for studying the impact of different combat rules on military performance.

Partial Differential Equations

Without accounting for variations due to movement in space as well as time, the ordinary differential equations forms of Lanchester can never represent such key elements of C2 such as maneuver and attrition on the move. To do so requires a partial differential equation formulation. Such work has been underway for some time at the Oak Ridge National Laboratory (Refs. 12, 24).

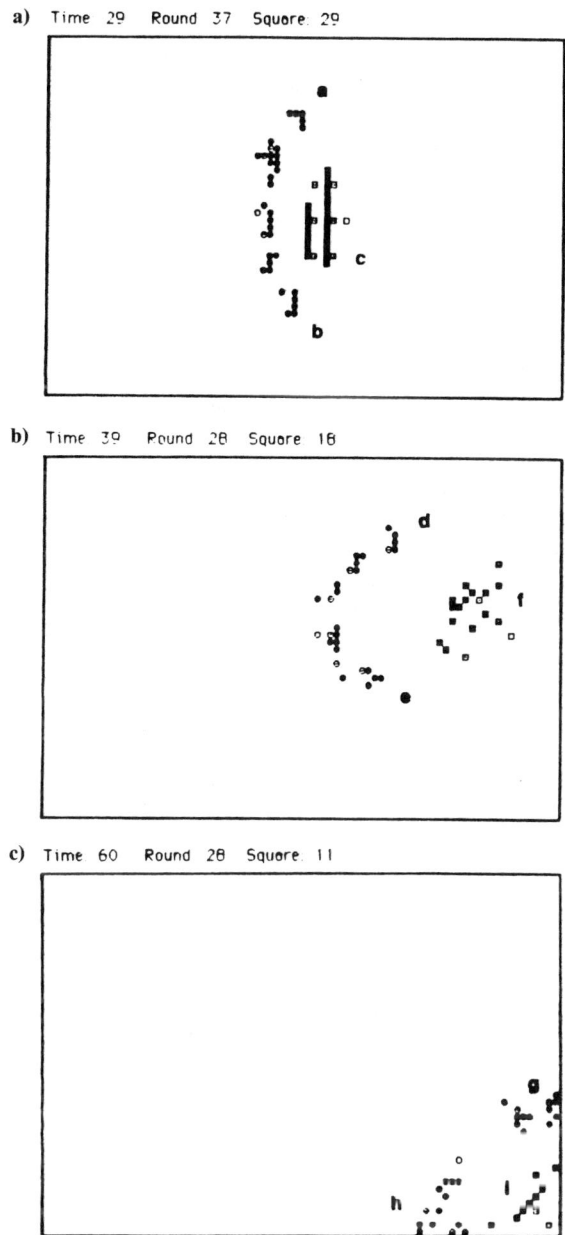

Figs. 12 Cellular automata-based representation of military flanking and envelopment maneuvers.

Beginning with two homogeneous equations which were expressed as $f(x,t)$, the work has now been expanded to encompass up to 16 heterogeneous equations which are $f(x,y,t)$. Moreover the results have been matched to actual field tests through the JANUS model comparisons. Evidence for chaos in the solutions has also been presented by Protopopescu, *et al.*[25].

The one-dimensional partial differential equation extension of the Lanchester equations takes the form shown below.

$$\partial u/\partial t = (D_1 u_x)_x + (C_1 u)_x + u[a_1 + b_1 u \\ + \int c_1 (x-y) v(y) dy] + d_1 v + e_1 \qquad (15)$$

$$\partial v/\partial t = (D_2 v_x)_x + (C_2 v)_x + v[a_2 + b_2 v \\ + \int c_2 (x-y) u(y) dy] + d_2 u + e_2 \qquad (16)$$

These equations include a diffusion term which expresses the natural tendency for force dispersion as a force moves and fights. There is also an advection term that represents the large-scale order flow of battle as opposed to the small-scale or local movement due to "diffusion," logistics, and resupply terms. Other terms include compensation for crowding, a saturation term, plus an interaction term which can be reduced to the classical Lanchester type of attrition equation under appropriate circumstances.

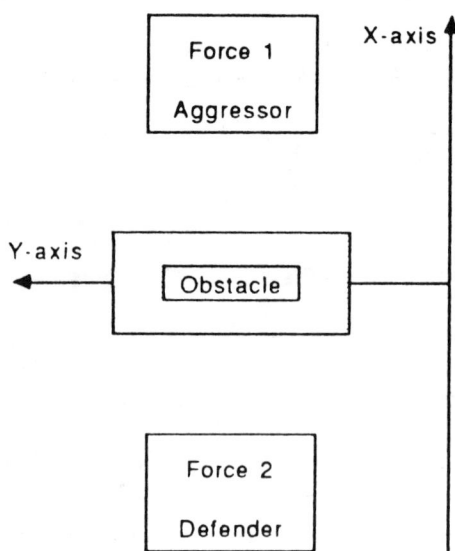

Fig. 13 Geometry for a partial differential equation-based analysis of combat behavior.

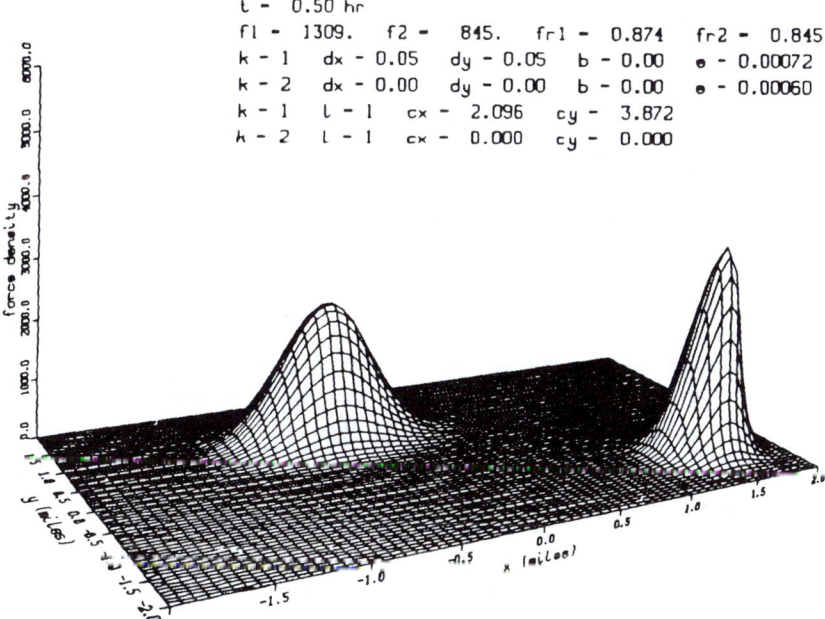

Fig. 14 Specific solution of a partial differential equation-based combat model.

Results obtained from the solution of such equations involving an aggressor and defender force maneuvering around an obstacle (Fig. 13) have been obtained with the aid of the Oak Ridge National Laboratory Cray XMP computer. One example of the output of such an analysis is shown in Fig. 14. These analyses have shown that extensions of the formal Lanchester modeling format can provide an analysis of spatial and temporal behavior during combat.

IV. Providing Structure

Overview

In this section we move beyond the catastrophe theory framework to consider mathematical tools and methods which directly address the question of providing a structure for C2 processes and an understanding of those processes. We also consider the special case of the commander's perceptions, where we again encounter the catastrophe theory framework, albeit from a different angle. Each of the following will be treated in turn: fuzzy sets, fractals, Q-analysis, casualty based entropy, relativistic information theory, and perceptual models.

Although space does not permit more than a fleeting reprise, two additional tools and methods which we borrow from other workers are Petri nets and the Langrangian formulation. The former are covered elsewhere in this volume. Although Petri nets lack the direct link to combat, nonetheless they provide a rather complete description of the traffic and protocols which occur in a functioning C2 system together with its ancillary communications systems. The Lagrangian treatment is due to Ingber[26] who has developed a representation of combat using strong physical analogies to the Lagrangian formulation of physics. From this he has been able to derive rigorous expressions for such concepts as combat momentum. We believe that there may exist theoretical connections between Ingber's formulation and our introduction of combat potentials based on catastrophe theory.

Fuzzy Sets

The fact of imprecision pervades the whole subject of C2 from the initial data input to the final operational orders. Essential applications to C2 problems are due to Dockery (Refs. 8-10 and 27). Basically, fuzzy sets replaces the conventional binary logic choice of "yes" and "no" with a continuum of possibilities that amount to the answer "maybe." Formally all that is mathematically required is to replace the characteristic function of set A defined so that

$$f(x) = 1 \ \forall \ x \in A \ \text{and} \ f(x) = 0 \ \forall \ x \notin A \qquad (17)$$

with the membership function

$$\mu(x) \in [0,1] \ \forall \ x \ \text{in} \ A \qquad (18)$$

The appropriate mathematical structure then becomes a lattice, and the appropriate combining operations are max and min although actually an entire hierarchy of operators which are based on the strength of the connection are possible.

A theory of C2 requires the definition of a structure, a characterization of transactions on that structure, and medium of exchange for the transactions. Fuzzy sets can be introduced to model the transactions involved in C2. They seem particularly appropriate to the task because military information and assessment and direction are often, and of necessity, fuzzy in character. Conventional communications theory can then enter to assess the degree to which the fuzzy information was transmitted. With respect to the functional C2 hierarchy already introduced in Table 3, one finds fuzzy sets are particularly appropriate for describing the traffic between linked nodes which drive the various C2 processes.

Fuzzy sets now form the basis of an extensive alternative to conventional control theory. With the advent of fuzzy logic chips, workers in Japan are revolutionizing some sectors of the commercial market, but as yet the idea of fuzzy logic seems to remain anathema to American designers. As a result the promise of fuzzy sets has not yet been translated into C2 hardware.

Fuzzy sets have been used to formulate a theoretical treatment of effectiveness measures by Dockery (Ref. 10). That work embeds measures of effectiveness in a category of concepts which is known to be isomorphic to the category containing fuzzy sets. The result is another example of a common thread running through candidate tools for a new understanding of C2. Tools in question deal either/both with situations which are context dependent or/and involve qualitative modeling.

Fractals and Combat

It has been increasingly clear to us that a proper metric for combat related phenomena is to be found in a study of fractal, or fractional dimensional entities. Details are described in Ref. 6. The original inspirations came from an observation that classical fortifications appeared to have boundaries which seemed similar to fractal patterns. Such was the case when Ft. Stanwix, a Revolutionary War fort in Rome, New York was compared to a fractal figure known as the Koch island fractal.

Inspection of the inner fortification line of Fort Stanwix (Fig. 15a) reveals that it can be inscribed within a square and that this fortification appears to be constructed from eight subunits (Fig. 15b). Iteration of the initial eight sub-unit pattern generates the pattern of the gunports of the fort. This process of substructural iteration is similar to the methods used to generate the Koch island outline.

A fractal fractal dimension of 1.07 has been computed for the inner fortification line of the fort. It implies to us that the defenders of Fort Stanwix would have a 7% advantage in projectable firepower over the attackers as a result of the impact of the structure of the fortification. The apex of such apparently fractal fortification construction seems to have been reached in the work of the French military architect Vauban.

We have also shown that such things as message arrival and command post movement obey fractal distributions. Moreover, the forward edge of the battle line for a World War II German offensive against Russia may be fit to a fractal power law relationship. The latter work uses arguments similar to those which find coastlines to be fractal entities. By using self-affine rather than self-similar fractal definitions, front-line deployments have been argued by Dockery, Anthony, and Woodcock to be fractal in nature. Assigning numerical dimensions to deployments permits a crucial coupling between what C2 controls and the algorithms used for such control.

Q-Analysis Methodology

This methodology is about connectivity *per se*. Thus, it is fundamental to any consideration of C2 since C2 can be considered to be the study of connectivity and relationships between combat entities. For example, "A 'reports' to B" or "A 'commands' B." The basic methodology is due to Atkin[28] and has been extensively explored by Johnson[29]. Military and C2 applications involving an air defense command and control network for a forward area have been explored by Dockery (Ref. 9).

Q-analysis yields a global structure from a consideration of pairwise relationships between two sets of entities. Q-analysis methodology can also be considered as a kind of cluster analysis. The pairwise relationships are displayed in a relational matrix with respective sets forming rows and columns. Such a structure is called a simplicial complex and is equivalent to a polyhedron. Connectivity is defined by how many polyhedral faces are shared by the entities

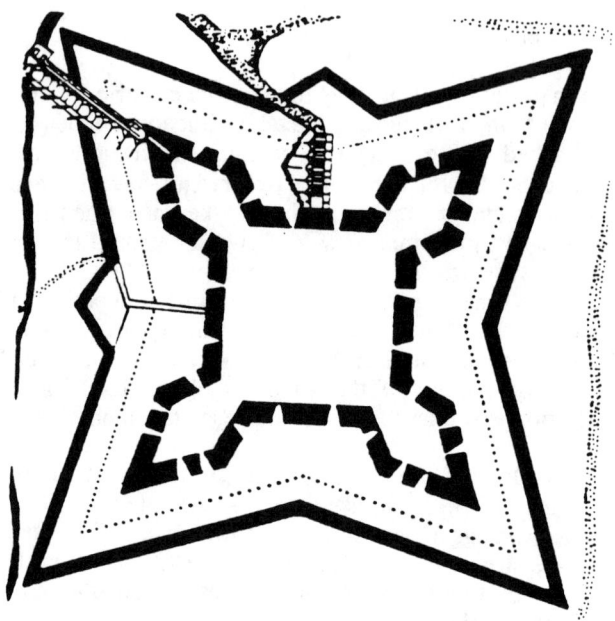

a) Ground Plan of Fort Stanwix, New York.

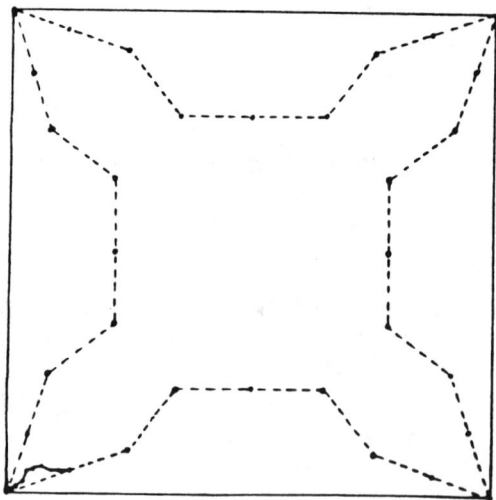

b) Fractal Structure of Inner Fortification.

Figs. 15 Fort Stanwix has a fractal inner fortification line.

in the respective sets. The "q" term in the methodology relates to the number of faces involved. The result is called a backcloth and such backcloths have been constructed for a range of combat-related situations.

Higher dimensional set entities can be related to each other through lower dimensional entities with which they share faces. In this manner a chain of relationships can be constructed. A quantity called the eccentricity is often computed. It measures the degree of isolation of any of the polyhedra from a comparison of the dimension at which the structure first appears with the dimension at which it is first connected to other elements in the structure. The number of pieces into which a structure falls at a given dimension, or q-level, may also be constructed and is known as the structure vector.

Some interesting things can be done with a backcloth. By noting the frequency of occurrence of a particular relationship, it serves as a framework against which traffic patterns may be measured. Changes to the backcloth which results from the addition and/or removal of relationships can be quantified through the calculation of properties called t-forces where t can have the dimensionality of force. The name derives from a formal mathematical analogy to physical forces when the methodology is applied to certain physics problems. By using the concept of t-forces it becomes possible to quantify such difficult concepts as the degree of stress placed upon C2 systems by electronic warfare jamming.

Work with q-analysis is computationally intensive. A program due to Johnson[29] called STARPACK can perform such computation. One should also note that the important concept of surprise has been expressed by Atkin[30] in q-analytical terms.

Entropy-Based Computations

Two applications of the mathematical tools and methods connected with the definition and use of entropy have been advanced. One of these is described in Ref. 7. in a contribution which inquires into the predictive capability of entropy definition based on battlefield casualties. The form of the equation is shown below.

$$H_S = (C_i/N_i)\ln[1/(C_i/N_i)] \tag{19}$$

where C_i represent casualties and N_i the original number of forces with i representing either the red or blue forces.

That work posits entropy as a fundamental parameter of systems analogous to something like mass or charge for physical systems. A case is made that a casualty-based entropy can be a predictor of combat success. The question is then raised as to whether or not C2 ought to be tracking the changes in battlefield entropy as well as the raw data from combat. As a further aid to design of C2, systems trajectories of battles that have been defined in an entropic state space have also been computed.

Relativistic information theory is a more speculative application of entropy which is based on the work of Jumarie[31] and Dockery and Woodcock[32]. Basically, Jumarie argues that entropy has four components which are equivalent

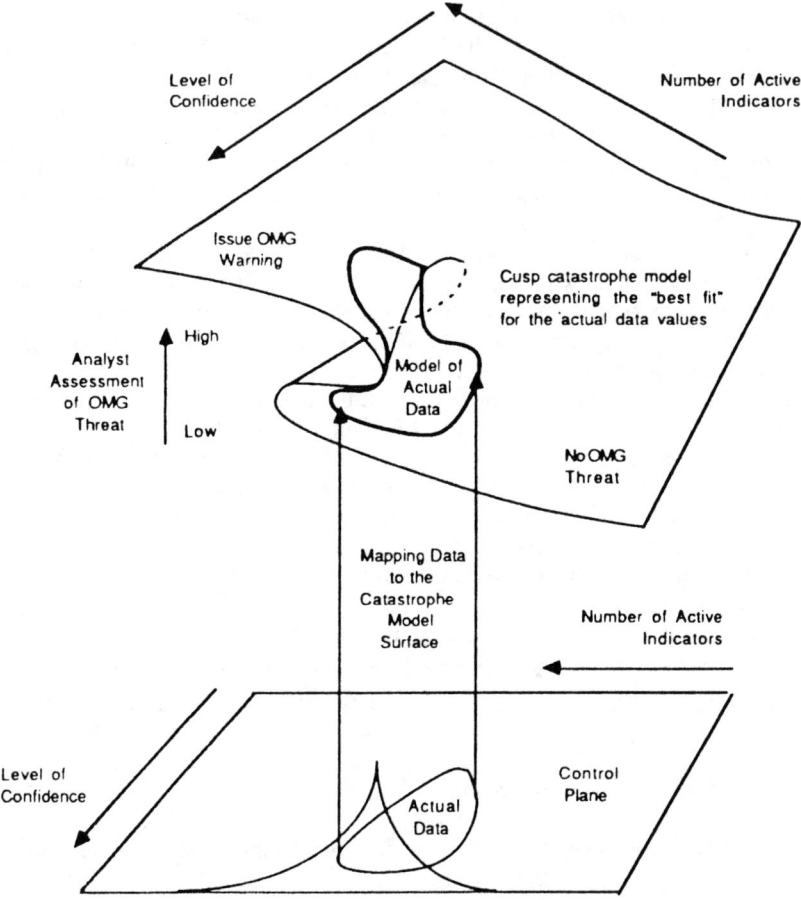

Fig. 16 Mapping data to the cusp model surface.

to a four vector that may be transformed like a Lorenz transformation familiar from special relativity. Something called the organizability of the systems plays the role of velocity permitting comparison between organizations in the manner in which they respond to data and external challenges.

The common observation that some (military) organizations "move" faster with respect to each other in terms of their internal structure is thus given mathematical expression. We find that except for extreme differences in "velocity" the differences are small but that it is nature of the transformation equations to accelerate differences leading to collapse of one or the other structures.

Perceptual Models

Many of the mathematical concepts come together in a consideration of the commander who we have asserted operates largely on perceptions. Software has been constructed to map the commanders perceptions onto a cusp manifold. Such a mapping graphically illustrates the phenomenon of perceptual ambiguity (Ref. 17).

The software tool has been used in experiments with intelligence analysts involving the simulated detection of Soviet Operational Maneuver Groups (OMGs). A set of unclassified notional indicators predicting the development of an OMG was developed and 10 specific settings of these indicators were presented to intelligence analysts, who were asked to assess the probability of OMG development.

Based on this analysis, it is possible to construct a cusp catastrophe model that is the best available "fit" for the intelligence analyst-derived data (Fig. 16, for example). The transformed data is located within the circle drawn on the control plane formed from transformations of the number of active indicators and level of confidence control factors. Some of these data lie inside, and the remainder lie outside, the region of bimodality or perceptual ambiguity on the control plane.

V. Work in Progress

With the framework of a theoretical structure in place, Woodcock[33-37] in a series of papers, has begun to explore other aspects of C2. Two questions in particular are being addressed. One concerns the inherent limits to achievement of combat objectives with even the best of C2 and intentions. The other has to do with providing coordinate systems appropriate to the new mathematical landscapes.

Charting the limits to combat involves a study of the computational stability of iterative equations representative of combat situations. Coefficients in, for example, population dynamics-based models are plotted *vs.* each other to chart regions for which military realizable solutions may exist. Woodcock calls these regions *possibility zones*. Linked to this work is a study of attractors. By working with the Jacobian of coupled iterative equations, boundaries may be plotted within which chaotic behavior is contained. In a sense a coordinate system for the chaos, which C2 paradoxically seeks to both create and confine, is thereby defined.

Additional work by Woodcock and Cobb[38-39] has investigated the dynamical aspects of combat with embedded command and control with the aid of such techniques as power spectral analysis. Results of such work can provide insights into the degree of time-responsiveness needed to provide adequate capabilities for command and control decision-making activities.

VI. Synthetic Combat Modeling Environment

Identifying requisite mathematical tools and methods turns out to have been only part of the task. Operationalizing them is the other part of the problem. Since many of the subjects discussed are not part of the usual designers lexicon, the need to build computer based tools through which to exercise the mathematical technology became a necessity. The key is make the details of the tool or method as transparent as possible to the user but still permit manipulation of concepts. What emerges from our activities is a kind of experimental mathematical approach to investigating C2.

Using a construct that they have called the *decision space* Woodcock, Cobb, and Dockery (Ref. 4) have constructed a computer-based modeling and analysis tool called the electronic workbench. The workbench incorporates aspects of the work involving a merging of catastrophe theory, selected Lanchester equations, and a cellular automata battlefield with decision models involving strategies and tactics. The workbench provides a representation of the strength of two opposing (blue and red) forces in terms of a position on a plane specified with reference to the two conflicting factor axes of a catastrophe function.

The workbench permits its users to define combat profiles and kill uncertainties for two opposing forces, select a range of different thresholds and one of three different combat termination rules, and study the impact of such combat parameters on combat outcomes. At the same time the same battlefield conditions can be studied from a cellular automata perspective from within the same software program.

VII. Summary and Conclusions

We have described a series of investigations in which candidate mathematical tools and methods appropriate for providing answers to the questions of C2 were identified. These in turn have resulted in the construction of a series of models which support the development of a theory of combat with embedded command and control. One group of models that we have developed describes stationary state aspects of combat behavior is based on catastrophe theory. Adding back layers of complexity to these models has produced other models based on aspects of catastrophe theory, population and systems dynamics, cellular automata, and partial differential equations that account for variations in combat properties as functions of time and space.

Other investigations are using fractal geometry, fuzzy sets, Q-analysis, and entropy-based information theory to describe aspects of the structure of the combat environment. The role of perception in the analysis of combat behavior has been considered and a combat modeling and analysis facility called the electronic workbench has been developed.

The framework provided by catastrophe theory which has emerged from our research appears to include the wider classes of dissipative and chaotic dynamical systems. One of the most important discoveries that we have made during our investigations is that simple combat models can generate elaborate patterns of behavior and this has implications for the role that C2 is playing, or should be playing, under such circumstances. We would end this chapter by stressing that without recourse to non-traditional mathematical tools, none of the foregoing modeling and analysis activities and the insights that this work has generated would have been possible.

The title of this chapter asserts our position in no uncertain terms that progress in building command and control theory is at an impasse absent the application of modern mathematical insights. The chapter itself catalogues our major efforts in the exploration of mathematical disciplines adequate for the job of letting some fresh air into the construction of command and control theory. Each candidate mathematical technology described in the chapter is tied to a previously identified requirement for some aspect of an eventual comprehensive theory of command and control.

References

[1] Woodcock, A. E. R. and Dockery, J. T., "Models of Combat with Embedded C^2 I: Catastrophe Theory and the Lanchester Equations," *International Command and Control, Communications and Information Systems Journal*, Vol. 2, No. 3, 1988, pp. 34-62. **

[2] Dockery, J. T. and Woodcock, A. E. R., "Models of Combat with Embedded C^2 II: Catastrophe Theory and Chaotic Behavior," *International Command and Control, Communications and Information Systems Journal*, Vol. 2, No. 4, 1988, pp. 17-51.

[3] Woodcock, A. E. R. and Dockery, J. T., "Models of Combat with Embedded C^2 III: Recruitment, Disaffection, and the Tactical Control of Insurgents," *International Command and Control, Communications and Information Systems Journal*, Vol. 3, No. 1, 1989, pp. 5-38. **

[4] Woodcock, A. E. R., Cobb, L., and Dockery, J. T, "Models of Combat with Embedded C^2 IV: The Decision Space," *International Command and Control, Communications and Information Systems Journal*, Vol. 3, No. 2, 1989, pp. 5-31. **

[5] Woodcock, A. E. R., Cobb, L., and Dockery, J. T., "Models of Combat with Embedded C^2 V: Cellular Automata," *International Command and Control, Communications and Information Systems Journal*, Vol. 3 No. 3, 1989, pp. 5-44.

[6] Dockery, J. T., Anthony, R., and Woodcock, A. E. R., "Fractals in Combat Data," *Proceedings of the 1991 Joint Directors of Laboratories/Basic Research Group Command and Control Research Symposium*, held at the National Defense University, Washington, D. C., 1991, McLean, VA., Science Applications International Corp. **

[7] Carvalho-Rodrigues, F., Dockery, J. T., and Woodcock, A. E. R., "Entropy in Combat Data," *Proceedings of the 1991 Joint Directors of Laboratories/Basic Research Group Command and Control Research Symposium*, held at the National Defense University, Washington, D. C., 1991, McLean, VA., Science Applications International Corp. **

** Note: The *International Command and Control, Communications and Information Systems Journal* has suspended publication due to the adverse business climate. Modified and updated versions of the papers referred to as 'in press' with that journal at the time that the first draft of this paper was written in 1990, as well as other selected materials by Dockery, Woodcock, Cobb, Carvalho-Rodrigues, Anthony, and Langendorf listed in this list of references, have appeared in an edited form in The *Military Landscape: Mathematical Models of Combat*, 1993, co-edited by Dockery, J. T. and Woodcock A. E. R., Abington Hall, Cambridge, Woodhead Publishing Ltd.

[8] Dockery, J. T., "Mathematics of Command and Control Analysis," *European Journal of Operational Research*, Vol. 21, 1984, pp. 172-188.

[9] Dockery, J. T., "Structure of Command and Control Analysis," *Proceedings of the Symposium on Modelling Defence Processes*, edited by R. Hüber, New York, Plenum Press, 1984.

[10] Dockery, J. T., 1984, *A Fuzzy Treatment of Effectiveness Measures*, SHAPE Technical Centre, TM-729, The Hague, The Netherlands, 1984.

[11] Wohl, J., "Force Management Decision Requirements for Air Force Tactical Command and Control." *I. E. E. E. Transactions on Systems, Man and Cybernetics*. Vol. SMC-11, pp. 610-639, 1981.

[12] Protopopescu, V., Santoro, R. T., Dockery, J. T., Cox, R. L., and Barnes, J. M., "Combat Modeling with Partial Differential Equations," Oak Ridge National Lab. Rept., ORNL/TM-10636, 1987.

[13] Dockery, J. T., and 'Project 80-5' Team Members, *Tools of the Trade for Command and Control Analysis*, SHAPE Technical Centre, TM 752 (NATO Confidential), 1985, The Hague, The Netherlands.

[14] Goodman, I. R., "Applications of an Exact Linearization-Gaussian Sum Technique to the Modeling of C^3 Nodes," *Proceedings of the 1990 Joint Directors of Laboratories/Basic Research Group Command and Control Research Symposium*, held at the Naval Postgraduate School, Monterey California, 1990, pp. 63-72, McLean, VA., Science Applications International Corp.

[15] Girard, P., "Stochastic Conditional Object Oriented System Analysis," *Proceedings of the 1990 Joint Directors of Laboratories/Basic Research Group Command and Control Research Symposium*, held at the Naval Postgraduate School, Monterey California, 1990, pp. 354-365, McLean, VA., Science Applications International Corp.

[16] Poston, T. and Stewart, I., *Catastrophe Theory and Its Applications*, London, Pitman, 1978.

[17] Woodcock, A. E. R., Cobb, L., and Langendorf, P. M., "A Catastrophe Theory-Based Analysis of C^3I Problems: Perceptual Hysteresis and the Resolution of Ambiguity." *International Command and Control, Communications and Information Systems Journal*, in press, 1990. **

[18] Dockery, J. T. and Chiatti, S., "Application of Catastrophe Theory to the Problems of Military Analysis," *European Journal of Operational Research*, Vol. 24, 1986, pp. 46-53.

[19] Cobb, L., *A Maximum Likelihood Computer Program to Fit a Statistical Cusp Hypothesis*, SHAPE Technical Centre, 1983, The Hague, the Netherlands.

[20] Taylor, J. G., *Lanchester Models of Warfare*, Vols. I and II, Arlington, VA., Operations Research Society of America, 1983.

[21] Wasserman, G, Stability of Unfoldings in Space and Time," *Acta Mathematica*, Vol., 135, 1975, pp. 57-128.

[22] Anon., *STELLA^{TM}*, A software product of High Performance Systems, Lyme, NH., 1990.

[23] Coyle, G., "Technical Elements of Systems Dynamics," *European Journal of Operational Research*, Vol. 14, 1981, pp. 359-370.

[24] Dockery, J. T. and Santoro, R. S., "Lanchester Revisited: Progress in Modeling C^2 in Combat." *Signal*, July, 1988, pp. 41-50.

[25] Protopopescu, V., Santoro, R., and Azmy, Y., "Recent Advances in Analytical Combat Simulation: From Modelling to Validation and Beyond," Presentation to Working Group 29, 58 Military Operations Research Society Symposium, Annapolis, MD., United States Naval Acad., 1990.

[26] Ingber, L., "Nonlinear Nonequlibrium Statistical Mechanics Approach to C^3I Systems," *Proceedings of the 9th. Massachusetts Institute of Technology/Office of Naval Research Workshop on C^3 Systems*, 1986, pp. 237-244, Cambridge, Massachusetts Institute of Technology.

[27] Dockery, J. T., "Fuzzy Design of Military Information Systems," *International Journal of Man-Machine Studies*, Vol. 16, 1982, pp. 1-38.

[28] Atkin, R., *Mathematical Structure in Human Affairs*, New York, Crane Russak, 1974.

[29] Johnson, G., *STARPACK*. Unpublished Contract Rpt. to the Joint Chiefs of Staff, The Pentagon, Washington, D. C., 1989.

[30] Atkin, R., "Theory of Surprises," *Environment and Planning B*, Vol. 8, 1981, pp. 359-365.

[31] Jumarie, G., *Subjectivity, Information, Systems*. New York, Gordon and Breach, 1986.

[32] Dockery, J. T. and Woodcock, A. E. R., "A Case Study in Selecting Relativistic Information Theory as a C^2 Analysis Tool," *Proceedings of the 1989 Joint Directors of Laboratories/Basic Research Group Command and Control Research Symposium* held at the National Defense University, Washington, D.C., 1989, McLean, VA., Science Applications International Corp. **

[33] Woodcock, A. E. R., "Toward a Theory of Combat with Embedded Command and Control," *Proceedings of the Third International Conference on Command, Control, Communications and Management Information Systems (C^3MIS)*, held in Bournemouth, England, 1989, pp. 127-138, London, Institution of Electrical Engineers. **

[34] Woodcock, A. E. R., "Trapping Chaos: Determining the Limits of Command and Control Systems," *International Command and Control, Communications and Information Systems Journal*, 1990, in press. **

[35] Woodcock, A. E. R., "The Limits of Combat I: Possibility Zones, Resource Mobilization, and Attrition Processes," *European Journal of Operational Research*, Vol. 67, 1993, pp. 111-125. **

[36] Woodcock, A. E. R., "The Limits of Combat II: "Application of Possibility Zones to the Analysis of Insurgent Recruitment and Coupled Logistic Dynamics," in preparation, 1990. **

[37] Woodcock, A. E. R., "The Limits of Combat III: "Power Spectral Analysis of Recruitment and Attrition Processes," in preparation, 1990. **

[38] Woodcock, A. E. R., and Cobb, L., "Power Spectral Analysis of Combat as a Chaotic Dynamical System," *International Command and Control, Communications and Information Systems Journal*, Vol. 4, No. 1, 1990, pp. 5-38. **

[39] Woodcock, A. E. R., and Cobb, L., "Power Spectral Analysis of Combat as a Chaotic Dynamical System II: The Impact of Lanchester-like Attrition Processes," *International Command and Control, Communications and Information Systems Journal*, 1990, in press. **

Formal Theory of C3 and Data Fusion

I. R. Goodman*

Naval Ocean Systems Center, San Diego, California 92152

Introduction

The history of C^3 analysis as an organized approach to defining the general military problem and, in particular, the command and control aspects, goes back several years. For a brief history of approaches based upon the MIT/ONR Workshop on C^3 systems, for many years the premier academic venue for C^3 analysis, see, e.g., Goodman.[1] Despite the large amount of literature produced on C^3 issues, whether it be from the Workshop or other government or private industry sources, a basic pattern emerges: little attention has been paid to the establishment of an overall C^3 model from a quantitative microscopic or "bottoms-up" point of view. Instead, much of the work has been devoted to either qualitatively-based analysis or quantitative analysis of bits and pieces of the whole C^3 panarama. This is obviously due to the great potential complexity involved in attempting to model the entire detailed process. A third approach to modeling C^3 systems utilizes a complete macroscopic or "top-down" viewpoint. Examples of the first two types of analysis are numerous. In fact, perusing through the last several issues of the Proceedings of the MIT/ONR Workshop and its subsequent successor, the Symposium on C^2 Research, one finds articles on command planning, fire control, tracking and filtering of targets, correlation and data association of multiple targets, surveillance, limited interacting multiple persons decision games, time studies, stochastic control problems, etc. Examples of the third men-

---

This paper is declared a work of the U.S. Government and is not subject to copyright protection in the United States.
 *Senior Mathematician, Command and Control Department.

tioned type of analysis are not as plentiful, but include papers on markovian models of C^3 systems relative to attrition and supply, variations of Lawson's macrothermodynamic analog, Lanchester's attrition equations and generalizations, use of general resource allocation principles, and applications of analogs with laws of behavior in economics and other large-scale systems.

Of course, the above mentioned examples certainly contribute toward the overall understanding of C^3 in general; however, they point up the lack of any attempt to model C^3 from a microscopic approach. It is the thesis here that it is not too early in the development of C^3 as a discipline to make this effort. Among the work directed previously to this goal, mention should be made of Ingber,[2,3] and Rubin and Mayk,[4] and Rubin et al.[5] Ingber utilizes the path integral principle from nonlinear nonequilibrium statistical mechanics to attempt a meso-macroscopic C^3 analogous model, while Rubin and Mayk's approach has a more microscopic flavor in extending the Lanchester equations. Finally, Refs. 6-8 should be noted. These are based upon a partial microscopic model of the SHOR paradigm concerning information throughput and transmittal relative to an overall organizational model. In a sense, the work of Levis and co-workers has influenced the author's thinking more than any other source with respect to modeling of C^3 systems.

Objectives and Approach

The long-range goal of this work is twofold.

1) Show tactical C^3 processes can be reasonably modeled within a game theory context, using a formal system of axioms which capture a minimal number of pertinent relations among the C^3 variables and operators.

2) Provide an outline for a feasible implementation of this program as an aid in the design of C^3 systems.

For present, we must be content with only the first goal; time will tell if the second goal can be achieved. In modeling C^3 processes one always must be aware of the tradeoff between the fidelity of theory and the complexity of practical implementations. With this in mind, a C^3 design game is proposed here based upon the outputs of a formal theory for the evolution of node states. This is predicated upon the asumption that a C^3 system as envisioned here is completely identified as a collection of such interacting node states, each operating according to the SHOR paradigm (S = Sense, H = Hy-

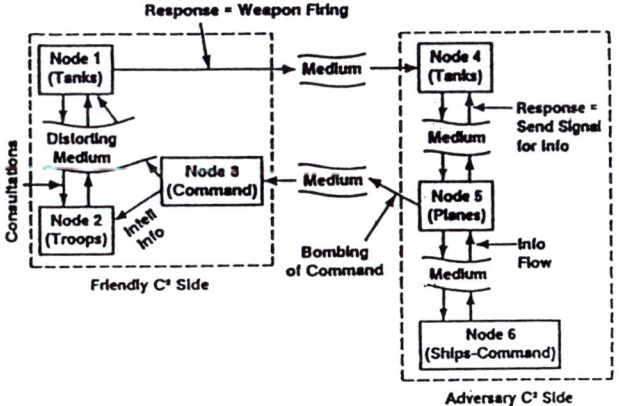

Fig.1 Qualitative aspects: external dynamics of C^3 processes simplified.

pothesize, O = Options available, R = Response). Externally, the model can be implemented via standard probability ideas, but internally, two factors involving nonstandard concepts are treated: incorporation of linguistic-based or narrative information and utilization of conditioned information, when the antecedents of the conditioning differ. More details of this will be presented in the following section.

Before proceeding to the development of the formal theory, a summary of the key ideas in describing and analyzing C^3 systems as viewed here is given in Figs. 1-5. Figure 1 illustrates a typical interaction of C^3 nodes. The SHOR paradigm is outlined in Fig. 2, with the basic evolution cycle of node signal processing shown in Fig. 3. Figure 4 outlines how knowledge flows, in general, in carrying out a formal theory; and, finally, Fig. 5 illustrates the decomposition of a C^3 node state into its proper and knowledge parts.

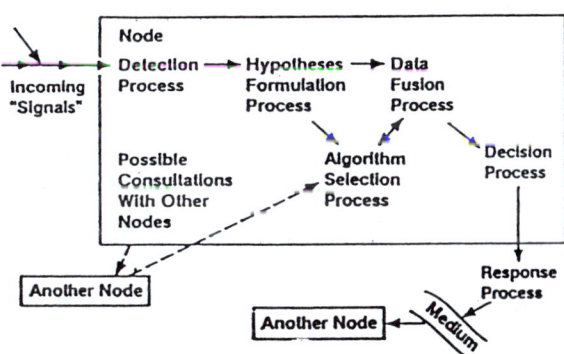

Fig.2 Qualitative aspects: internal dynamics of C^3 processes simplified.

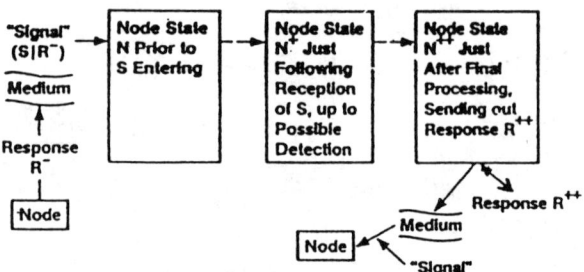

Fig.3 Qualitative aspects: basic evolution cycle of a node due to "signal" processing and response.

Rationale for Use of Formal Theory

Basically, all scientific disciplines are concerned with developing theories as comprehensive as possible. This is both to explain past empirically obtained data and to predict as accurately as possible likely future performance or behavior. Given sufficient specialization and localization, these goals have been realized to varying degrees of success in many areas comprising the "hard" sciences. These include: mechanics, biology, and chemistry; the more abstract-rooted, but related fields of statistical communications and information; and the yet more abstract general fields of mathematics and logic. On the other hand, much less success has been achieved in developing explanatory theories for the "soft" sciences related to human thought and relations, including natural language, cognition, psychology, sociology, and law.

At the outset, it must be pointed out that any attempt at describing systematically C^3 ought to span both soft and hard sciences. This is due to the interdependencies of the following three factors: 1) necessary physical actions and

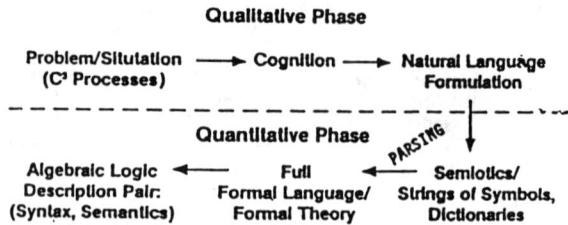

Fig.4 Quantitative aspects: knowledge flow in describing the information present.

FORMAL THEORY C3

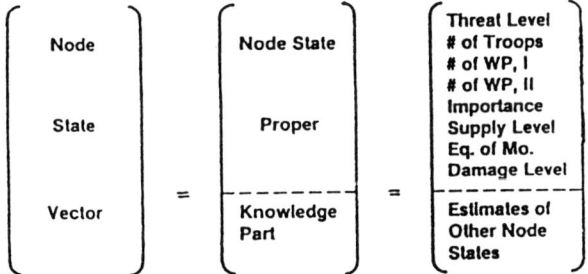

NODE = (NODE STATE , NODE STRUCTURE)

Fig.5 Quantitative aspects: components of C^3 node states.

effects (possibly deadly) involved in moving about men and supplies relative to the execution of weapons and resulting damage given and received; 2) decision-making schemes used in carrying out all aspects of part 1; and 3) information content present upon which part 2 operates (in this case both sensor-oriented (stochastic) and human-oriented (linguistic) information may be present).

In synthesizing the above mentioned concepts into a coherent whole, it is reasonable to attempt a full formal approach, drawing from previous more localized situation-specific analyses of C^3 systems. Such a comprehensive formal theory of C^3 can help in the long run to place it more within the realm of science, rather than just art and empiricism. Most importantly, such a framework is relatively easily amenable to changes, such as for finer tuning or modifications of relations, when deemed necessary, and, as well, exhibits the basic logical relations and algebraic structure for all the variables.

Examples of formal theory abound. To illustrate this point, see, e.g., the work of Woodger[9] in biology, Carnap[10] for aspects of sociology and law, and Jammer[11] for a good survey of the several attempts to axiomatize quantum mechanics. Also, by using a formal theory, C^3 systems can be analyzed from a more universal mathematical-logical viewpoint. In particular, the newly developed calculus of conditional events [12,13] becomes readily applicable to more consistent modeling of combination of conditional evidence, as part of data fusion.

The theory outlined in the subsequent sections is based upon a distillation of the work found in Refs. 14-18.

C^3 Variables

In building the formal theory, one must first scope out the relevant variables describing the system. Generally, these variables are indicated by the end of the alphabet letters in italics as X, Y, Z. Particular variables are denoted by other letters such as R denoting response of a node, N for the entire node state, and ALG for the algorithm selected for a given node following "signa;" reception (the quotes later to be explained), etc. Two specifically designated variables are actually constants: Ω for the universal or always true event or action and $\emptyset$ for the null or always false event or action.

Each variable, where necessary, indicates through appropriate subscripting or superscripting, the time, the hostile vs. friendly status of the C^3 system, as well as the identification number. In addition, each variable X has associated with it a natural domain of values that the variable can achieve. Call this $\text{dom}(X)$. Depending on the nature of the variable X, $\text{dom}(X)$ consists of, usually, a collection of subsets of a given set, or, in particular, of a collection of singleton "points" making up the parent set $\text{dom}(X)$.

It should also be noted that all axioms involving variables can be converted to corresponding ones with any domain value substituted for the corresponding variable. For convenience, variables are divided into two basic types: intranodal and internodal, with further subdivisions where warranted.

Intranodal Variables

These designated variables describe the functioning of a typical C^3 decision node. Three subdivisions arise: node state proper, knowledge aspect, and node structure.

1) N denotes the ensemble of node state proper variables for a typical node. Some examples include: TRP, the number of troops present, and EQ, the true equations of motion of the entire node, such as straight line constant velocity, second degree motion in a parabolic path, and circular tangential motion, etc. Also, WP6 indicates number of weapons of type 6 present in the node and DAM damage level to the node so far. Thus, one can write typically,

$$N = (\ldots, TRP, EQ, \ldots, WP6, \ldots, DAM, \ldots,) \qquad (1)$$

filling in the appropriate variables.

2) K denotes the collection of knowledge-related variables for the node of interest. Generally, this is taken here to be the estimates of the variables belonging to all other nodes, friendly or hostile. In many cases, this will be vacuous from lack of pertinent information. Thus, e.g., one might have

FORMAL THEORY C3

$\widehat{WP6}_{i,j}$, the hat indicating node i's estimate of WP6 relative to node j. A typical example of K can be

$$K_3 = (\ldots, \widehat{EQ}_{3,7}, \ldots, \widehat{WP4}_{3,2}, \ldots,) \qquad (2)$$

3) T denotes the collection of variables describing the actual functioning of the node. These include: Det, detection; HYP, hypotheses formulation; ALG, algorithm selection; FUS, data fusion; and DEC, decisions; all based upon incoming "signal" S. The quotes around S (deleted subsequently) refer to the fact that S could be a signal in the classical sense or an incoming weapon about to explode, or any other physical or sensory interaction between nodes. Thus, one could write

$$T = (DET, HYP, ALG, FUS, DEC, \ldots,) \qquad (3)$$

Internodal Variables

The second type of C^3 variable is the internodal or between-node type. These variables describe the factors present that affect and relate one decision node with another. These include R, the node response following all data/signal processing of signal S and S itself.

The fundamental relationship between an outgoing node response becoming eventually a signal relative to another node or nodes is determined by the intervening environment or medium which can distort and/or produce "additive" (in some algebraic sense such as ordinary arithmetic addition or multiplication) noise to the original response. Symbolically, one has the general regression relation

$$S = G(R) \oplus Q \qquad (4)$$

where internodal variable G is actually a numerical valued (vector or scalar) function which is possibly nonlinear in R, and Q represents additive error. (Other relations among internodal and intranodal variables will be considered in the next sections.) Thus, the internodal variables can be displayed as, say

$$J = (S, R, G, Q) \qquad (5)$$

with, of course, suitable time and node identifier indices.

Examples of Domains of Variables and Transforms of Linguistic Information to Stochastic

Some examples of domains are:

$$\text{dom(TRP)} = \{0,1,2,\ldots,6000\}, \quad \text{dom(EQ)} = \{e(s,v,a): s \in S_0, v \in V_0, a \in A_0\}$$

where $e(s,v,a)$ indicates constant acceleration, two-dimensional motion with initial position s, initial velocity v, and constant acceleration (possibly 0) a, where S_0, V_0, A_0 are suitably chosen sets of 2×1 real vectors.

Some additional examples worth noting:

$$\text{dom(ALG)} = \{\text{ALG1, ALG2},\ldots,\text{ALG31}\}$$

where ALG1 is a piecewise linear Kalman filter, ALG2 is an alpha-beta filter,..., ALG17 a hypotheses tester which assumes the general linear regression model, ALG18 a hypotheses tester based upon AI procedures,...

$$\text{dom(DET)} = \{\text{no detect, detect}\}, \quad S = \{S_1, S_2, \ldots, S_{23}\}$$

where each S_i is a linguistic or stochastic variable. The following are examples of linguistic variables:

S_1 = ship appears short, maybe under 300 ft long

S_2 = ship appears to be of medium length, maybe in the neighborhood of 200-400 ft long

S_3 = ship appears to be very wide and in fog a reddish flag was spotted

On the other hand, $S_4, S_5, \ldots, S_{23}$ can represent stochastic variables, such as

$$S_5 = \text{ship}_1 \text{ is 4.8 miles from ship}_2$$

where the above outcome is assumed to come from an exponential distribution with parameters $\lambda = 0.7$ miles and $\sigma = 2.1$ miles, so that S_5 represents the outcome of a random variable with a well-defined distribution which is known.

In the case of S_5 and other linguistic-based descriptions, one can utilize a technique (see Ref. 19) which converts first the linguistic description to a fuzzy set or possibilistic form and then to a random set structure, or equivalently, to a cdf (cumulative distribution function). For example, S_3 can be stated as

$$S_3 = [ht(ship) \varepsilon\, very(long)] \cdot [col(flag) \varepsilon\, red \mid weath\, \varepsilon\, fog]$$

where the symbols ε and $\cdot$ above refer to formal attribute membership and conjunction, respectively, and where the domain of values is, e.g.,

$$dom(S_3) = \underbrace{[0',1000']}_{= A_{3,1}} \times \underbrace{[degrees\ of\ redness}_{= A_{3,2}}\ for\ a\ scale]$$

The symbol $\mid$ refers to conditioning. (See the next section for further explication.) Here, S_3 corresponds to the fuzzy set (membership function) $g_3 : dom(S_3) \to [0,1]$ in the compound form

$$g_3(x,y) = g_{3,1}(x) \odot g_{3,2}(y);\ x = ht(ship),\ y = col(flag)$$

where functions $g_{3,1} : A_{3,1} \to [0,1]$ and $g_{3,2} : A_{3,2} \to [0,1]$ are both obtained from expert prior advice and intelligence information. The range values of the $g_{3,i}$ are possibilities: in general, representing overlapping compound events, and, hence, not necessarily disjoint probabilities. The operator $\odot$ is not necessarily multiplication and is obtained following the specification of the stochastic interpretation. Each $g_{3,i}$ can be identified with the one point coverage probability of random set $g_{3,i}^{-1}[U_{3,i},1]$, or, equivalently, as the probability of

$$(U_{3,i} \leq g_{3,i}(x))_{x \in A_{3,i}}$$

or, equivalently,

$$([U_{3,i}^{-1}[0, g_{3,i}(x)]])_{x \in A_{3,i}}$$

Each $U_{3,i}$ is a random variable uniformly distributed over the unit interval $[0,1]$ and the joint distribution of $U_{3,1}$ and $U_{3,2}$, as well as with other similarly introduced uniform $[0,1]$ random variables, is determined by experts or from previous knowledge. In particular, one extreme case is where the $U_{3,i}$ are all identical; another is where they are the negation (unity minus the value) of each other; an intermediate case is where they are all statistically independent, among an infinity of other possible levels of correlation. All of this corresponds to choices of the operator $\odot$, called a *copula* in the literature. (See Ref. 19, section 2.3.6.)

In summary, all C^3 variables can be expressed as states of random variables, or in a related form, as collections of such descriptions, indexed by the points in the associated domains when the variables are linguistic in nature.

Unconditional Logical Operators / Relations

Following the determination of all variables and the appropriate transforms and domains of variables, logical operations are considered next. These are merely formal counterparts for the ordinary set and classical logic operators: · ("and", conjunction, etc.), v ("or", disjunction, etc.), ( )' ("not", negation, complement, etc.). As usual, these operators obey the laws of boolean algebra relative to any variables (or their domain values). Thus, if X,Y,Z are any C^3 variables, provided it is meaningful to apply any of these operators throughout a given relation, one has[20]:

$$X*(Y*Z) = (X*Y)*Z \qquad \text{associativity} \qquad (6)$$

$$X*X = X \qquad \text{idempotency} \qquad (7)$$

$$X*Y = Y*X \qquad \text{commutativity} \qquad (8)$$

for $* = \cdot, v$.

$$\emptyset v X = X = X v \emptyset$$
$$\Omega \cdot X = X = X \cdot \Omega \qquad \text{identities} \qquad (9)$$

$$X \cdot (Y v Z) = (X \cdot Y) v (X \cdot Z); \; X v (Y \cdot Z) = (X v Y) \cdot (X v Z) \qquad (10)$$
$$\text{distributivity}$$

$$(X \cdot Y)' = X' v Y'; (X v Y)' = X' \cdot Y' \qquad \text{deMorgan} \qquad (11)$$

$$X'' = X \qquad \text{involution} \qquad (12)$$

$$\emptyset' = \Omega \; ; \; \Omega' = \emptyset \qquad \text{zero-unity dual} \qquad (13)$$

$$X \cdot X' = \emptyset \; ; \; X v X' = \Omega \qquad \text{orthocomplements} \qquad (14)$$

$$X v (X \cdot Y) = X = X \cdot (X v Y) \qquad \text{absorption} \qquad (15)$$

noting that all of the above axioms for boolean relations are not independent of one another, but are presented for purpose of completeness and clarity.

In addition, one has the basic partial (lattice) order (corresponding to subset inclusion)

$$X \leq Y \quad \text{iff} \quad X = X \cdot Y \quad \text{iff} \quad Y = X v Y \qquad (16)$$

where the term "iff" means "if and only if", i.e., logical equivalence. Strict order < (corresponding to proper subset inclusion) holds when ≤ holds, but = does not, i.e.,

$$X < Y \quad \text{iff} \quad (X \leq Y \text{ and } X \neq Y) \qquad (17)$$

FORMAL THEORY C3

Finally, unless otherwise indicated, the normalization axiom will be assumed here for all variables of interest

$$(\ldots x \vee x \vee \ldots) = \vee_{x \in \mathrm{dom}(X)} x = \Omega \tag{18}$$

Thus, the domain of X is a possibly overlapping but exhaustive covering of Ω.

Conditional Logical Operators / Relations

Although many readers of this work will be familiar with the axioms characterizing boolean algebra of unconditional classical logical relations presented in the previous section, few will recognize the following extension to *conditional logic* operators and relations; yet such conditioning plays a key role in many of the problems arising in C^3 and elsewhere. Because of historical reasons a gap has existed between conditioning in probability and that in classical logic. In Refs. 12,13 this is rectified through the rigorous derivation of a sound, and practical to implement, calculus of operators and relations. For example, if one wishes to evaluate the expression $p(s)$, where p is an ordinary probability evaluation and

$$s = \text{if event } b \text{ occurs, then } a \text{ happens or, if } d \text{ occurs, then so does } c \tag{19}$$

where, e.g.,

  a = enemy resupplies sector A ,
  b = enemy has increased men at sector C
  c = enemy will advance against us
  d = enemy has increased supply at sector B

no current standard probability procedure exists at present for dealing with the modeling of s that is both mathematically sound and efficient and that is also compatible with the usual interpretations for conditional probability evaluations

$$\begin{aligned} p(\text{if } b, \text{ then } a) &= p(a \mid b) = p(a \cdot b)/p(b) \\ p(\text{if } d, \text{ then } c) &= p(c \mid d) = p(c \cdot d)/p(d) \end{aligned} \tag{20}$$

provided that $p(b), p(d) > 0$.

On the other hand, the new development permits the full evaluation of $p(s)$, where s is as in Eq.(19), as

$$p(s) = p[(a \cdot b) \vee (c \cdot d) \mid (a \cdot b) \vee (c \cdot d) \vee (b \cdot d)] \tag{21}$$

thus obtainable through the usual evaluation of conditional probabilities as well as unconditional ones.

The above problem holds because of the appearance of *differing antecedents* in the conditional information present. In any case, the new axioms or laws governing the behavior of conditional events of the form $(X|Y)$, read as "if Y, then X" or "X given Y", are, for all (unconditional) X,Y,Z,W,

$$p[(X|Y)] = p(X|Y) \quad \text{evaluations} \quad (22)$$

$$(X|\Omega) = X \; ; \; (X \cdot Y|Y) = (X|Y) \quad \text{fundamentals} \quad (23)$$

$$(X|Y) = (W|Z) \text{ iff } [X \cdot Y = W \cdot Z \text{ and } Y = Z] \text{ pairwise equiv.} \quad (24)$$

$$[(X|Y)|(W|Z)] = (X \cdot Y \cdot W \cdot Z \mid Y \cdot [(W \cdot Z) v (X' \cdot Z')]) \quad (25)$$

$$\text{higher order cond.}$$

$$(X|Y)' = (X'|Y) \quad \text{negation} \quad (26)$$

$$(X|Y) \cdot (W|Z) = [X \cdot Y \cdot W \cdot Z \mid (X' \cdot Y) v (W' \cdot Z) v (Y \cdot Z)] \quad (27)$$

$$\text{conjunction}$$

$$(X|Y) v (W|Z) = [(X \cdot Y) v (W \cdot Z) \mid (X \cdot Y) v (W \cdot Z) v (Y \cdot Z)] \quad (28)$$

$$\text{disjunction}$$

$$(X|Y) \times (W|Z) = (X \times \Omega_1 \mid Y \times \Omega_1) \cdot (\Omega_2 \times W \mid \Omega_2 \times Z)$$

$$= [(X \cdot Y) \times (W \cdot Z) \mid ((X' \cdot Y) \times \Omega_2) v (\Omega_1 \times (W' \cdot Z)) v (Y \times Z)] \quad (29)$$

$$\text{cartesian product}$$

Partial ordering over boolean algebras is extended here and characterized as

$$(X|Y) \leq (W|Z) \quad \text{iff} \quad (X|Y) = (X|Y) \cdot (W|Z)$$

$$\text{iff} \quad (W|Z) = (X|Y) v (W|Z)$$

$$\text{iff} \quad X \cdot Y \leq W \cdot Z \quad \text{and} \quad W' \cdot Z \leq X' \cdot Y \quad (30)$$

with a natural restriction for strict order $<$, analogous to Eq.(17).

All of the above leads to an algebraic structure for the set of all conditional events $(X|Y)$, that though not quite boolean, is a relatively pseudocomplemented lattice which is also a Stone algebra with additional properties. (Again, see Refs. 12,13 for further properties. Author's note: Since the original writing of this paper, a new conditional event algebra has been developed based upon product probability space considerations which possesses more desirable properties than the one proposed in the above references, but requires more implementation time. One of the desirable properties of the newly proposed algebra is that it is actually boolean. See Ref. 21 for more details.)

Some Specific SHOR Paradigm Relations as Axioms

With the general logical structure of variable relations established, the remaining axioms required to specify the formal C^3 theory fully are now given. These relations essentially divide up into two types: weak sufficiency axioms and strong sufficiency axioms. The weak correspond to the classical sufficiency conditions in probability, and hence are dependent on the specification of particular families of cdf's. For example, when processing information, if the regression relation introduced earlier becomes a linear one and if noise Q and structure variable T are jointly distributed gaussianly where p, as before, indicates appropriate probability evaluation, and the regression relation is

$$S = B \cdot R + Q \qquad (31)$$

where B is a constant m×k real matrix of rank k, Q is m×1 S is m×1, then one has the relation

$$p(T|S) = p(T|\hat{R}); \quad \hat{R} = (B^T \cdot \text{Cov}(Q)^{-1} \cdot B)^{-1} B^T \cdot \text{Cov}(Q)^{-1} S \qquad (32)$$

Here, $(\ )^T$ denotes matrix transpose. $\hat{R}$ is the standard least squares (weighted) estimator of R through S.[22]

However, when the above assumptions do not hold, then the corresponding sufficiency condition is invalid. On the other hand, independent of the probability evaluation chosen and the specific function forms involved, the following strong sufficiency conditions hold relative to conditional event form, the +,- superscripts referring to relative times:

$$(N^{++}|R^{++} \cdot T^+ \cdot N^+ \cdot S \cdot R^- \cdot N) = (N^{++}|R^{++} \cdot N^+) \qquad (33)$$

$$(R^{++}|T^+ \cdot N^+ \cdot S \cdot R^- \cdot N) = (R^{++}|\text{DEC}^+ \cdot N^+) \qquad (34)$$

$$(T^+|N^+ \cdot S \cdot R^- \cdot N) = (T^+|N^+) \qquad (35)$$

etc., where all of the above are derived as reasonable fits to the sequence of data processing occurring within a typical node during the SHOR paradigm (see Figs. 2,3). A longer list of strong sufficiency relations can be found in Ref. 17, p. 97. Further subdivisions of variables such as for T and S

can lead to additional relations, e.g. where

$$(DEC|S \cdot DET \cdot HYP \cdot ALG) = (DEC|S) \vee (DEC|ALG) \qquad (36)$$

to reflect possible man-over-ride relative to use of algorithms available for incoming signals.

Theorems Deduced from the Formal Theory

In summary, the formal theory of C^3 consists of the usual alphabet with appropriate sub- and superscripts to indicate time and node identification: Eqs. (1-5) and (33-35) [with additional axioms representing further subdivisions of relations such as in Eq.(36)] representing C^3 proper relations; Eqs.(6-18) representing the unconditional classical logical operators and relations constituting boolean algebra; Eqs. (20-30) representing the conditional extension of logical operators and relations; and finally, the evaluations and interpretations furnished in the examples given in Eqs.(19), (31-32), and in other parts. In addition, the preliminaries to implementing the theory include the evaluation of specific domains of variables and the replacement of linguistic descriptions by stochastic ones, as detailed earlier.

Next, a simple list of results is given in the form of Theorems 1-3, leading, in turn, to the chief results provided in Theorems 4,5, where the data processing cycle of a typical node according to the SHOR paradigm is quantified recursively.

Theorem 1. Equal antecedent case for combining conditionals.

For any C^3 variables $X_1,\ldots,X_n,Y$ and logical operators, such as $\cdot$ and $\vee$, or any well-defined combination of them, indicated by $*$,

$$(X_1|Y) * \cdots * (X_n|Y) = (X_1 * \cdots * X_n | Y)$$

Proof: Use conditional event algebra axioms specialized to the equal antecedent case [see Eqs.(26-28) with $Y = Z$]. ∎

Theorem 2. Conditional forms in expanded disjunctive expressions of auxiliary variables.

For any C^3 variables X,Y and any auxiliary C^3 variables chosen for convenience, say, $Z_1,\ldots,Z_n$, assuming normalization of the Z_i,

$$(X|Y) = \bigvee_{\substack{\text{all } z_i \,\in\, \text{dom}(Z_i), \\ i=1,\ldots,m}} (X \cdot Z_1 \cdots Z_m | Y)$$

Proof: Combine Theorem 1 with normalization (18), associativity and identity extended to conditionals (as a result of applications of assumptions (22-28)). ∎

Theorem 3. Fundamental chaining relation for conditionals.

For any C^3 variables $X, Y, Z_1, \ldots, Z_m$, the following relation always holds

$$(X \cdot Z_1 \cdot \ldots \cdot Z_m | Y) = (X | Z_1 \cdot \ldots \cdot Z_m \cdot Y) \cdot (Z_1 | Z_2 \cdot \ldots \cdot Z_m \cdot Y) \cdot \ldots$$
$$\cdot (Z_{m-1} | Z_m \cdot Y) \cdot (Z_m | Y)$$

Proof: Apply iteratively the conjunction axiom (27) to the right-hand side above. ∎

Theorem 4. Formal recursive expansion of evolving node states in simplified form.

$$(N^{++} | N^+) = \bigvee_{\begin{bmatrix} \text{all}: R^{++} \varepsilon \text{ dom}(R^{++}), \\ DEC^+ \varepsilon \text{ dom}(DEC^+), \\ \ldots\ldots\ldots\ldots\ldots ; \\ DET^+ \varepsilon \text{ dom}(DET^+) \end{bmatrix}} F(N^{++}, N^+; R^{++}, DEC^+, \ldots, DET^+)$$

where

$$F(N^{++}, N^+; R^{++}, DEC^+, \ldots, DET^+) = (N^{++} | R^{++} \cdot N) \cdot (R^{++} | DEC^+ \cdot N^+)$$
$$\cdot (DEC^+ | FUS^+ \cdot HYP^+ \cdot ALG^+ \cdot DET^+ \cdot N^+)$$
$$\ldots\ldots\ldots\ldots\ldots\ldots\ldots\ldots\ldots\ldots\ldots$$
$$\cdot (ALG^+ | DET^+ \cdot N^+) \cdot (DET^+ | N^+)$$

Each of the above factors can be decomposed into further chains where required.

Proof: Combine Theorems 2 and 3 and use, e.g., Eqs. (33-35). ∎

Theorem 5. Probability evaluation of evolving node states for SHOR paradigm.

Let p be any probability measure. Then, with the usual assumptions,

$$p(N^{++} | N^+) = \sum_{\begin{bmatrix} \text{all}: R^{++} \varepsilon \text{ dom}(R^{++}), \\ DEC^+ \varepsilon \text{ dom}(DEC^+), \\ \ldots\ldots\ldots\ldots\ldots ; \\ DET^+ \varepsilon \text{ dom}(DET^+) \end{bmatrix}} p[F(N^{++}, N^+; R^{++}, DEC^+, \ldots, DET^+)]$$

where

$$p[F(N^{++},N^+;R^{++},DEC^+,\ldots,DET^+)] = p(N^{++}|R^{++}\cdot N^+)\cdot p(R^{++}|DEC^+\cdot N^+)$$
$$\cdot p(DEC^+|FUS^+\cdot HYP^+\cdot ALG^+\cdot DET^+\cdot N^+)$$
$$\cdots\cdots\cdots\cdots\cdots\cdots\cdots\cdots\cdots\cdots\cdots$$
$$\cdot p(ALG^+|DET^+\cdot N^+)\cdot p(DET^+|N^+)$$

Proof and Important Remark: The above result follows immediately from use of Eq.(22) and the basic properties of conditionals and conditional probabilities. Although Theorem 5 could be obtained rather easily as a standard application of the total probability theorem in conjunction with probabilistic chaining, the point here is that within the chaining factors above one can expand at will -by non-chaining forms - various probabilities of the conditionals using the conditional event algebra relations given in Eqs.(23-29). For example, note that Eq.(36) or a related form can be used to evaluate the factor $p(DEC^+|FUS^+\cdot HYP^+\cdot ALG^+\cdot DET^+\cdot N^+)$. (Or, see Eqs.(19,28,22).) ∎

Implementation of the Theory and the C^3 Design Game

Applying the above outlined theory to a particular C^3 setting requires specification of all appropriate variables and their possible distributions.

Because of the microscopic nature of this approach, an exponential growth can be expected, in general, for the computations involved as the number of variables is increased for fidelity of modeling. One technique for possible reduction of this load is outlined in Ref. 23, where a combination of an "exact" linearization procedure is utilized with gaussian sum expansions of distributions. Another is the judicious use of key relations and the omission or simplification of others. In Ref. 24, Girard outlines such an implementation of theory for a reduced version of the Naval outer-inner air battle, where a blue fighter engages an orange bomber in the outer zone. The full-scale implementation of this is yet to be developed which will include modeling of missile launches, counterattacks, and maneuvering. See also Ref. 23 for an outline of an implementation scheme related to the outer-inner air battle and Ref.5 for the basic presentation of the numerically-oriented Markov process type of approach of Rubin and Mayk.

It is intended that the outputs of the model as developed here be used in producing a full C^3 design game. Here, the adversary and friendly forces are identified with the possible choices one can make subject to the constraints of terrain,

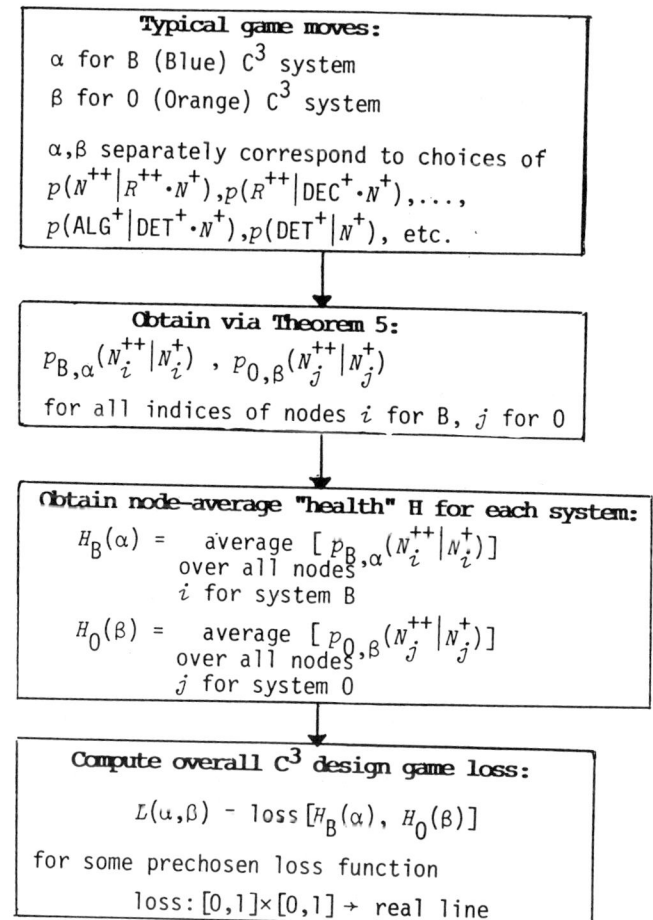

Fig.6 Outline of C^3 design game.

politics, resources, etc., for the functional forms of the various conditional probability distributions (or cdf's) that can be chosen among the C^3 variables. A summary of a generic C^3 design game following these ideas is presented in Fig. 6.

References

[1] Goodman, I.R., "Combination of Evidence in C^3 Systems", *Proc. 8th MIT/ONR Workshop on C^3 Systems*, Laboratory for Information Decision Systems, Massachusetts Inst. of Tech., Cambridge, MA, Dec. 1985, pp. 161-166.

[2] Ingber, L., "Nonlinear Nonequilibrium Statistical Mechanics Approach to C^3 Systems", *Proc. 9th MIT/ONR Workshop on C^3 Systems*,

Laboratory for Information Decision Systems, Massachusetts Inst. of Tech., Cambridge, MA, Dec. 1986, pp. 237-244.

[3]Ingber,L., "C^3 Decision Aids:Statistical Mechanics Application of Biological Intelligence", *Proc. 1987 Symposium on C^2 Research,* Science Applications Intern. Corp., McLean, VA, Sept.1987, pp. 49-57.

[4]Rubin, I. and Mayk, I., "Markovian Modeling of Canonical C^3 Systems Components", *Proc. 8th MIT/ONR Workshop on C^3 Systems,* Laboratory for Information Decision Systems, Massachusetts Inst. of Tech., Cambridge, MA, Dec., 1985, pp. 15-23.

[5]Rubin, I., Baker, J., and Mayk, I., "A Stochastic Model for the Naval Multi-Phase Outer Air and Inner Air Battles", *Proc.1989 Symposium on C^2 Research,* Science Applications Intern. Corp., McLean, VA, Aug. 1989, pp. 48-56.

[6]Levis, A.H., "Information Processing and Decisionmaking Organizations:A Mathematical Description", *Proc.6th MIT/ONR Workshop on C^3 Systems,* Laboratory for Information Decision Systems, Massachusetts Inst. of Tech., Cambridge, MA, Dec. 1983, pp. 30-38.

[7]Tomovic, M.M. and Levis, A.H., "On the Design of Organizational Structures for Command and Control", *Proc.7th MIT/ONR Workshop on C^3 Systems,* Laboratory for Information Decision Systems, Massachusetts Inst. of Tech., Cambridge, MA, Dec. 1984, pp. 131-138.

[8]Cothier, P. and Levis, A.H., "Assessment of Timeliness in Command and Control", *Proc.8th MIT/ONR Workshop on C^3 Systems,* Laboratory for Information Decision Systems, Massachusetts Inst. of Tech. Cambridge, MA, Dec.1985, pp. 39-48.

[9]Woodger, J.H., "The Technique of Theory Construction", *Encyclopedia of Unified Science,* Vol.2, No.5, Univ. of Chicago Press,1956.

[10]Carnap, R., *Introduction to Symbolic Logic and Its Applications,* Dover, New York, 1958.

[11]Jammer, M., *Philosophy of Quantum Mechanics,* Wiley, New York, 1974, sections 8.4, 8.5.

[12]Goodman, I.R., "A Measure-Free Approach to Conditioning", *Proc. 3rd Workshop on Uncertainty in Artificial Intelligence,* Univ. of Washington, Seattle, WA, July, 1987, pp. 270-277.

[13]Goodman, I.R., Nguyen, H.T., and Walker, E.A., *Conditional Inference and Logic for Intelligent Systems: A Theory of Measure-Free Conditioning,* North-Holland, Amsterdam, Neth., 1991.

[14]Goodman, I.R., "A General Theory for the Fusion of Data", *Proc. First Tri-Service Data Fusion Symposium,* Applied Physics Lab., Johns Hopkins Univ., Laurel, MD, Dec.1987, pp. 254-270.

[15]Goodman, I.R., "Applications of a Conditional Event Algebra to Data Fusion", *Proc. 2nd Tri-Service Data Fusion Symposium,* Applied Physics Lab., Johns Hopkins Univ., Laurel, MD, Dec. 1988, pp.179-187.

[16]Goodman, I.R., "Applications of a Conditional Event Algebra to Data Fusion: Part 2", *Proc. 3rd Tri-Service Data Fusion Symposium,* Applied Physics Lab., Johns Hopkins Univ., Laurel, MD, Dec. 1989, pp. 181-193.

[17]Goodman, I.R., "Toward a General Theory of C^3 Processes", *Proc. 1988 Symposium on C^2 Research,* Science Applications Inter.Corp.,McLean, VA, Dec. 1988, pp. 92-105.

[18]Goodman, I.R., "Toward a General Theory of C^3 Processes: Part 2", *Proc. 1989 Symposium on C^2 Research,* Science Applications Inter.Corp., McLean, VA, Dec. 1989, pp. 57-67.

[19]Goodman, I.R. and Nguyen, H.T., *Uncertainty Models for Knowledge-Based Systems,* North-Holland, Amsterdam, Neth., 1985.

[20]Mendelson, E., *Boolean Algebra and Switching Circuits*, Schaum Outline Series, McGraw-Hill, New York, 1970.

[21]Goodman, I.R. and Nguyen, H.T., "A Theory of Conditional Information for Probabilistic Inference in Intelligent Systems: Part 2, Product Space Approach", to appear in *Information Sciences*.

[22]Rao, C.R., *Linear Statistical Inference and Its Applications*, Wiley, New York, 1973.

[23]Goodman, I.R., "Applications of an Exact Linearization-Gaussian Sum Technique to the Modeling of C^3 Nodes", *Proc. 1990 Symposium on C^2 Research*, Science Applications Intern. Corp., McLean, VA, Sept. 1990, pp. 63-72.

[24]Girard, P.E., "A Combat and Decision Model Based on Conditional Probability Logic", *Proc. 1989 Symposium on C^2 Research*, Science Applications Intern. Corp., McLean, VA, Aug. 1989, pp. 37-47.

Statistical Mechanics of Combat and Extensions

Lester Ingber*
Lester Ingber Research, McLean, Virginia 22101

I. Introduction: C^2 in Training and Computer Models

A. Necessity of Comparing Computer Models to Exercise Data

This project addresses the utility of establishing a mathematical approach to compare exercise data to large scale computer models whose underlying microscopic interactions among men and machines are driven by the natural laws of physics. In this study, the focus is to compare the Janus(T) wargame to National Training Center (NTC) data, since both systems then take into account human interactions.

It also should be noted that "large scale" here refers to battalion level. (Army systems scale by factors of 3–5, from company to battalion to brigade to division to corps to army.) If these battalion level computer models can be favorably compared, and if consistency can be achieved between the hierarchy of large scale battalion level, larger scale corps level, and largest scale theater level computer models, then these higher echelon computer models also can be favorably compared. This could only enhance the value of training on these higher echelon computer models.[1]

The requirement of depending more and more on combat computer models (including simulations and wargames) has been brought into sharper focus because of many circumstances, e.g.: (1) the nonexistence of ample data from previous wars and training operations, (2) the rapidly shortening time scale on which tactical decisions must be made, (3) the rapidly increasing scale at which men and machines are to be deployed, (4) the increasing awareness of new scenarios that are fundamentally different from historical experiences, (5) and the rapidly increasing expense of conducting training exercises.

Furthermore, such computer models could be used to augment training. We now spend several million dollars to cycle each battalion through NTC. The

---

Copyright © 1993 by the American Institute of Aeronautics and Astronautics, Inc. All rights reserved.
*President, P.O. Box 857; *EMail:* ingber@alumni.caltech.edu.

training of these commanders could be greatly enhanced if inexpensive pre and post training wargames were provided that statistically replicate their training missions. Even, or rather especially, for the development of such training aids, proper analysis and modeling is required to quantitatively demonstrate that the computer models are good statistical representations of the training mission.

However, the level of acceptance of computer models in major military battle-management and procurement decisions appears to be similar to the level of acceptance of computer simulations in physics in the 1960s. In physics, prior to the 1960s, theory and experiment formed a close bond to serve to understand nature. In the 1960s, academicians were fascinated with evolving computer technology, but very few people seriously accepted results from computer simulations as being on a par with good theory and good experiment. Now, of course, the situation is quite different. The requirement of understanding truly complex systems has placed computer simulation, together with theory and experiment, as an equal leg of a tripod of techniques used to investigate physical nature.

The requirements necessary to bring combat computer models to their needed level of importance are fairly obvious. To have confidence in computer-model data, responsible decision makers must be convinced that computer models model reality, not metaphors of reality, models of models, or models of models of models, etc. Many people feel that not much progress has been made in the last decade[2,3] with regard to this issue, despite a general awareness of the problem.

If a reasonable confidence level in computer models of large scale combat scenarios could be obtained, there are several immediate payoffs to be gained. More objective data could be presented for procurement decisions, e.g., provided by sensitivity analyses of sets of computer models differing in specific weapons characteristics. To give proper weight to these differing characteristics, their influence within the global context of full combat scenarios would be tested.

B. Need for Data

With the present development of high quality simulators (essentially computer simulations coupled with hardware to simulate close to actual machinery and communication) to augment training in the field, e.g., SIMNET, it should become increasingly important to "validate" the simulator system to some reality, even if only exercise reality. For example, several hundred simulators are likely to be driven in the context of tens of thousands of units being simulated by software similar to present-day computer simulations. In fact, many people see the necessity of integrating simulations and simulators to properly augment training. Thus, the semi-automated forces (SAF) driving the simulators must be validated. We believe we offer here the appropriate methodology to carry out this program.

As we will see, proper high quality validation requires "high quality" data, recorded in "timely" fashion. The proper time scale required for validation of battalion- to brigade level combat appears to be on the order of minutes. High quality typically means accurate "killer victim" scoreboards (KVS) as a function of time. This timely and high quality data is currently only available from such exercise arenas as NTC, certainly not from records of actual combat. However, qualifying NTC data is laborious, and in these days of diminishing research and development funding, it might soon become a vanishing art form. This does not diminish the requirement for gathering more high quality data for analysis, if analysis is to properly serve training; nor does it excuse the expenditure of billions of dollars on hardware without spending at least millions on validation of software driving this machinery.

C. Large Scale C^2 and Need for Mathematical Modeling

Modeling phenomena is as much a cornerstone of 20th century science as is collection of empirical data.[4] In practically all fields of science, mathematical models of the real world become tested by fitting some parameters to empirical data. Since systems in the real world are often nonlinear and stochastic, it is not surprising that often this fitting process must involve fitting statistical, nonlinear, functional (algebraic) forms to data. They are nonlinear because typically their alternative outputs can be complicated functions of their inputs and feedback. They are stochastic because typically they have many constituents that are generally treated as aggregate entities, and/or the specification of these entities requires statistical judgment about past or future performance.

As in other fields of science, in the context of modeling combat, reductionist doctrine is simply inadequate to fully understand large scale systems. For example, a threshold is quickly reached at a level of any large system, be it physical, biological or social, when a "language" shift is required for effective command and control. A high level commander cannot use a grease board to track individual units, albeit he might periodically sample his units, but he/she must rather look at the overall systematics, e.g., aggregated measures of force (MOF) or effectiveness (MOE), attrition, resupply, etc. At this level we properly require command and control (C^2), rather than "supra-battle-management" from commanders. At this level we denote the system as large scale.

This issue of utilizing MOFs and MOEs, e.g., starting at approximately battalion level of combat, is relevant to computer models as well actual combat. Merely aggregating data to form MOFs or MOEs does not determine if results from one mission (combat or computer-model scenario) are comparable to another mission. For example, small differences in tempo or in spatial distribution of forward line of own troops (FLOT), or forward edge of the battle area (FEBA), may cause tables of numbers to appear quite different.

Mathematical models of aggregated data should be expected to uncover "mechanisms" of combat, e.g., line firing or area firing in simple Lanchester theory. More complex missions plausibly will contain more subtle mechanisms as well as weighted contributions of more basic mechanisms, e.g., quantification of dynamic "force," "stress," "strain," "momentum," "energy," etc., as is possible with this approach. Using this as hindsight, in some systems it may then be possible to specify a figure of merit, some simple set of numbers to capsulate the influence of these mechanisms.

It must be emphasized that this approach requires an evolution of knowledge. This project is developing models suitable to describe the statistical nature of selected force-on-force battalion and brigade scenarios. It is expected that the accumulation of models of many types of scenarios will lead to a better fundamental understanding of combat with direct operational applications.

D. Models Versus Reality

It must be stated that there are still many problems faced by all computer models of combat which must be solved before they can be accepted as models of reality. For example, a very basic problem exists in the quality of acquisition algorithms, i.e., how to construct an algorithm that realistically portrays human attention (preattentive as well as selective) and perception, under various combat and weather conditions, night vs day, etc. The influence of attention and perception on complex physical[5-7] and mental tasks[8,9] has received considerable attention by the author. Currently, the best combat computer models treat acquisition as serial and logical processes, whereas the human brain acquires data by parallel and associative processes. Therefore, the inclusion of human players in multiple runs of similar scenarios is essential, if a probabilistic mathematical model is to be developed to model exercise data such as that obtained from NTC.

Now, line of sight (LOS) algorithms seem to be the most costly time factor in running Janus(T) computer models. Even if more realistic acquisition algorithms are developed, they must be tailored to the needs of real time computer models if they are to be used in wargaming and in training.

E. Outline of Paper

Section II describes the present day empirical approximate of the real world suitable for gathering data, e.g., NTC. We must model for reasons given above. Good physical theory always must interact with the real world, and so we must see what observed data is available.

Section III demonstrates that the natural evolution of mathematical methodology often starts out in a strikingly simple fashion describing some region of appropriate reality. Given the form that data is collected at NTC and from Janus(T), i.e., KVSs, a natural formal structure is intuited for this beginning mathematical description.

However, as is true for many physical systems, simple equations describing one patch of reality often can require quite abstract and sophisticated transformations and algorithms to faithfully calculate other empirical observables in some other region of reality. In fact, the degree to which theory can extrapolate or at least interpolate from one region of reality to another is the practical scientific method of testing theories. Therein lies much of the utility of theories, to present patterns of information to human minds, more significant than enumeration of tables of statistics and empirical data. Complex reality therefore, not surprisingly, requires complex theoretical structures, and such complex structures require good quality data to fit a theory to one region and to test the theory in other regions. In this case, nonlinear stochastic combat leads to nonlinear stochastic equations, requiring state-of-the-art methodologies much more advanced than available even only a decade ago.

Section IV presents the numerical algorithms required to faithfully carry our the calculations intended by the mathematics derived in Sec. III. The numerical algorithms required likewise are state-of-the-art. Without these numerical techniques, the mathematical formalism of Sec. III would have to be drastically undercut or shelved, leading to abuses of modeling of actual NTC and Janus data, more typical (and then excusable) of pioneering work done decades ago without such formalism and numerical techniques.

Section V gives a short outline of the relevance of our work to finding mathematical "chaos" in battalion level combat. The presence of chaos simply is not supported by the facts, tentative as they are because of sparse data.

Section VI describes an ambitious but reasonable approach to more explicitly model human factors in combat, using techniques developed for describing command and control of teleoperated vehicles.

Section VII describes extension of our statistical mechanics of combat (SMC) methodology to other scenarios. This work was done with several officer students of the author.

II. Janus Computer Simulation of National Training Center

A. Individual Performance

Clearly individual performance is extremely important in combat,[10] ranging in scale from battle management of the commander, to battle leadership of sub-commanders, to the degree of participation of individual units, to the more subtle degradation of units performing critical tasks.

Our analyses of NTC data concludes that the quality of data collected to date is not sufficient to accurately statistically judge individual performance across these scales. However, we do believe that this data is sufficient to analyze battle management and perhaps battle leadership at the company or platoon level, in some cases reflecting the influence of a human commander.

It is important to recognize and emphasize the requirement of improving data collection at NTC, to permit complementary analyses of human factors at finer scales than our statistical approach permits.

Therefore, as understood from experience in simulating physics systems, many trajectories of the "same" stochastic system must be aggregated before a sensible resolution of averages and fluctuations can be ascertained. Given two scenarios that differ in one parameter, and given a sufficient number of trajectories of each scenario, then the sensitivity to changes of a "reasonable" algebraic function to this parameter can offer some analytic input into decisions involving the use of this parameter in combat scenarios. NTC is the best source of such data, albeit it is sparse.

B. Description of National Training Center

The U.S. Army National Training Center (NTC) is located at Fort Irwin, just outside Barstow, California. As of 1989, there have been about 1/4 million soldiers in 80 brigade rotations at NTC, at the level of two battalion task forces (typically about 3500 soldiers and a battalion of 15 attack helicopters), which train against two opposing force (OPFOR) battalions resident at NTC. NTC comprises about 2500 km^2, but the current battlefield scenarios range over about 5 km linear spread, with a maximum lethality range of about 3 km. NTC is gearing up for full brigade level exercises.

Observer-controllers (OC) are present at various levels of command down to about platoon level. A rotation will have three force-on-force missions and one live-fire mission. OPFOR platoon and company level personnel are trained as US Army soldiers; higher commanders practice Soviet doctrine and tactics. An OPFOR force typically has ~100 red armored personnel carriers (BMPs) and ~40 red tanks (T72s).

The primary purpose of data collection during an NTC mission is to patch together an after action review (AAR) within a few hours after completion of a mission, giving feedback to a commander who typically must lead another mission soon afterward. Data from the field, i.e., multiple integrated laser engagement system (MILES) devices, audio communications, OCs, and stationary and mobile video cameras, is sent via relay stations back to a central command center where this all can be recorded, correlated and abstracted for the AAR. Within a couple of weeks afterwards, a written review is sent to commanders, as part of their NTC take home package. It now costs about 4×10^6 dollars per NTC rotation, 1 million of which goes for this computer support.

There are 460 MILES transponders available for tanks for each battle. The B units have transponders, but most do not have transmitters to enable complete pairings of kills targets to be made. (New MILES devices being implemented have transmitters which code their system identification, thereby greatly increasing the number or recordings of pairings.) Thus, MILESs without transmitters cannot be tracked. Man packs with B units enable these men to be tracked, but one man pack can represent an aggregate of as many as five people.

B units send data to A stations (was 48, though 68 can be accommodated), then collected by two C stations atop mountains, and sent through cables to central VAXs forming a core instrumentation system (CIS). There is a present limitation of 400 nodes in computer history for video tracking (but 500 nodes can be kept on tape). Therefore, about 200 blue and 200 OPFOR units are tracked.

By varying the laser intensity and focusing parameters, a maximum laserbeam spread is achieved at the nominal range specified by the Army. A much narrower beam can reach as far as the maximum range. Focusing and attenuation properties of the laser beam makes these nominal and maximum ranges quite sharp, with resolution supposedly considerably less than several hundred meters under ideal environmental conditions. For example, a weapon might send out a code of 8 words (spaced apart by ns), 2 of which must register on a target to trigger the Monte Carlo routine to calculate a probability of kill (PK). Attenuation of the beam past its preset range means that it rapidly becomes unlikely that 2 words will survive to reach the target.

With increasing demands to make training more realistic, the MILES devices need to be upgraded. For example, degradation of the laser beam under conditions of moderate to heavy smoke and dust might be at least partially offset by sending fewer words per message. New sensor abilities to encode specific shooters will also greatly aid data collection.

It should be understood that present training problems at NTC, e.g., training commanders—especially at Company level—to handle synchronization of more than three tasks, misuse of weapons systems, etc., overshadow any problems inherent in the MILES systems. We repeatedly have expressed this view for well over a year, after going to NTC several times; but only at a meeting at Carlisle Barracks, PA, on May 17, 1989, when various school commanders briefed Gen. Maxwell Thurman, TRADOC Commander, was this view broadly accepted.

Therefore, to the degree possible in this project, our wargaming efforts strive to place commanders under these constraints of current interest, e.g., under requirements to synchronize the timing of the movement or repositioning of forces, request for supporting fires (artillery, air strike, etc.), initiation of fires

C. Qualification Process

Missing unit movements and initial force structures were completed in the NTC database, often making "educated guesses" by combining information on the CIS tapes and the written portion of the take-home package.

This project effectively could not have proceeded if we had not been able to automate transfers of data between different databases and computer operating systems. One of the author's students, Bowman,[11] wrote a thesis on the management of the many information-processing tasks associated with this project. He has coordinated and integrated data from NTC, Training and Doctrine Command (TRADOC) Analysis Command (TRAC) at White Sands Missile Range, New Mexico (TRAC-WSMR) and at Monterey, California (TRAC-MTRY) for Janus(T) wargaming at TRAC-MTRY, and for use at at Lawrence Livermore National Laboratory (LLNL) Division B, and for Janus(T) and NTC modeling.

D. Description of Janus(T)

Janus(T) is an interactive, two-sided, closed, stochastic, ground combat computer simulation. As discussed below, we have expanded Janus(T) to include air and naval combat, in several projects with the author's previous thesis students at the Naval Postgraduate School (NPS).

Interactive refers to the fact that military analysts (players and controllers) make key complex decisions during the simulation, and directly react to key actions of the simulated combat forces. Two-sided (hence the name Janus of the Greek two-headed god) means that there are two opposing forces simultaneously being directed by two set of players. Closed means that the disposition of the enemy force is not completely known to the friendly forces. Stochastic means that certain events, e.g., the result of a weapon being fired or the impact of an artillery volley, occur according to laws of chance [random number generators and tables of probabilities of detection (PD), acquisition (PA), hit (PH), kill (PK), etc.]. The principle modeling focus is on those military systems that participate in maneuver and artillery operations. In addition to conventional direct fire and artillery operations, Janus(T) models precision guided munitions, minefield employment and breaching, heat stress casualties, suppression, etc.

Throughout the development of Janus(T), and its Janus precursor at Lawrence Livermore National Laboratory, extensive efforts have been made to make the model "user friendly," thereby enabling us to bring in commanders with combat experience, but with little computer experience, to be effective wargamers. There is now a new version, Janus(A), bringing together the strengths of these predecessors.

III. Mathematical Formalism

A. Model Development

Consider a scenario taken from our NTC study: two red systems, red T-72 tanks (RT) and red armored personnel carriers ($RBMP$), and three blue systems, blue M1A1 and M60 tanks (BT), blue armored personnel carriers ($BAPC$), and blue tube-launched optically-tracked wire-guided missiles (*BTOW*), where *RT* specifies the number of red tanks at a given time t, etc. Consider the kills suffered by *BT*, ΔBT, e.g., within a time epoch $\Delta t \approx 5$ min

$$\Delta BT/\Delta t \equiv \dot{BT} = x^{BT}_{RT} RT + y^{BT}_{RT} RT\, BT + x^{BT}_{RBMP} RBMP + y^{BT}_{RBMP} RBMP\, BT \tag{1}$$

Here, the x terms represent attrition owing to point fire; the y terms represent attrition owing to area fire. Note that the algebraic forms chosen are consistent with current perceptions of aggregated large scale combat.

Now consider sources of noise, e.g., that at least arise from PD, PA, PH, PK, etc. Furthermore, such noise likely has its own functional dependencies, e.g., possibly being proportional to the numbers of units involved in the combat. Now we write

$$\begin{aligned}\frac{\Delta BT}{\Delta t} \equiv \dot{BT} &= x^{BT}_{RT} RT + y^{BT}_{RT} RT\, BT + x^{BT}_{RBMP} RBMP + y^{BT}_{RBMP} RBMP\, BT \\ &+ z^{BT}_{BT} BT \eta^{BT}_{BT} + z^{BT}_{RT} \eta^{BT}_{RT} + z^{BT}_{RBMP} \eta^{BT}_{RBMP}\end{aligned} \tag{2}$$

where the η represent sources of (white) noise (in the Itô prepoint discretization discussed below). The noise terms are taken to be log normal (multiplicative) noise for the diagonal terms and additive noise for the off-diagonal terms. The diagonal z term (z^{BT}_{BT}) represents uncertainty associated with the *target BT*, and the off-diagonal z terms represent uncertainty associated with the *shooters RT* and *RBMP*. The x and y are constrained such that each term is bounded by the mean of the KVS, averaged over all time and trajectories of similar scenarios; similarly, each z term is constrained to be bounded by the variance of the KVS. The methodology presented here can accommodate any other nonlinear functional forms, and any other variables that can be reasonably represented by such rate equations, e.g., expenditures of ammunition or bytes of communication.[12] Variables that cannot be so represented, e.g., terrain, C^3, weather, etc., must be considered as "super-variables" that specify the overall context for the above set of rate equations.

Equations similar to the $\dot{BT}$ equation are also written for *RT*, *RBMP*, *BAPC*, and *BTOW*. Only x and y that reflect possible nonzero entries in the KVS are free to be used for the fitting procedure. For example, since Janus(T) does not permit direct-fire fratricide, such terms are set to zero. In most NTC scenarios, fratricide typically is negligible. Nondiagonal noise terms give rise to correlations in the covariance matrix. Thus, we have

$$\begin{aligned} M^G &= \{RT, RBMP, BT, BAPC, BTOW\} \\ \dot{M}^G &= g^G + \sum_i \hat{g}^G_i \eta^i \\ \hat{g}_i &= \begin{cases} z^G_i M^G, & i = G \\ z^G_i, & i \neq G \end{cases} \end{aligned} \tag{3}$$

B. Problems in Lanchester Theory

Quasilinear deterministic mathematical modeling is not only a popular theoretical occupation, but many wargames, e.g., joint Theater Level Simulation

(JTLS), use such equations as the primary algorithm to drive the interactions between opposing forces.

In its simplest form, this kind of mathematical modeling is known as Lanchester theory:
$$\dot{r} = dr/dt = x_r b + y_r rb$$
$$\dot{b} = db/dt = x_b r + y_b br \qquad (4)$$

where r and b represent red and blue variables, and the x and y are parameters that somehow should be fit to actual data.

It is well known, or should be well known, that it is notoriously difficult, if not impossible, to use the simple Lanchester equations to mathematically model any real data with any reasonable degree of precision. These equations perhaps are useful to discuss some gross systematics, but it is discouraging to accept that, for example, a procurement decision involving billions of dollars of national resources could hinge on mathematical models dependent on Lanchester theory.

Some investigators have gone further, and amassed historical data to claim that there is absolutely no foundation for believing that Lanchester theory has anything to do with reality.[13] However, although there is some truth to the above criticisms, the above conclusions do not sit comfortably with other vast stores of human experience. Indeed, this controversy is just one example that supports the necessity of having human intervention in the best of C^2 plans, no matter how (seemingly) sophisticated analysis supports conclusions contrary to human judgment.[10] That is, when dealing with a dynamic complex system, intuition and analysis must join together to forge acceptable solutions. The purpose of good theory and good data should be to ease the burden placed on the human decision maker, and to enable better decisions to be made.

We need better nonlinear stochastic theory than provided by Lanchester theory.[14] Just as important, as we have numerically detailed in this project, data to test such models must be "time dense'," i.e., available on time scales of minutes, not days or weeks as is typically given by actual combat records. Also, the kind of data required, KVSs, is hard enough to extract from laser-recorded exercises, e.g., at NTC; getting such data from combat journals is unrealistic.

C. Nonlinear Stochastic Processes

Aggregation problems in such nonlinear nonequilibrium systems typically are "solved" (accommodated) by having new entities/languages developed at these disparate scales to efficiently pass information back and forth.[12,15,16] This is quite different from the nature of quasi-equilibrium quasi-linear systems, where thermodynamic or cybernetic approaches are possible; these approaches typically fail for nonequilibrium nonlinear systems.

In the late 1970's, mathematical physicists discovered that they could develop statistical mechanical theories from algebraic functional forms

$$\dot{r} = f_r(r,b) + \sum_i \hat{g}_r^i(r,b)\eta_i$$
$$\dot{b} = f_b(b,r) + \sum_i \hat{g}_b^i(b,r)\eta_i \qquad (5)$$

where the $\hat{g}$ and f are general nonlinear algebraic functions of the variables r and b.[17-22] The f are referred to as the (deterministic) drifts, and the square of the $\hat{g}$ are related to the diffusions (fluctuations). In fact, the statistical mechanics can be developed for any number of variables, not just two. The η are sources of Gaussian-Markovian noise, often referred to as "white noise." The inclusion of the $\hat{g}$, called "multiplicative" noise, has been shown to very well model mathematically and physically other forms of noise, e.g., shot noise, colored noise, dichotomic noise.[23-25]

The ability to include many variables also permits a "field theory" to be developed, e.g., to have sets of (r, b) variables (and their rate equations) at many grid points, thereby permitting the exploration of spatial-temporal patterns in r and b variables. This gives the possibility of mathematically modeling the dynamic interactions across a large terrain. Modern computer capabilities are daily brought to bear on similar problems of this magnitude.

These new methods of nonlinear statistical mechanics have been applied to complex large scale physical problems, demonstrating that empirical data can be described by of these algebraic functional forms. Success was gained for large scale systems in neuroscience, in a series of papers on statistical mechanics of neocortical interactions (SMNI),[26-40] in nuclear physics,[41,42] and in financial markets.[43-45] These have been proposed for problems in C^3.[12,15,46-48]

Thus, now we can investigate various choices of f and $\hat{g}$ to see if algebraic functional forms close to the Lanchester forms can actually fit the data. In physics, this is the standard phenomenological approach to discovering and encoding knowledge and empirical data, i.e., fitting algebraic functional forms that lend themselves to physical interpretation. This gives more confidence when extrapolating to new scenarios, exactly the issue in building confidence in combat computer models.

The utility of these algebraic functional forms goes further beyond their being able to fit sets of data. There is an equivalent representation to these stochastic differential equations, called a "path integral" representation for the long-time probability distribution of the variables. This short-time conditional probability distribution is driven by a "Lagrangian," that can be thought of as a dynamic algebraic "cost" function. The path integral representation for the long-time distribution possesses a variational principle, which means that simple graphs of the algebraic cost-function give a correct intuitive view of the most likely states of the variables, and of their statistical moments, e.g., heights being first moments (likely states) and widths being second moments (uncertainties). Like a ball bouncing about a terrain of hills and valleys, one can quickly visualize the nature of dynamically unfolding r and b states.

Especially because we are trying to mathematically model sparse and poor data, different drift and diffusion algebraic functions can give approximately the same algebraic cost-function when fitting short-time probability distributions to data. The calculation of long-time distributions permits a better choice of the best algebraic functions, i.e., those which best follow the data through a predetermined epoch of battle. Thus, dynamic physical mechanisms, beyond simple "line" and "area" firing terms, can be identified. Afterwards, if there are closely competitive algebraic functions, they can be more precisely assessed by calculating higher algebraic correlation functions from the probability distribution.

It must be clearly stated that, like any other theory applied to a complex system, these methods have their limitations, and they are not a panacea for all systems. For example, probability theory itself is not a complete description when applied to categories of subjective "possibilities" of information.[49,50] Other non-stochastic issues are likely appropriate for determining other types of causal relationships, e.g., the importance of reconnaissance to success of missions.[51]

These statistical mechanical methods appear to be appropriate for describing stochastic large scale combat systems. The details of our studies will help to determine the correctness of this premise.

As discussed above, the mathematical representation most familiar to other modelers is a system of stochastic rate equations, often referred to as Langevin equations. From the Langevin equations, other models may be derived, such as the times-series model and the Kalman filter method of control theory. However, in the process of this transformation, the Markovian description typically is lost by projection onto a smaller state space.[52,53] This work only considers multiplicative Gaussian noise, including the limit of weak colored noise.[24] These methods are not conveniently used for other sources of noise, e.g., Poisson processes or Bernoulli processes. It remains to be seen if multiplicative noise can emulate these processes in the empirical ranges of interest, in some reasonable limits.[25] At this time, certainly the proper inclusion of multiplicative noise, using parameters fit to data to model general sources of noise, is preferable to improper inclusion or exclusion of any noise.

D. Algebraic Complexity Yields Simple Intuitive Results

Consider a multivariate system, but with the multivariate variance a general nonlinear function of the variables. The Einstein summation convention helps to compact the equations, whereby repeated indices in factors are to be summed over.

The Itô (prepoint) discretization for a system of stochastic differential equations is defined by

$$\bar{t}_s \in [t_s, t_s + \Delta t]$$
$$M(\bar{t}_s) = M(t_s)$$
$$\dot{M}(\bar{t}_s) = M(t_{s+1}) - M(t_s) \qquad (6)$$

The stochastic equations are then written as

$$\dot{M}^G = f^G + \hat{g}_i^G \eta^i$$
$$i = 1, \cdots, \Xi$$
$$G = 1, \cdots, \Theta \qquad (7)$$

The operator ordering (of the $\partial/\partial M^G$ operators) in the Fokker-Planck equation corresponding to this discretization is

$$\frac{\partial P}{\partial t} - VP + \frac{\partial(-g^G P)}{\partial M^G} + \frac{1}{2} \frac{\partial^2(g^{GG'} P)}{\partial M^G \partial M^{G'}}$$
$$g^G = f^G + \frac{1}{2} \hat{g}_i^{G'} \frac{\partial \hat{g}_i^G}{\partial M^{G'}}$$
$$g^{GG'} = \hat{g}_i^G \hat{g}_i^{G'} \qquad (8)$$

The Lagrangian corresponding to this Fokker-Planck and set of Langevin equations may be written in the Stratonovich (midpoint) representation, corresponding to

$$M(\bar{t}_s) = \frac{1}{2}[M(t_{s+1}) + M(t_s)] \qquad (9)$$

This discretization can be used to define a Feynman Lagrangian L that possesses a variational principle, and which explicitly portrays the underlying Riemannian geometry induced by the metric tensor $g_{GG'}$, calculated to be the inverse of the covariance matrix.

$$P = \int \cdots \int DM \exp(-\sum_{s=0}^{u} \Delta t L_s)$$

$$DM = g_{0_+}^{1/2}(2\pi\Delta t)^{-1/2} \prod_{s=1}^{u} g_{s_+}^{1/2} \prod_{G=1}^{\Theta} (2\pi\Delta t)^{-1/2} dM_s^G$$

$$\int dM_s^G \to \sum_{\alpha=1}^{N^G} \Delta M_{\alpha s}^G, \quad M_0^G = M_{t_0}^G, \quad M_{u+1}^G = M_t^G$$

$$L = \frac{1}{2}(\dot{M}^G - h^G)g_{GG'}(\dot{M}^{G'} - h^{G'}) + \frac{1}{2}h^G_{;G} + R/6 - V$$

$$[\cdots]_{,G} = \frac{\partial[\cdots]}{\partial M^G}$$

$$h^G = g^G - \frac{1}{2}g^{-1/2}(g^{1/2}g^{GG'})_{,G'}$$

$$g_{GG'} = (g^{GG'})^{-1}$$

$$g_s[M^G(\bar{t}_s), \bar{t}_s] = \det(g_{GG'})_s, \quad g_{s_+} = g_s[M_{s+1}^G, \bar{t}_s]$$

$$h^G_{;G} = h^G_{,G} + \Gamma^F_{GF}h^G = g^{-1/2}(g^{1/2}h^G)_{,G}$$

$$\Gamma^F_{JK} \equiv g^{LF}[JK, L] = g^{LF}(g_{JL,K} + g_{KL,J} - g_{JK,L})$$

$$R = g^{JL}R_{JL} = g^{JL}g^{JK}R_{FJKL}$$

$$R_{FJKL} = \frac{1}{2}(g_{FK,JL} - g_{JK,FL} - g_{FL,JK} + g_{JL,FK}) + g_{MN}(\Gamma^M_{FK}\Gamma^N_{JL} - \Gamma^M_{FL}\Gamma^N_{JK})$$

(10)

A "potential" term V is included, e.g., which might arise to simulate boundary conditions.

Because of the presence of multiplicative noise, the Langevin system differs in its Itô (prepoint) and Stratonovich (midpoint) discretizations. The midpoint-discretized covariant description, in terms of the Feynman Lagrangian, is defined such that (arbitrary) fluctuations occur about solutions to the Euler-Lagrange variational equations. In contrast, the usual Itô and corresponding Stratonovich discretizations are defined such that the path integral reduces to the Fokker-Planck equation in the weak-noise limit. The term $R/6$ in the Feynman Lagrangian includes a contribution of $R/12$ from the WKBJ approximation[54] (named after Wentzel, Kramers, Brillouin, and Jefferys) to the same order of $(\Delta t)^{3/2}$.[21]

Now, consider the generalization to many cells. In the absence of any further information about the system, this increases the number of variables, from the set $\{G\}$ to the set $\{G, v\}$.

A different prepoint discretization for the same probability distribution P, gives a much simpler algebraic form, but the Lagrangian L' so specified does not satisfy a variational principle useful for moderate to large noise.

$$L' = \frac{1}{2}(\dot{M}^G - g^G)g_{GG'}(\dot{M}^{G'} - g^{G'}) \tag{11}$$

Still, this prepoint-discretized form has been quite useful in all systems examined thus far, simply requiring a somewhat finer numerical mesh.

It must be emphasized that the output need not be confined to complex algebraic forms or tables of numbers. Because L possesses a variational principle, sets of contour graphs, at different long-time epochs of the path integral of

P over its r and b variables at all intermediate times, give a visually intuitive and accurate decision-aid to view the dynamic evolution of the scenario.

This Lagrangian approach to combat dynamics permits a quantitative assessment of concepts previously only loosely defined.

Momentum:
$$\Pi^G = \frac{\partial L}{\partial(\partial M^G/\partial t)} \qquad (12)$$

Mass:
$$g_{GG'} = \frac{\partial L}{\partial(\partial M^G/\partial t)\partial(\partial M^{G'}/\partial t)} \qquad (13)$$

Force:
$$\frac{\partial L}{\partial M^G} \qquad (14)$$

$F = ma$:
$$\delta L = 0 = \frac{\partial L}{\partial M^G} - \frac{\partial}{\partial t}\frac{\partial L}{\partial(\partial M^G/\partial t)} \qquad (15)$$

where M^G are the variables and L is the Lagrangian. These relationships are derived and are valid at each spatial-temporal point of M^G. Reduction to other mathematical and physics modeling can be achieved after fitting realistic exercise and/or simulation data.

These physical entities provide another form of intuitive, but quantitatively precise, presentation of these analyses. A visual example is given below.

IV. Numerical Implementation

A. Fitting Parameters

The five coupled stochastic differential equations, for variables $M^G = \{RT, RBMP, BT, BAPC, BTOW\}$, can be represented equivalently by a short-time conditional probability distribution P in terms of a Lagrangian L:

$$P(R\cdot, B\cdot; t + \Delta t | R\cdot, B\cdot; t) = \frac{1}{(2\pi\Delta t)^{5/2}\sigma^{1/2}} \exp(-L\Delta t) \qquad (16)$$

where σ is the determinant of the inverse of the covariance matrix, the metric matrix of this space, "$R \cdot$" represents $\{RT, RBMP\}$, and "$B \cdot$" represents $\{BT, BAPC, BTOW\}$. (Here, the prepoint discretization is used, which hides the Riemannian corrections explicit in the midpoint discretized Feynman Lagrangian; only the latter representation possesses a variational principle useful for arbitrary noise.)

This defines a scalar "dynamic cost function," $C(x, y, z)$,

$$C(x, y, z) = L\Delta t + \frac{5}{2}\ln(2\pi\Delta t) + \frac{1}{2}\ln \sigma \qquad (17)$$

which can be used with the very fast simulated reannealing (VFSR) algorithm[55] further discussed below, to find the (statistically) best fit of $\{x, y, z\}$ to the data.

The form for the Lagrangian L and the determinant of the metric σ to be used for the cost function C is

$$L = \sum_G \sum_{G'} \frac{(\dot{M}^G - g^G)(\dot{M}^{G'} - g^{G'})}{2 g^{GG'}}$$
$$\sigma = \det(g_{GG'})$$
$$(g_{GG'}) = (g^{GG'})^{-1}$$
$$g^{GG'} = \sum_i \hat{g}_i^G \hat{g}_i^{G'} \tag{18}$$

Generated choices for $\{x, y, z\}$ are constrained by empirical KVSs (taken from exercises or from computer simulations of these exercises)

$$g^G(t) \leq n^G < \Delta M^G(t) >$$
$$\hat{g}_i^G(t) \leq n_i^G [< (\Delta M^G(t))^2 >]^{1/2} \tag{19}$$

where n^G and n_i^G are the number of terms in g^G and $\hat{g}_i^G$, respectively, and averages $< \cdot >$ are taken over all time epochs and trajectories of similar scenarios.

If there are competing mathematical forms, then it is advantageous to use the path integral to calculate the long-time evolution of P.[12] Experience has demonstrated that, since P is exponentially sensitive to changes in L, the long-time correlations derived from theory, measured against the empirical data, is a viable and expedient way of rejecting models not in accord with empirical evidence.

Note that the use of the path integral is a posteriori to the short-time fitting process, and is a subsidiary physical constraint on the mathematical models to judge their internal soundness and suitability for attempts to extrapolate to other scenarios.

B. Combat Power Scores

After the $\{x, y, z,\}$ are fit to the data and a mathematical model is selected, another fit can be superimposed to find the effective "combat scores," defined here as the relative contribution of each system to the specific class of scenarios in question. Using a fundamental property of probability distributions, a probability distribution $P_A(q)$ of aggregated variables $q_1 + q_2$ can be obtained from the probability distribution for $P(q_1, q_2)$

$$P_A(q = q_1 + q_2) = \int dq_1 \, dq_2 \, P(q_1, q_2) \delta(q - q_1 - q_2) \tag{20}$$

where $\delta(\cdot)$ is the Dirac delta function.

Thus, we calculate the aggregated conditional probability
$$P_A(r, b; t + \Delta t | R\cdot, B\cdot; t)$$
$$= \int dRT \, dRBMP \, dBT \, dBAPC \, dBTOW \; P(R\cdot, B\cdot; t + \Delta t | R\cdot, B\cdot; t)$$
$$\times \delta(r - w_{RT}^r RT - w_{RBMP}^r RBMP)$$
$$\times \delta(b - w_{RT}^b BT - w_{BAPC}^b BAPC - w_{BTOW}^b BTOW) \tag{21}$$

where the w represent the desired combat scores. After the $\{x, y, z\}$ have been fitted, the new parameters $\{w\}$ can be fit to the data by maximizing the cost function $C'(w)$ using VFSR,

$$C'(w) = -\ln P_A \tag{22}$$

Note that the simple linear aggregation by systems above can be generalized to nonlinear functions, thereby taking into account synergistic interactions among systems that contribute to overall combat effectiveness.

We will be able to explore the possibility of developing human factors combat power scores, since we will be similarly including human-factors variables in such equations, as discussed below.

C. Very Fast Simulated Reannealing

Two major computer codes have been developed, which are key tools for use of this approach to mathematically model combat data. The first code, very fast simulated reannealing (VFSR),[55] fits short-time probability distributions to empirical data, using a most-likelihood technique on the Lagrangian. This algorithm has been developed to fit empirical data to a theoretical cost function over a D dimensional parameter space,[55] adapting for varying sensitivities of parameters during the fit. The annealing schedule for the "temperatures" (artificial fluctuation parameters) T_i decrease exponentially in "time" (cycle-number of iterative process) k, i.e., $T_i = T_{i0} \exp(-c_i k^{1/D})$.

Heuristic arguments have been developed to demonstrate that this algorithm is faster than the fast Cauchy annealing:[56] $T_i = T_0/k$, and much faster than Boltzmann annealing:[57] $T_i = T_0/\ln k$. To be more specific, the kth estimate of parameter α^i,

$$\alpha_k^i \in [A_i, B_i] \tag{23}$$

is used with the random variable x^i to get the $k + 1$th estimate,

$$\alpha_{k+1}^i = \alpha_k^i + x^i(B_i - A_i)$$
$$x^i \in [-1, 1] \tag{24}$$

Define the generating function

$$g_T(x) = \prod_{i=1}^{D} \frac{1}{2\ln(1 + 1/T_i)(|x^i| + T_i)} \equiv \prod_{i=1}^{D} g_T^i(x^i)$$
$$T_i = T_{i0}\exp(-c_i k^{1/D}) \tag{25}$$

The cost-functions $\underline{L}$ we are exploring are of the form

$$h(M; \alpha) = \exp(-\underline{L}/T)$$
$$\underline{L} = L\Delta t + \frac{1}{2}\ln(2\pi\Delta t g_t^2) \tag{26}$$

where L is a Lagrangian with dynamic variables $M(t)$, and parameter coefficients α to be fit to data. The g_t is the determinant of the metric. Note that the use of $\underline{L}$ is *not* equivalent to doing a simple least squares fit on $M(t + \Delta t)$.

Recently, a comparison of VFSR was made with another popular approach, Genetic Algorithms (GA).[58] GA previously has been demonstrated to be competitive with other standard Boltzmann-type simulated annealing techniques. Presenting a suite of six standard test functions to GA and VFSR codes from previous studies, without any additional fine tuning, strongly suggests that VFSR can be expected to be orders of magnitude more efficient than GA. Other studies have shown VSFR to be superior to other simulated annealing techniques.[59,60] A new algorithm has been outlined, combining the strengths with other powerful algorithms.[39] Since VFSR was recoded and made publicly available, many groups worldwide have made it the algorithm of choice for complex systems.[61]

D. Path Integral

The second code develops the long-time probability distribution from the Lagrangian fit by the first code. A robust and accurate histogram-based (non-Monte Carlo) path integral algorithm to calculate the long-time probability distribution has been developed to handle nonlinear Lagrangians,[15,62-64] including a two variable code for additive and multiplicative cases.

The histogram procedure recognizes that the distribution can be numerically approximated to a high degree of accuracy as the sum of rectangles at points M_i of height P_i and width ΔM_i. For convenience, just consider a one dimensional system. The above path integral representation can be rewritten, for each of its intermediate integrals, as

$$P(M; t + \Delta t) = \int dM' [g_s^{1/2}(2\pi\Delta t)^{-1/2} \exp(-L_s \Delta t)] P(M'; t)$$
$$\equiv \int dM' G(M, M'; \Delta t) P(M'; t)$$
$$P(M; t) = \sum_{i=1}^{N} \pi(M - M_i) P_i(t)$$

$$\pi(M - M_i) = \begin{cases} 1, & (M_i - \frac{1}{2}\Delta M_{i-1}) \leq M \leq (M_i + \frac{1}{2}\Delta M_i) \\ 0, & \text{otherwise} \end{cases} \quad (27)$$

which yields

$$P_i(t + \Delta t) = T_{ij}(\Delta t) P_j(t)$$
$$T_{ij}(\Delta t) = \frac{2}{\Delta M_{i-1} + \Delta M_i} \int_{M_i - \Delta M_{i-1}/2}^{M_i + \Delta M_i/2} dM \int_{M_j - \Delta M_{j-1}/2}^{M_j + \Delta M_j/2} dM' G(M, M'; \Delta t)$$
$$(28)$$

T_{ij} is a banded matrix representing the Gaussian nature of the short-time probability centered about the (varying) drift.

This histogram procedure has been extended to two dimensions for this combat analysis, that is, using a matrix T_{ijkl},[15] e.g., similar to the use of the A matrix in the previous section. Explicit dependence of L on time t also can be included without complications. We see no problems in extending it to other dimensions, other than care must be used in developing the mesh in ΔM, which is dependent on the diffusion matrix.

Fitting data with the short-time probability distribution, effectively using an integral over this epoch, permits the use of coarser meshes than the corresponding stochastic differential equation. The coarser resolution is appropriate, typically required, for numerical solution of the time dependent path integral: By considering the contributions to the first and second moments of ΔM^G for small time slices θ, conditions on the time and variable meshes can be derived.[62] The time slice is determined by $\theta \leq \bar{L}^{-1}$, where $\bar{L}$ is the "static" Lagrangian with dM^G/dt = 0, throughout the ranges of M^G giving the most important contributions to the probability distribution P. The variable mesh, a function of M^G, is optimally chosen such that ΔM^G is measured by the covariance $g^{GG'}$, or $\Delta M^G \sim (g^{GG}\theta)^{1/2}$.

As is true for many systems described by such stochastic equations, the equations themselves are but a part of the mathematical model, being complemented by boundary conditions. These are usually difficult to implement in differential equations. That is, codes that the author has seen for Lanchester-type systems do not include such conditions as bounds on numbers of units available. However, in the path integral formalism typically these boundary conditions can be readily and properly included,[64] and we have done so.

E. Modeling of National Training Center

The "kills" attrition data from NTC and our Janus(T)/NTC simulation at once looks strikingly similar during the force-on-force part of the combat (Fig. 1). Note that we are fitting (only half) the middle part of the engagement, where the slope of attrition is very steep (and almost linear on the given scale), i.e., the "force on force" part of the engagement. The second half of the data must be predicted by our models.

From the single NTC trajectory qualified to date, seven 5-min intervals in the middle of the battle were selected. From six Janus(T) runs, similar force-on-

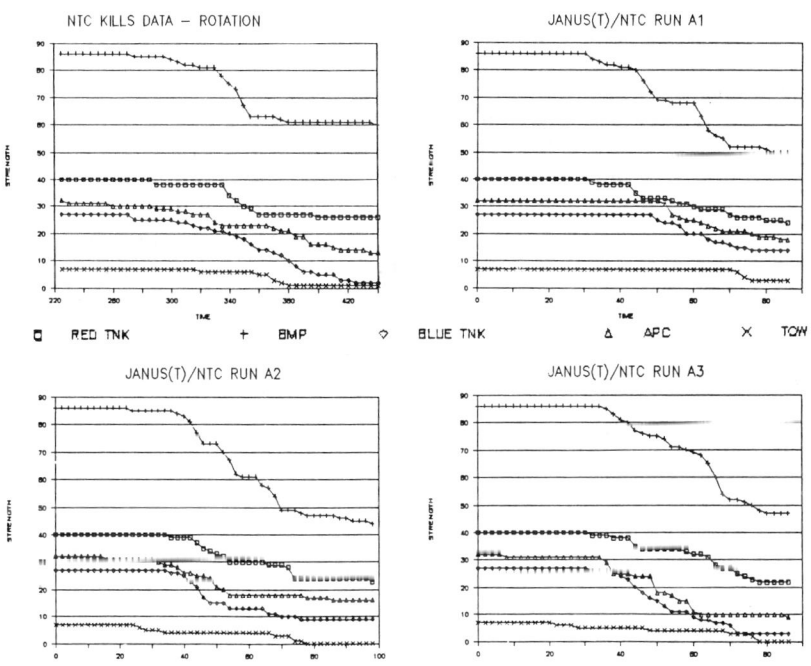

Fig 1 NTC vs Janus(T), attrition ("kills") data for an NTC mission (upper left box) and for three Janus(T) runs using the NTC qualified database.

force time epochs were identified, for a total of 42 data points. In the following fits, r represents red tanks, and b represents blue tanks.

Fitting NTC data to an additive noise model, a cost function of 2.08 gave

$$\dot{r} = -2.49 \times 10^{-5}b - 4.97 \times 10^{-4}br + 0.320\eta_r$$
$$\dot{b} = -2.28 \times 10^{-3}r - 3.23 \times 10^{-4}rb + 0.303\eta_b \quad (29)$$

Fitting NTC data to a multiplicative noise model, a cost function of 2.16 gave

$$\dot{r} = -5.69 \times 10^{-5}b - 4.70 \times 10^{-4}br + 1.06 \times 10^{-2}(1+r)\eta_r$$
$$\dot{b} = -5.70 \times 10^{-4}r - 4.17 \times 10^{-4}rb + 1.73 \times 10^{-2}(1+b)\eta_b \quad (30)$$

Fitting Janus(T) data to an additive noise model, a cost function of 3.53 gave

$$\dot{r} = -2.15 \times 10^{-5}b - 5.13 \times 10^{-4}br + 0.530\eta_r$$
$$\dot{b} = -5.65 \times 10^{-3}r - 3.98 \times 10^{-4}rb + 0.784\eta_b \quad (31)$$

Fitting Janus(T) data to a multiplicative noise model, a cost function of 3.42 gave

$$\dot{r} = -2.81 \times 10^{-4}b - 5.04 \times 10^{-4}br + 1.58 \times 10^{-2}(1+r)\eta_r$$
$$\dot{b} = -3.90 \times 10^{-3}r - 5.04 \times 10^{-4}rb + 3.58 \times 10^{-2}(1+b)\eta_b \quad (32)$$

This comparison illustrates that two different models about equally fit the short-time distribution. The multiplicative noise model shows that approximately a factor of 100 of the noise might be "divided out," or understood in terms of the physical log normal mechanism.

To discern which model best fits the data, we turn to the path integral calculation of the long-time distribution, to see which model best follows the actual data. Figure 2 presents the long-time probability of finding values of these forces. In general, the probability will be a highly nonlinear algebraic function, and there will be multiple peaks and valleys.

Figures 3 and 4 give the means and variances of tank attrition from the Janus(T) and NTC databases. Since we now have only one NTC mission qualified, the variance of deviation from the mean is not really significant; it is given only to illustrate our approach that will be applied to more NTC missions as they are qualified and aggregated. Note that only the blue Janus(T) variances of the additive noise model are consistent with the NTC data.

F. Discussion of Study

Data from 35 to 70 minutes was used for the short-time fit. The path integral code was used to calculate the long-time evolution of this fitted short-time (5-min) distribution from 35 to beyond 70 min. This serves to compare long-time correlations in the mathematical model vs the data, and to help judge extrapolation past the data used for the short-time fits. More data and work are required to find a better (or best?) algebraic form. The resulting form is required for input into higher echelon models. As more NTC data becomes available (other NTC missions are in the process of being qualified, wargamed, and analyzed), we will be able to judge the best models with respect to how well they extrapolate across slightly different combat missions.

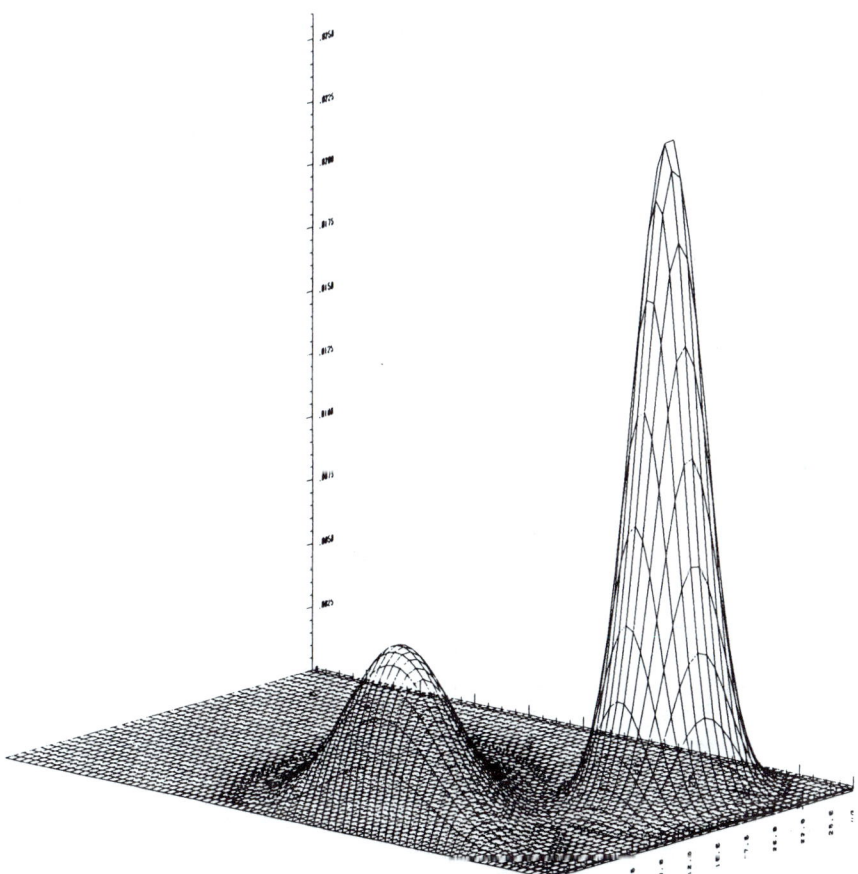

Fig 2 Path integral calculation of long-time distribution: The horizontal axes represent red and blue forces. For this Janus(T) additive noise case, two time slices are superimposed. Taking the initial onset of the engagement as 35 minutes, these peaks represent 50 and 100 minutes.

V. Chaos or Noise?

Given the context of current studies in complex nonlinear systems, the question can be asked: What if combat has chaotic mechanisms that overshadow the above stochastic considerations? The real issue is whether the scatter in data can be distinguished from that due to noise or chaos. Several studies have been proposed with regard to comparing chaos to simple filtered (colored) noise.[65-67]

The combat analysis was possible only because now we had data on combat exercises from the NTC of sufficient temporal density to attempt dynamical mathematical modeling. The criteria used to (not) determine chaos in this dynamical system is the nature of propagation of uncertainty, i.e., the variance. For example, following by now standard arguments, propagation of uncertainty may be considered as (1) diminishing, (2) increasing additively, (3) or increasing

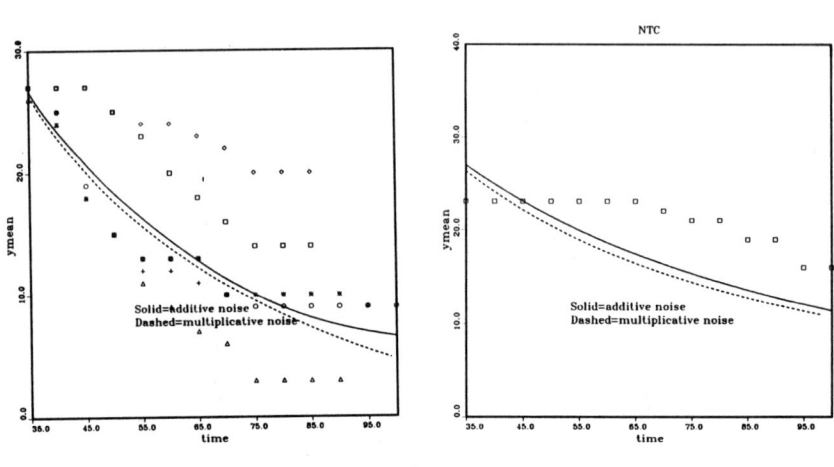

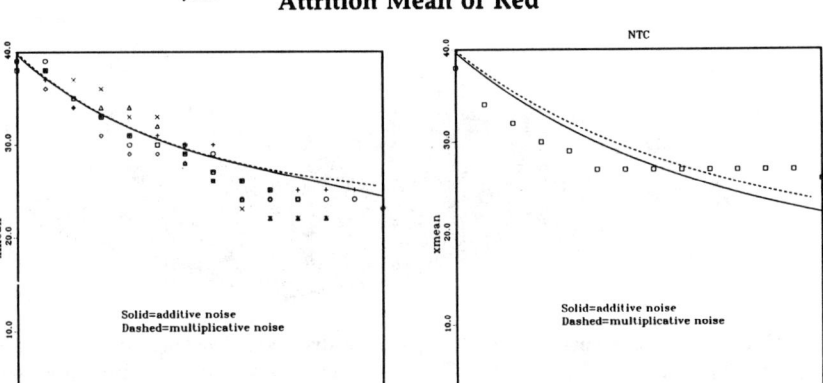

Fig 3 Attrition means: The left two boxes are blue; the right two boxes are red. The left-hand-side box of each pair represents Janus(T); the right-hand-side box represents NTC; solid lines the additive noise model; dotted lines the multiplicative noise model; small circles in the means' boxes empirical data.

multiplicatively. An example of (1) is the evolution of a system to an attractor, e.g., a book dropped onto the floor from various heights reaches the same point no matter what the spread in initial conditions. An example of (2) is the propagation of error in a clock, a cyclic system. Examples of (3) are chaotic systems, of which very few real systems have been shown to belong. An example of (3) is the scattering of a particle in a box whose center contains a sphere boundary: When a spread of initial conditions is considered for the particle to scatter from the sphere, when its trajectories are aligned to strike the sphere at a distance

Attrition Variance of Blue

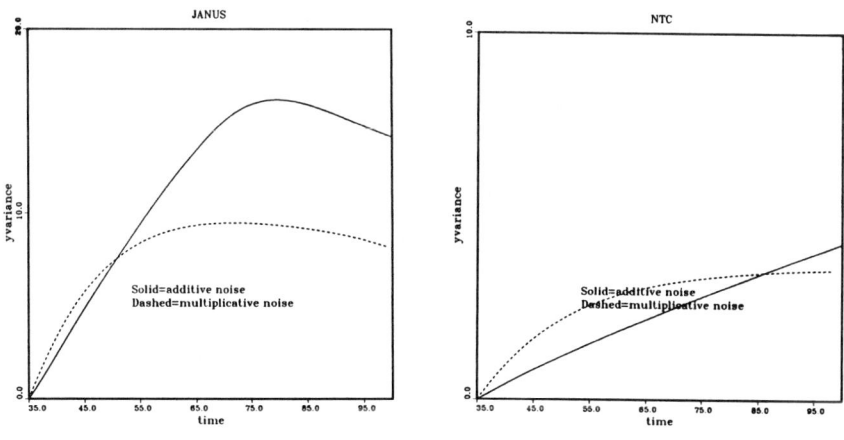

Attrition Variance of Red

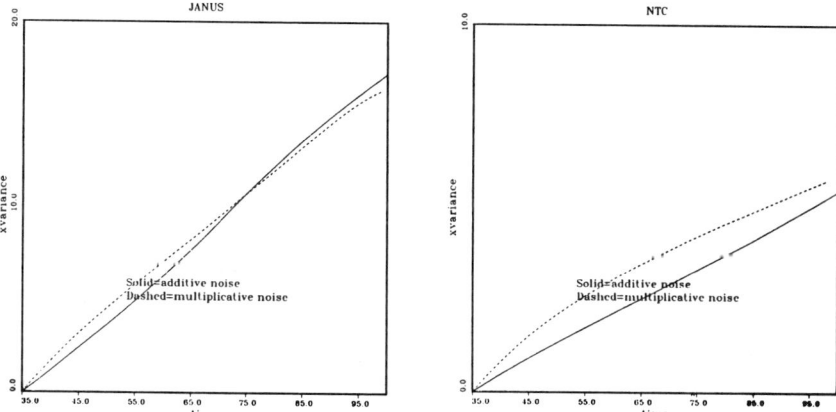

Fig 4 Attrition variances.

from its center greater than the diameter, the spread in scattering is a factor of about three greater than the initial spread.

In our analysis of NTC data, we were able to fit the short-time attrition epochs (determined to be about 5 min from mesh considerations determined by the nature of the Lagrangian) with short-time nonlinear Gaussian-Markovian probability distributions with a resolution comparable to the spread in data. When we computed the long-time path integral from some point (spread) at the beginning of the battle, we found that we could readily find a form of the Lagrangian that made physical sense and that also fit the multivariate variances as well as the means at each point in time of the rest of the combat interval; i.e., there was not any degree of sensitivity to initial conditions that prevented us from "predicting" the long-time means and variances of the system. Of course, since the system is dissipative, there is a strong tendency for all moments to diminish in time, but in fact this combat is of sufficiently modest duration (typically 1–2 h) that variances do increase somewhat during the middle of the battle.

In summary, this battalion-regiment scale of battle does not seem to possess chaos. Similar to serious work undertaken in several fields,[68,69] here too, the impulse to cry "chaos!" in combat appears to have been shown to be premature. It is not supported by the facts, tentative as they are because of sparse data. Of course, some other combat conditions might show some elements of chaos in some spatial-temporal domain, and then the resolution of the analysis would determine the influence of that chaos on that combat scenario.

A more purposeful project is to compare stochastic with deterministic models of combat data. Today much attention is turning to the use of deterministic chaotic models for short-time predictions of systems. For example, if only short-time predictions are required, and if a deterministic chaotic model could well describe stochastic data within these epochs, then this model might be more computationally efficient instead of a more "correct" stochastic model that would be necessary for long-time predictions. The scales of time involved are of course system dependent, and the deterministic chaotic modeling of noisy data is still in its infancy.[70]

VI. Statistical Mechanics of Combat with Human Factors

A. Rationale

In many complex activities involving interactions between human and inanimate systems, e.g., modern combat, the nonlinear synergies capable between these systems make it impossible to separate their influences from the total scenario. However, their relationships and functional dependencies still might be amenable to explicit scientific description. Working with Sworder at University of California at San Diego, we are developing the approach below.[47]

For example, if $h(t)$ could be determined to be a (time dependent) human factor, and if $x(t)$ could be determined to be an inanimate factor, then one could imagine that a "cost function" C fitting data from a specific class of combat scenarios could be fit by a probability distribution $P[C]$ (emphasizing the uncertainty and "noise" in both systems). For specificity, consider the completely arbitrary distribution selected only for purposes of illustration.

$$P[C] = N \exp(-C)$$
$$C(x, h; t) = \frac{(dx/dt + 3.04xh^2 + 0.21x^2)^2}{(2.84 + 0.133h^2x^2)} + \frac{(dh/dt + 5.10x^2 + 1.13hx^2)^2}{(4.21 + 0.071h^2x^2)} \tag{33}$$

where $N(x, h; t)$ is a normalization factor for this distribution.

As discussed above,[15] we have derived similar stochastic nonlinear forms, in terms of nonlinear stochastic time dependent inanimate combat variables (tank attrition, etc.), that fit quite well to data from exercises at NTC. We propose here to more explicitly include the human-factor variables relevant to decision making processes at NTC. The determination of such a cost function permits the accurate derivation of graphical aids to visualize the sensitivity of combat macrovariables (force, mass, momentum) as a function of the human decision making process.[46]

It is perhaps just as important for us to clearly state what we are not proposing. We are not proposing that the human-factors variables we will derive, e.g., in the sense of h, will be explicit representations of cognitive activity, such as attentional processes. Rather, these variables are to be considered metavariables representing the behavioral characteristics of human decision making,[71] in the context of specific NTC scenarios. The above sensitivity measures of the decision making process perhaps come closest with this methodology to explicit identification of human factors.

We believe that we can deliver normative (probabilistic) standards of a class(es) of NTC scenarios, by which specific unit decision making performance can be gauged within this context. Furthermore, our explicit representations of these human factors permits these equations to be used directly in many combat computer models, thereby increasing the utility of these computer models for training and analysis. The lack of human-factors algorithms in combat computer models is notorious.[72-74]

Our major thrust will be to identify and to interpret reasonable functional forms much more extensive and detailed than above. This project is new research territory and it will require extensive and intensive interaction between accumulating critical analyses and accumulating experience with operations at NTC.

The inclusion of human factors in a single equation is too naive to capture the essence of human decision making, even if we generalize h and x to include many variables from each opponent, e.g., $\vec{h}$ and $\vec{x}$ vectors. To include sharp bifurcations, e.g., alternative branching scenarios because of the perceptions of commanders and actions thereby taken at critical times in combat, we plan to fit a "tree" of distributions, each branch representing an alternative scenario.

Our rationale for this attempt, to generalize our previous NTC fits of inanimate variables,[46] is based on our other work modeling the human decision maker controlling teleoperated robotic vehicles, a decision making process that is conceptually similar to the role of the commander in combat.[71,75] We should thereby gain greater fidelity in our fits to NTC data by more explicitly including human factors.

A major thrust of our research will be to expand the linearized theory beyond that currently formulated,[71] to include more robust nonlinear features of the underlying theory of human decision making. This approach is now possible because of the spin-off work in the previous project,[46] i.e., developing the VFSR methodology to fit such nonlinear multivariate stochastic models.[55]

B. Significant Aspects to be Studied

We believe we are addressing the following issues:

1) Human factors, especially in combat, are nonlinear. Nonlinearity arises for many reasons, ranging from synergies of human factors with physical systems, to multivalued decision trees depicting future states. The inclusion of realistic movements on realistic terrain typically presents a nonlinear spatial-temporal surface on which the variables evolve.

2) Human factors are stochastic. There are relatively separable influences on decision making, e.g., probabilities associated with detections, acquisitions,

hits, and kills. Furthermore, especially in a given complex situation, not only will different people often make different decisions at different times, but, given the same opportunities, the same person often will make different decisions at different times. Therefore, we need multiple runs of similar situations in order to deduce these distributions. Such sets of data, albeit not ideal, are present at NTC.

3) Human factors are typically observed as "metavariables" of human performance. Especially because information possessed by decision makers is often incomplete or known to be at least partially incorrect, decision makers must make their decisions based on their current perceptions, their extrapolations to future perceptions, their perceptions of their opponent's perceptions, etc. These possibilities give rise to alternate behavioral states, in part contributing the nonlinearities and stochasticities discussed above.

4) Human factors are very context and domain dependent. Other approaches to human factors, e.g., in the field of artificial intelligence, are also converging on this realization. Analogical reasoning is often more efficient than logical deduction.

5) Even combat models with reasonable combat algorithms do not have reasonable human factors algorithms. Especially because of real time constraints, these computer models require relatively simple functional relationships if they are to include human factors at all. Our robust stochastic nonlinear approach permits us to identify (multiple) ranges of likely probable states, that then can be approximated by quasi-linear algebraic forms in each range. This forms the basis of an "expert" system that derives knowledge from objective fits of theoretical models to empirical data.

It is relevant to this paper to note that the projects discussed here, modeling NTC and Janus(T) as described above, and modeling teleoperated vehicles, both have brought powerful mathematical machinery to bear to the stages of numerical specificity with state-of-the-art successful description of realistic empirical data.

VII. Extensions to Other Systems

A. Amphibious Model

One of the author's students wrote a nontechnical thesis on the mathematical methodology.[76] Upton is looking at amphibious models, filling the gap in the spatial scales now using Air Force, Army, and Navy systems.

B. Joint Janus Model

Gallagher, another of the author's students, wrote a thesis documenting a Mideast Army-Navy joint scenario using a Battleship Battle Group with Tomahawk missiles supporting air-land combat.[77] Balaconis, another of the author's students, wrote a thesis documenting the extension of this joint concept to a NATO scenario, including studies of competitive strategies and integrated strike warfare, using two Carrier Battle Groups with Tomahawk and SLAM missiles, F-14 and A-6 tactical air support, and remotely piloted vehicles (Fig. 5).[78]

These projects have established a direction for further study (Fig. 6). We especially are aware of the necessity to include more Air Force systems.

C. Issues of Higher Echelon Extrapolation

After fitting data from microscopic unit interactions to mesoscopic equations at battalion-regiment level, these equations can be used to drive higher level macroscopic scenarios at corps and theater levels (Fig. 7). This mathematical aggregation is required for interpretation at multiple scales.

However, there are many issues yet to be resolved in using this approach. This requires approximately company-fidelity combat data from the unit interactions, e.g., barely the level obtained from NTC. It may be possible soon to obtain similar fidelity at division level, as NTC gears up for this scale of play.

Currently there are four main approaches to modeling theater level combat. 1) Distribution of combat scenarios: The approach in this paper uses stochastic trajectories of high-fidelity interactions and develops stochastic distributions of lower-echelon scenarios. Linear MOFs are derived and battle nodes are coordinated for theater combat. 2) Distribution of system-system interactions: This approach, e.g., in COSAGE used at the Army Concepts Analysis Agency (CAA), uses statistical distributions of representative variables (including terrain and LOS) and distributions of system KVSs to develop attrition model for theater KVSs. 3) Deterministic combat scenarios: This approach, e.g.,

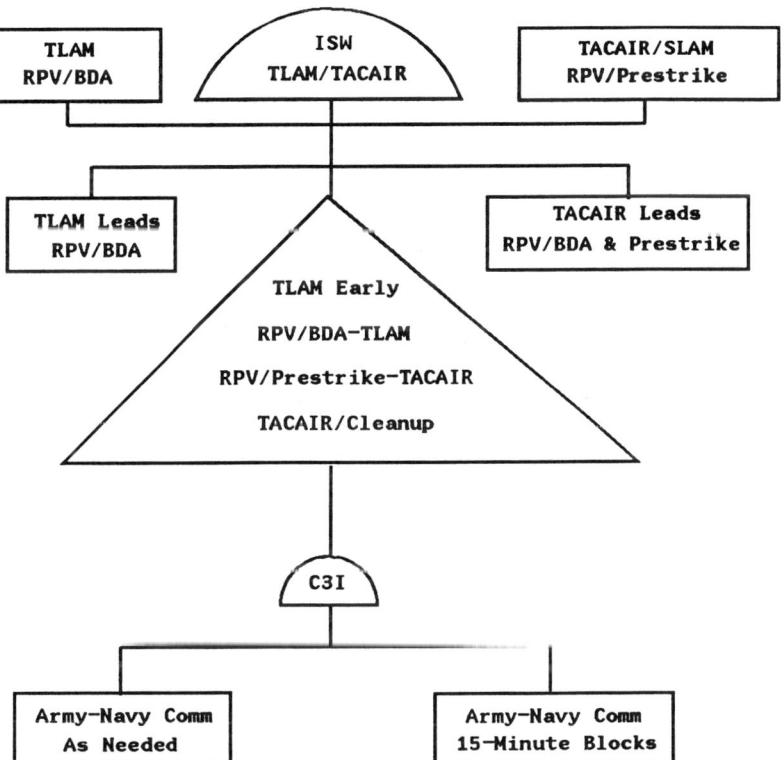

Fig 5 Integrated strike warfare scenarios; rectangles represent the six scenarios that were wargamed.

Increase Air-to-Ground weapons capabilities of Strike aircraft to include "TACIT RAINBOW" and "HARM" anti-radiation weapons	Increase "air" dimension in the AirLand scenario development; introduce F-14 naval fighter aircraft. Build an "air defense" system on a "flyer" with direct fire weapon systems.	Introduce Air Force F-15, F-16, and ATF expanding the "jointness" and usage of joint-JANUS (T) model.	Expand naval usage by modeling more anti-surface weapon systems (i.e., TASM, Harpoon)
Aircraft Carrier Battle Group (CVBG) with AEGIS projecting Integrated Strike Warfare	U.S. Navy Attack Aircraft (F/A-18 and A-6)	Introduction of the Standoff Land Attack Missile (SLAM), RPVs, and TLAM-D (GPS)	Coastal/Near Coast Scenario on NATO's Central Front (Northern Flank)
Development of TLAM-C Weapon System with the BBBG in a Persian Gulf Scenario			
Validation of JANUS (T) Through Mathematical Comparison of JANUS (T) to National Training Center Data. Establish Credibility Based On Correspondence to the real (exercise) world			

Fig 6 Ongoing development of joint Janus(T).

used in VIC at Training and Doctrine Command (TRADOC) Analysis Command (TRAC), White Sands Missile Range, New Mexico (TRAC-WSMR), develops KVSs from lower echelon scenarios and uses system KVSs for theater models. 4) Theater stochastic high fidelity model: This approach requires no aggregation, and studies all spatial-temporal scales simultaneously. This approach has regularly failed because of the huge computer resources required. Furthermore, aggregation is really required anyway, to simulate MOFs, MOEs, etc. required for cognitive decisions at various levels of command.

The important issues are: 1) sensitivity of theater models to different approaches, 2) inclusion/absorption of human factors into variables/parameters, 3) fidelity of representation of modern systems, e.g., cruise missiles (possessing short reaction times, large spatial coverage, and C^3I at multiple scales), 4) statistical comparison of approaches, and 5) baselining of these approaches to some reality.

Response-surface methodologies and central composite design have been quite useful in a number of disciplines, effectively fitting (usually quadratic) algebraic forms to judicially selected instances of the appropriately scaled vari-

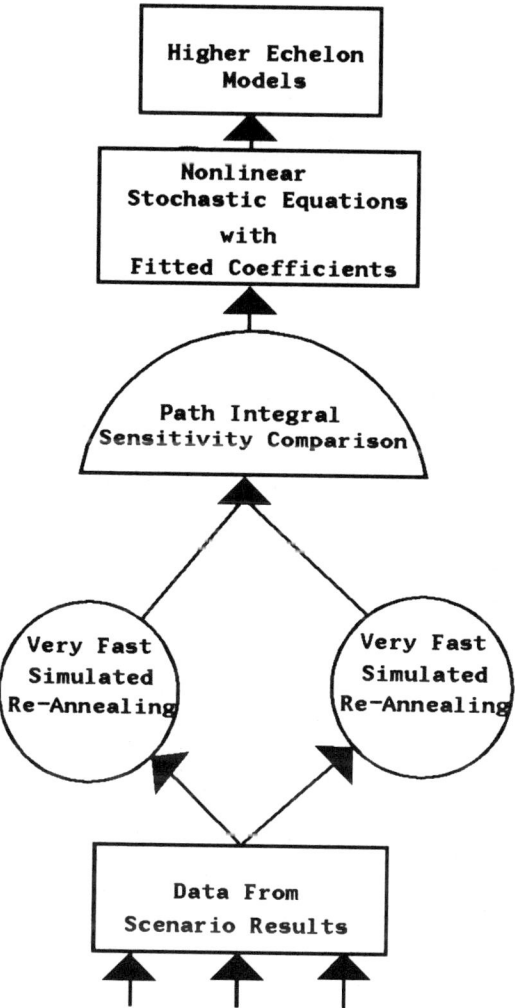

Fig 7 Support of Higher-Echelon Models.

ables. Of course, the phenomenology underlying this approach might not work well for combat systems, but it should be tried; i.e., the nonlinear nature of combat might extend to super-variable space, requiring more sophisticated analysis than the typical regression on quadratic surfaces afforded by current implementations of this approach. Here, the methods of VFSR[55,58] can be quite useful. At the least, this approach should be viewed as: given a fixed constraint of limited funds and time for a study, then there is a tradeoff between the depth (fidelity, etc.) of each scenario run vs breadth (number of runs, etc.).

We have suggested implementation these ideas, developing a mesoscopic filter of a joint model of models (JMOM). JMOM would take microscopic high fidelity combat information up from individual simulations/wargames into macroscopic quick response joint theater level simulations/wargames. This capability could greatly enhance the fidelity and validity of many high level analyses, e.g., of net assessments.

D. Applications to Process Aggregated Information

This statistical mechanics approach represents the mesoscale as a pattern processing computer. The underlying mathematical theory, i.e., the path integral approach, specifies a parallel-processing algorithm that statistically finds those parameter regions that contribute most to the overall probability distribution. This theoretical approach would not be very useful if we could not perform the calculations afforded by VFSR.

This theory represents a kind of "intuitive" algorithm, globally searching a large multivariate database to find parameter regions deserving more detailed local information processing. The derived probability distribution can be thought of as a filter, or processor, of incoming patterns of information; and this filter can be adapted, or updated, as it interacts with previously stored patterns of information.

As an example of how we can develop "intuitive" measures of performance, Figure 8 shows the derived force acting within the NTC system. These figures compare the NTC additive and multiplicative noise cases above, where we have graphed the force of red minus the force on blue. The difference between the red force and the blue force is portrayed as a function of number of vehicles. The solution of the evolution of the probability distribution gives the most likely number of vehicles existing at any time and their rates of attrition. Such figures at least serve to demonstrate that different models of combat can present quite different visual decision aids representing the nature of a particular scenario. This dramatizes the necessity for developing more sensitive algorithms to determine the proper models driving simulations and combat decision aids.

Two of the author's students, Connell,[79] and Yost,[80] have written theses examining multiple scales of interaction in large scale systems, including combat systems. These mathematical methods are quite general, and they have been applied to neuroscience, referenced previously as the SMNI papers, detailing properties of short-term memory derived from neuronal synaptic interactions, and calculating most likely frequencies observed in electroencephalographic (EEG) data and velocities of propagation of information across neocortex. We have detailed applications of this methodology to understand multiple scales of contributions to EEG data, developing software to perform accurate correlations between human behavioral states and EEG data.[37,38]

Many systems require the processing of large sets of data, so large that it is generally conceded that it seems unlikely that even projected computer architectures will be adequate to handle these demands.[81] Among these approaches offering some glimmer of hope are those that attempt to model the human information processing system, neocortex. "Neural nets" do not seem adequate.[82] However, "neural nets" also have not demonstrated that they well model neocortex. Therefore, more investigations into the nature of neocortex is certainly still a plausible approach.

Recently, the SMNI methodology has been used to define an algorithm to construct a mesoscopic neural net (MNN), based on realistic neocortical processes and parameters, to record patterns of brain activity and to compute the

STATISTICAL MECHANICS OF COMBAT

NTC Additive Noise Model

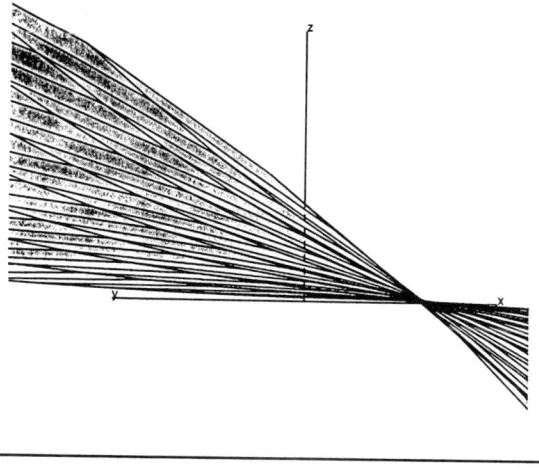

NTC Multiplicative Noise Model

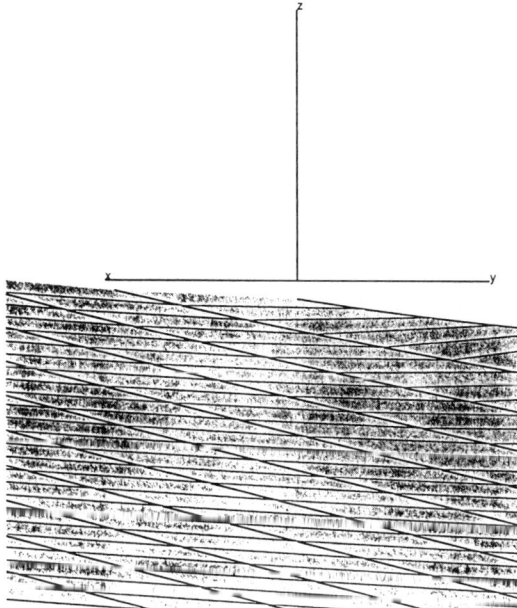

Fig 8 Comparison of NTC Models. The first graph shows the NTC additive noise model. The second graph shows the NTC multiplicative noise model with the same view and scale.

evolution of this system.[39] Furthermore, this new algorithm is quite generic, and can be used to similarly process information in other systems amenable to modeling by mathematical physics techniques alternatively described by path integral Lagrangians, Fokker-Planck equations, or Langevin rate equations, e.g., such as the combat analyses presented here. This methodology is made possible and practical by a confluence of techniques drawn from SMNI itself, modern methods of functional stochastic calculus defining nonlinear Lagrangians, VFSR, and parallel-processing computation.

It should not be too surprising that our present project is at a similar stage to where we began: The issues presented in modeling and understanding combat are quite similar, mathematically and operationally, to the issues presented in modeling and understanding the human brain!

References

[1] Bartlett, G.T., "Battle Command Training Program," *Phalanx*, Vol. 21, 1988, pp. 18-20.

[2] Comptroller General, "Models, Data, and War: A Critique of the Foundation for Defense Analyses," U.S. General Accounting Office, Washington, DC, Report No. PAD-80-21, 1980.

[3] Comptroller General, "DOD Simulations: Improved Assessment Procedures Would Increase the Credibility of Results," U.S. General Accounting Office, Washington, DC, Report No. GAO/PEMD-88-3, 1987.

[4] Jammer, M., *The Philosophy of Quantum Mechanics*, Wiley & Sons, New York, NY, 1974.

[5] Ingber, L., *The Karate Instructor's Handbook*, Physical Studies Institute-Institute for the Study of Attention, Solana Beach, CA, 1976.

[6] Ingber, L., *Karate: Kinematics and Dynamics*, Unique, Hollywood, CA, 1981.

[7] Ingber, L., *Elements of Advanced Karate*, Ohara, Burbank, CA, 1985.

[8] Ingber, L., "Editorial: Learning to Learn," *Explore*, Vol. 7, 1972, pp. 5-8.

[9] Ingber, L., "Attention, Physics and Teaching," *Journal Social Biological Structures*, Vol. 4, 1981, pp. 225-235.

[10] van Creveld, M., *Command in War*, Harvard University Press, Cambridge, MA, 1985.

[11] Bowman, M., "Integration of the NTC Tactical Database and JANUS(T) Towards a Combat Decision Support System," Naval Postgraduate School, Monterey, CA, M.S. Thesis, 1989.

[12] Ingber, L., "Mathematical Comparison of JANUS(T) Simulation to National Training Center," *The Science of Command and Control: Part II, Coping With Complexity*, edited by S.E. Johnson and A.H. Levis, Armed Forces Communications and Electronics Association International, Washington, DC, 1989, pp. 165-176.

[13] Dupuy, T.N., "Can We Rely on Computer Combat Simulations?," *Armed Forces Journal*, Vol. August, 1987, pp. 58-63.

[14] Dockery, J.T. and Santoro, R.T., "Lanchester Revisited: Progress in Modeling C^2 in Combat," *Signal*, Vol. July, 1988, pp. 41-48.

[15] Ingber, L., Fujio, H., and Wehner, M.F., "Mathematical Comparison of Combat Computer Models to Exercise Data," *Mathematical and Computer Modelling*, Vol. 15, Series 1, 1991, pp. 65-90.

[16] Ingber, L., "Mathematical Comparison of Computer Models to Exercise Data," *Symposium on Data Efficiency Using Pre-Processing*, edited by T.K. Gardenier, Turkan Kumbaraci Gardenier Consultants, Vienna, VA, 1989, pp. 72-115.

[17] Dekker, H., "Quantization in Curved Spaces," *Functional Integration: Theory and Applications*, edited by J.P Antoine and E. Tirapegui, Plenum, New York, 1980, pp. 207-224.

[18] Grabert, H. and Green, M.S., "Fluctuations and Nonlinear Irreversible Processes," *Physical Review A*, Vol. 19, 1979, pp. 1747-1756.

[19] Graham, R., "Covariant Formulation of Non-Equilibrium Statistical Thermodynamics," *Zeitschrift für Physik*, Vol. B26, 1977, pp. 397-405.

[20] Graham, R., "Lagrangian for Diffusion in Curved Phase Space," *Physical Review Letters*, Vol. 38, 1977, pp. 51-53.

[21] Langouche, F., Roekaerts, D., and Tirapegui, E., *Functional Integration and Semiclassical Expansions*, Reidel, Dordrecht, The Netherlands, 1982.

[22] Schulman, L.S., *Techniques and Applications of Path Integration*, J. Wiley & Sons, New York, 1981.

[23] Ingber, L., "Nonlinear Nonequilibrium Statistical Mechanics Approach to C^3 Systems," *9th MIT/ONR Workshop on C^3 Systems: Naval Postgraduate School, Monterey, CA, 2-5 June 1986*, MIT, Cambridge, MA, 1986, pp. 237-244.

[24] Fox, R.F., "Uniform Convergence to an Effective Fokker-Planck Equation for Weakly Colored Noise," *Physical Review A*, Vol. 34, 1986, pp. 4525-4527.

[25] van der Broeck, C., "On the Relation Between White Shot Noise, Gaussian White Noise, and the Dichotomic Markov Process," *Journal Statistical Physics*, Vol. 31, 1983, pp. 467-483.

[26] Ingber, L., "Towards a Unified Brain Theory," *Journal Social Biological Structures*, Vol. 4, 1981, pp. 211-224.

[27] Ingber, L., "Statistical Mechanics of Neocortical Interactions. I. Basic Formulation," *Physica D*, Vol. 5, 1982, pp. 83-107.

[28] Ingber, L., "Statistical Mechanics of Neocortical Interactions. Dynamics of Synaptic Modification," *Physical Review A*, Vol. 28, 1983, pp. 395-416.

[29] Ingber, L., "Statistical Mechanics of Neocortical Interactions. Derivation of Short-Term-Memory Capacity," *Physical Review A*, Vol. 29, 1984, pp. 3346-3358.

[30] Ingber, L., "Statistical Mechanics of Neocortical Interactions. EEG Dispersion Relations," *IEEE Transactions Biomedical Engineering*, Vol. 32, 1985, pp. 91-94.

[31] Ingber, L., "Statistical Mechanics of Neocortical Interactions: Stability and Duration of the 7 ± 2 Rule of Short-Term-Memory Capacity," *Physical Review A*, Vol. 31, 1985, pp. 1183-1186.

[32] Ingber, L., "Towards Clinical Applications of Statistical Mechanics of Neocortical Interactions," *Innov. Tech. Biol. Med.*, Vol. 6, 1985, pp. 753-758.

[33] Ingber, L., "Statistical Mechanics of Neocortical Interactions," *Bulletin of the American Physical Society*, Vol. 31, 1986, pp. 868.

[34] Ingber, L., "Applications of Biological Intelligence to Command, Control and Communications," *Computer Simulation in Brain Science: Proceedings, University of Copenhagen, 20-22 August 1986*, edited by R. Cotterill, Cambridge University Press, London, 1988, pp. 513-533.

[35] Ingber, L., "Statistical Mechanics of Mesoscales in Neocortex and in Command, Control and Communications (C^3): Proceedings, Sixth International Conference, St. Louis, MO, 4-7 August 1987," *Mathematical and Computer Modelling*, Vol. 11, 1988, pp. 457-463.

[36] Ingber, L., "Mesoscales in Neocortex and in Command, Control and Communications (C^3) Systems," *Systems with Learning and Memory Abilities: Proceedings, University of Paris 15-19 June 1987*, edited by J. Delacour and J.C.S. Levy, Elsevier, Amsterdam, 1988, pp. 387-409.

[37] Ingber, L. and Nunez, P.L., "Multiple Scales of Statistical Physics of Neocortex: Application to Electroencephalography," *Mathematical and Computer Modelling*, Vol. 13, Series 7, 1990, pp. 83-95.

[38] Ingber, L., "Statistical Mechanics of Neocortical Interactions: A Scaling Paradigm Applied to Electroencephalography," *Physical Review A*, Vol. 44, Series 6, 1991, pp. 4017-4060.

[39] Ingber, L., "Generic Mesoscopic Neural Networks Based on Statistical Mechanics of Neocortical Interactions," *Physical Review A*, Vol. 45, Series 4, 1992, pp. R2183-R2186.

[40] Ingber, L., "Statistical Mechanics of Multiple Scales of Neocortical Interactions," *Neocortical Dynamics and Human EEG Rhythms*, edited by P. Nunez, Oxford University Press, New York, NY, 1993, pp. (to be published).

[41] Ingber, L., "Path-integral Riemannian Contributions to Nuclear Schrödinger Equation," *Physical Review D*, Vol. 29, 1984, pp. 1171-1174.

[42] Ingber, L., "Riemannian Contributions to Short-Ranged Velocity-Dependent Nucleon-Nucleon Interactions," *Physical Review D*, Vol. 33, 1986, pp. 3781-3784.

[43] Ingber, L., "Statistical Mechanics of Nonlinear Nonequilibrium Financial Markets," *Mathematical Modelling*, Vol. 5, Series 6, 1984, pp. 343-361.

[44] Ingber, L., "Statistical Mechanical Aids to Calculating Term Structure Models," *Physical Review A*, Vol. 42, Series 12, 1990, pp. 7057-7064.

[45] Ingber, L., Wehner, M.F., Jabbour, G.M., and Barnhill, T.M., "Application of Statistical Mechanics Methodology to Term-Structure Bond-Pricing Models," *Mathematical and Computer Modelling*, Vol. 15, Series 11, 1991, pp. 77-98.

[46] Ingber, L., "Mathematical Comparison of Computer Models to Exercise Data," *1989 JDL C^2 Symposium: National Defense University, Washington, DC, 27-29 June 1989*, Science Applications International Corporation, McLean, VA, 1989, pp. 169-192.

[47] Ingber, L. and Sworder, D.D., "Statistical Mechanics of Combat with Human Factors," *Mathematical and Computer Modelling*, Vol. 15, Series 11, 1991, pp. 99-127.

[48] Ingber, L., "Statistical Mechanical Measures of Performance of Combat," *Proceedings of the 1991 Summer Computer Simulation Conference 22-24 July 1991, Baltimore, MD*, edited by D. Pace, Society for Computer Simulation, San Diego, CA, 1991, pp. 940-945.

[49] Zadeh, L., "A Computational Theory of Dispositions," *International Journal Intelligent Systems*, Vol. 2, 1987, pp. 39-63.

[50] Goodman, I.R., "A Probabilistic/Possibilistic Approach to Modeling C^3 Systems Part II," *1987 Symposium on C^3 Research at National Defense University, Washington, DC*, National Defense University, Washington, DC, 1988, pp. 41-48.

[51] Goldsmith, M. and Hodges, J., "Applying the National Training Center Experience: Tactical Reconnaissance," RAND, Santa Monica, CA, Report No. N-2628-A, 1987.

[52] Kishida, K., "Physical Langevin Model and the Time-Series Model In Systems Far From Equilibrium," *Physical Review A*, Vol. 25, 1982, pp. 496-507.

[53] Kishida, K., "Equivalent Random Force and Time-Series Model in Systems Far From Equilibrium," *Journal of Mathematical Physics*, Vol. 25, 1984, pp. 1308-1313.

[54] Mathews, J. and Walker, R.L., *Mathematical Methods of Physics*, 2nd ed., Benjamin, New York, NY, 1970.

[55] Ingber, L., "Very Fast Simulated Re-Annealing," *Mathematical and Computer Modelling*, Vol. 12, Series 8, 1989, pp. 967-973.

[56] Szu, H. and Hartley, R., "Fast Simulated Annealing," *Physics Letters A*, Vol. 122, Series 3-4, 1987, pp. 157-162.

[57] Kirkpatrick, S., Gelatt, C.D., Jr., and Vecchi, M.P., "Optimization by Simulated Annealing," *Science*, Vol. 220, Series 4598, 1983, pp. 671-680.

[58] Ingber, L. and Rosen, B., "Genetic Algorithms and Very Fast Simulated Reannealing: A Comparison," *Mathematical and Computer Modelling*, Vol. 16, Series 11, 1992, pp. 87-100.

[59] Ingber, L., "Simulated Annealing: Practice Versus Theory," *Statistics Comput.*, 1993, pp. (to be published).

⁶⁰ Rosen, B., "Function Optimization Based on Advanced Simulated Annealing," *IEEE Workshop on Physics and Computation - PhysComp '92*, , 1992, pp. 289-293.

⁶¹ Ingber, L. and Rosen, B., "Very Fast Simulated Reannealing (VFSR)," AT&T Bell Labs, Murray Hill, NJ, [ftp research.att.com: /netlib/opt/vfsr.Z] or [ftp lib.stat.cmu.edu: /general/vfsr], 1992.

⁶² Wehner, M.F. and Wolfer, W.G., "Numerical Evaluation of Path-Integral Solutions to Fokker Planck Equations. I.," *Physical Review A*, Vol. 27, 1983, pp. 2663-2670.

⁶³ Wehner, M.F. and Wolfer, W.G., "Numerical Evaluation of Path-Integral Solutions to Fokker-Planck Equations. II. Restricted Stochastic Processes," *Physical Review A*, Vol. 28, 1983, pp. 3003-3011.

⁶⁴ Wehner, M.F. and Wolfer, W.G., "Numerical Evaluation of Path Integral Solutions to Fokker-Planck Equations. III. Time and Functionally Dependent Coefficients," *Physical Review A*, Vol. 35, 1987, pp. 1795-1801.

⁶⁵ Theiler, J., "Correlation Dimension of Filtered Noise," UC San Diego, La Jolla, CA, Report 6/29/1988, 1988.

⁶⁶ Pool, R., "Is It Chaos, or Is It Just Noise?," *Science*, Vol. 243, 1989, pp. 25-28.

⁶⁷ Rapp, P.E., Albano, A.M., Schmah, T.I., and Farwell, L.A., "Filtered Noise Can Mimic Low Dimensional Chaotic Attractors," *Physical Review E*, , 1993, pp. (to be published).

⁶⁸ Brock, W.A., "Distinguishing Random and Deterministic Systems: Abridged Version," *Journal of Economic Theory*, Vol. 40, 1986, pp. 168-195.

⁶⁹ Grassberger, P., "Do Climatic Attractors Exist?," *Nature*, Vol. 323, 1986, pp. 609-612.

⁷⁰ Abarbanel, H.D.I., Brown, R., and Kadtke, J.B., "Prediction in Chaotic Nonlinear Systems: Methods for Time Series with Broadband Fourier Spectra," *Physical Review A*, Vol. 41, 1990, pp. 1782-1807.

⁷¹ Sworder, D.D. and Clapp, G.A., "Supervisory Control of C³ Systems," *1989 JDL C² Symposium at National Defense University, Washington, DC*, Science Applications International Corporation, McLean, VA, 1989, pp. 478-482.

⁷² Miller, G.J. and Bonder, S., "Human Factors Representations for Combat Models," Defense Technical Information Center, Alexandria, VA, Vector Research Report AD-A133351, 1982.

⁷³ Van Nostrand, S., "Including the Soldier in Combat Models," Industrial College of the Armed Forces, Fort McNair, Washington, DC, Thesis S73, 1988.

⁷⁴ Dupuy, T.N., *Numbers, Predictions & War*, Bobbs-Merrill, Indianapolis, MN, 1979.

⁷⁵ Sworder, D.D. and Clapp, G.A., "Quantifying Uncertainty in C³ Decision Makers," *1990 JDL C² Symposium: Naval Postgraduate School, Monterey, CA, June 1990*, Science Applications International Corporation, McLean, VA, 1990, pp. 348-353.

⁷⁶ Upton, S.C., "A Statistical Mechanics Model of Combat," Naval Postgraduate School, Monterey, CA, M.S. Thesis, 1987.

⁷⁷ Gallagher, J.F., "A Joint Army-Navy Combat Model Using TLAM C/D," Naval Postgraduate School, Monterey, CA, M.S. Thesis, 1988.

⁷⁸ Balaconis, R.J., "Integrated Strike Warfare High Fidelity Simulation: Cruise Missile and TACAIR Support of AirLand Battle," Naval Postgraduate School, Monterey, CA, M.S. Thesis, 1989.

⁷⁹ Connell, J.C., Jr., "Memory Efficient Evaluation of Nonlinear Stochastic Equations and C³ Applications," Naval Postgraduate School, Monterey, CA, M.S. Thesis, 1987.

⁸⁰ Yost, C.P., "A Virtual Statistical Mechanical Neural Computer," Naval Postgraduate School, Monterey, CA, M.S. Thesis, 1987.

⁸¹ Richardson, W.E., Miller, J.R., and Murphy, C.G., "Command, Control, and Communications (C³) Technology Projection and Assessment Advanced Computer Architectures," Defense Communications Agency, Reston, VA, Technical Report, 1988.

⁸² Pennisi, E., "The New Neural Network Tussle: 'Top Down' or 'Bottom Up'," *The Scientist*, Vol. 2 (21), 1988, pp. 1,4-5.

Impact of Organizational Structure on Team Performance: Experimental Findings

Victoria Y. Jin*
AT&T Bell Laboratories, Holmdel, New Jersey 07733
and
Alexander H. Levis†
George Mason University, Fairfax, Virginia 22030

Introduction

Experiments involving several human decisionmakers and computer simulations are generally complex and difficult to design and control. One of the difficulties is that a large number of parameters is involved; another is the determination of which parameters should be varied and over what range. Also, since human decisionmakers participate in the experiment, a large number of trials is not feasible. Although there is an extensive bibliography on experiments with single decisionmakers, no useful guidelines for model driven experiments with decisionmaking teams were found.

To design a controllable experiment in a complicated environment, a model is necessary for determining appropriate variables that ought to be controlled or measured. In the physical sciences and in engineering, procedures have been developed over the years for using models to design experiments. For example, to address the problem of many parameters and the problem of physical scale, dimensional analysis has been developed and is routinely used in mechanical, aeronautical, and astronautical engineering[1]. This well-established technique from the physical sciences has been extended to include the cognitive aspects of the distributed decisionmaking environment[2].

The model of the interacting decisionmaker developed by Boettcher and Levis[3] was used to represent the members of the organization in the experiment, predict organizational performance, and from those predictions formulate hypotheses that could be tested experimentally.

The experimental design starts with the posing of the problems and issues for which data are to be collected. First, one or more organizational forms are

---
Copyright © 1993 by the American Institute of Aeronautics and Astronautics, Inc. All rights reserved.
* Member of Technical Staff, Network Planning and Technology.
† Professor of Electrical, Computer, and Systems Engineering, School of Information Technology and Engineering.

designed using algorithms and procedures already available (e.g., the lattice algorithm developed by Remy and Levis[4]. Then, the specific procedures for each individual organization member are specified and the protocols for interaction between organization members that instantiate the organizational structure are designed. Finally, the set of normative and prescriptive strategies is formulated which determines the set of admissible strategies (the strategy space). Dimensional analysis is applied to the mathematical model of the team to select controlled and measured variables and construct dimensionless groups. These dimensionless groups are the basis for determining the variables to be measured and the ranges of values to be used for the controlled variables. Running a small-scale pilot experiment may be necessary at this stage to determine the ranges for certain time variables for which no theoretically derived quantitative estimate is available. For example, it may not be possible to predict for a particular cognitive task the minimum amount of time necessary to carry it out correctly. A separate single person experiment may be run for this type of task to obtain an estimate of the minimum time; this estimate is then used in the team model where members carry out similar cognitive tasks. Finally, both simulations and analysis are used to predict team performance for various values of the controlled variables. These predictions lead then to the generation of hypotheses to be tested experimentally. One key consideration in the elucidated approach is the need for the assumptions embedded in the model to be consistent with those underlying the experiment.

This engineering-based methodology is feasible and practical and was shown to be applicable to the design of model driven experiments in team decisionmaking; it has led to the formulation of hypotheses and to the design of an experiment and has guided the collection and analysis of data for proving or disproving the hypotheses. A special class of organizations was considered: a team of well-trained decisionmakers repetitively executing a set of well-defined cognitive tasks under severe time pressure. The cognitive limitations of decisionmakers (DMs) imposed a constraint on organizational performance. Performance, in this case, was assumed to depend mainly on the time available to perform a task and on the cognitive workload associated with the task. When the time available to perform a task is very short (time pressure is very high), decisionmakers are likely to make mistakes (human error) so that performance will be degraded.

The experimental results showed, as predicted, that the accuracy of the response decreases as the available time to do a task is reduced. The variation in performance is less between different teams than between different individual DMs within a team, which means that *organizational performance is more predictable than individual performance*. It has also been found that degradation of accuracy as a function of available time is less abrupt for organizations than for individuals. *Interaction among DMs in an organization compensates for differences in individual performance characteristics*. These results are consistent with the predictions from the theoretical model. Furthermore, *the critical value of the ratio of response time to available time for doing a task is an observable measure of the bounded rationality constraint*. Therefore, this ratio, which is observable from simple experiments, can be used in development of future organization designs as a key design parameter. Finally, organizational design is not unique. In the early stages of the design, there may be several or many structures that seem to be suitable for the task. Which one to choose to proceed with in the detailed design is an important question. The models supporting this work and the insights obtained into organizational behavior from the experimental investigation provide useful guidance for addressing that question.

Illustrative Application

The approach described in the introduction has been used to investigate the effect of organizational structure on performance. The particular problem chosen to illustrate the approach is the Naval outer air battle. The objective of a naval outer air battle is to monitor incoming enemy aircraft and deploy interceptors to engage threats so as to prevent the enemy from entering the range where missiles can be fired at ships in the battle group. A highly abstracted and simplified version of that battle was used (see Fig. 1). The carrier is at the center of the circles. Airborne warning radar aircraft (E2C) patrol the area at a distance R_p from the carrier. Each E2C commands several squadrons of interceptors, which can intercept the threats. The E2Cs are equipped with passive radar (ESM) and active radar. The ESM receives the radar transmission of other aircraft, while the active radar receives the reflection of its transmission from other objects. The ESM has a range of R_0 and the active radar has a range of R_a ($R_0 > R_a$). The range of the enemy's missiles is R_m.

ESM has a larger range for detecting incoming threats, but provides less specific data than the active radar. It provides only the presence and bearing (direction to) of the threats. The active radar provides more detailed data such as the position and speed of a threat. The signature of an aircraft is provided by ESM when the threat is closer. An emitter signature indicates the existence of an aircraft with its corresponding emitter.

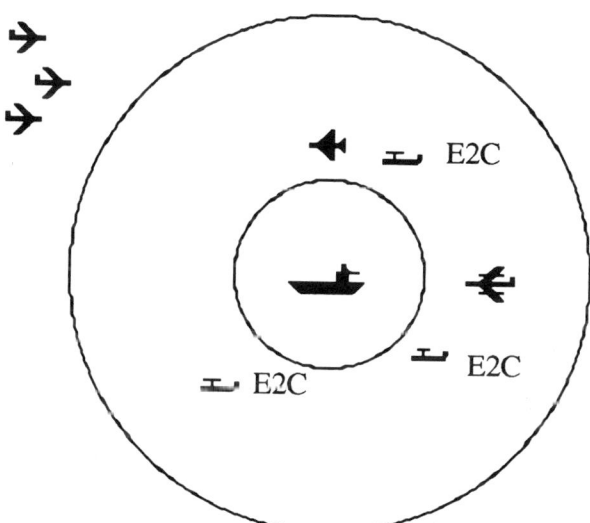

Fig. 1 Naval outer air battle environment.

Defense against air threats is to be performed by two small (three DM) decisionmaking organizations. Specifically, the task of the DMs is to detect incoming enemy aircraft ("threats"); find out the type and the number of threats; then allocate their own aircraft ("resources") to intercept the threats.

The E2C initially operates only ESM to avoid being detected by the enemy's radar. When enemy aircraft approach the E2C and are within a range R_a, the E2C turns on the active radar. When all information (speed, emitter signature, and so on) about a threat is available, the enemy aircraft can be identified. Correlation between the emitter signature and the speed of the aircraft can be used to classify the type of the aircraft with some level of certainty. Based on the assessment of incoming threats, the E2C mission commanders allocate resources to intercept the enemy aircraft. The resources are Tomcat fighter aircraft (F14), Hornet fighter/attack aircraft (F18), and Prowler aircraft (EA-6B).

There are situations in which uncertainty and conflict exist. For example, there may be a threat detected by more than one E2C. Then, the question becomes one of determining who is going to deal with it. In a situation like this, coordination between organization team members is necessary. The coordination is done through communication. In addition, the E2C mission commanders may have to communicate with the antiair warfare commander (AAWC) on the carrier to report the situation or to ask him to launch more interceptors. The protocol for communication is different for different organizational structures.

In this simple example, it has been assumed that when a threat reaches the point where the carrier is within its missile range (R_m) without having been engaged by interceptors, the nature of the task with respect to that particular threat shifts to the inner air battle and is of no further concern to the outer air battle team. Therefore, if enemy aircraft are not intercepted before they enter the inner air battle region, the outer air battle defense for these threats is considered to have failed. Note that the team members have to defend against a number of threats and that each threat may consist of one or more aircraft.

Modeling and Analysis

To perform the task described above, two three-DM organizational structures are considered: a parallel one and a hierarchical one. In the parallel structure (Fig. 2), all DMs are at the same level of authority. They are working together in coordinated fashion. In the hierarchical structure (Fig. 3), authority varies with the rank of a DM, that is, the position a DM holds in the organization. DM2 in Fig. 3 plays a supervisory role in coordinating the other two DMs. In both structures, the task is the same and members of the organization have to act as a team to perform the task.

The task has three stages: situation assessment, fusion, and response selection. When contacts are detected, the situation has to be assessed. The situation assessment (SA) function provides information such as the number of contacts, the position and speed of the contacts, the type and number of aircraft represented by the contact, and their classification as friendly or hostile. Note that a single contact may represent one or more aircraft operating together as a unit (a simplification used in both the model and the experiment). Situation assessment involves data gathering and processing because some of the information can be directly obtained from the observed data, while other information is available only after the raw data are processed. Depending on the particular situation,

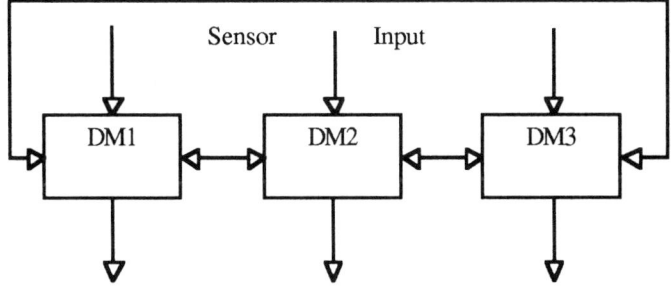

Fig. 2 Parallel organization.

communication may be required after the situation assessment. The results of the communication are processed in the information fusion stage for the parallel structure and in the command interpretation for the hierarchical structure. The last stage is the response selection stage. On the basis of the fusion data or command interpretation results, resources can be allocated to counter the threats. The command decision task is completed after resources are allocated.

Parallel Organization

In the parallel organization, the defense area is divided into three equal sectors (120 deg) as shown in Fig. 4. Each DM is an E2C mission commander and is responsible for operations in one sector, that is, this DM is responsible for selecting a response for all threats in the sector and only this sector. In Fig. 4, the solid straight lines in the radar screen define the three sectors or areas of operations. However, the area of awareness or observation of each E2C commander is larger than his area of operations. For example, the area of operations of one commander is the unshaded area bounded by the two solid

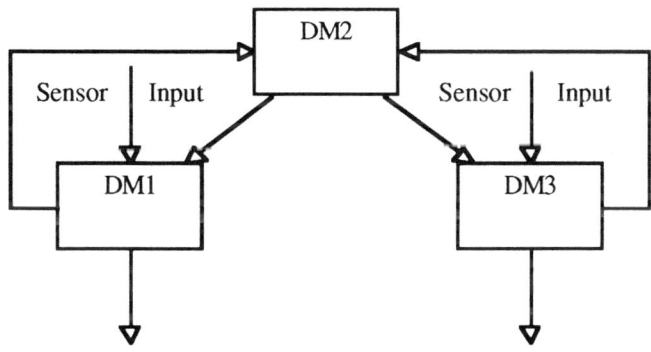

Fig. 3 Hierarchical organization.

lines. His area of awareness is the whole unshaded sector. The dashed lines within the unshaded sector show the parts of his area of operations that fall within the areas of awareness of the other two commanders. In the three-DM case, each DM can see a part of the other two sectors.

It should be clear that there are three types of sectors for each DM in which the responsibilities are different. One is his area of operations. The second is the observation sector, the unshaded area in Fig. 4, which includes his area of operations and the areas that he can observe in the other commanders' areas of operations. The third one consists of the two sectors within his area of operations (unshaded, bounded by solid and dashed lines) that can be observed by the other DMs.

The threats which are in his operations area and can be observed only by him can be processed without exchanging information with the adjacent DMs. For the threats in the operations area that can be observed by another DM, partial information is received; therefore, coordination with other DMs in the team is necessary. Coordination takes place through communication. In the real environment of an outer air battle, the procedure for coordination is quite complex. In this experiment, the procedure is simplified so that it is controllable and serves the purpose of the experiment.

Hierarchical Organization

In this organization, the defense area is divided into two action sectors. Two DMs are E2C mission commanders and play the role of subordinates. Only subordinates can observe the defended area directly. The third DM is the antiair warfare commander on the carrier; he performs a supervisory role and coordinates

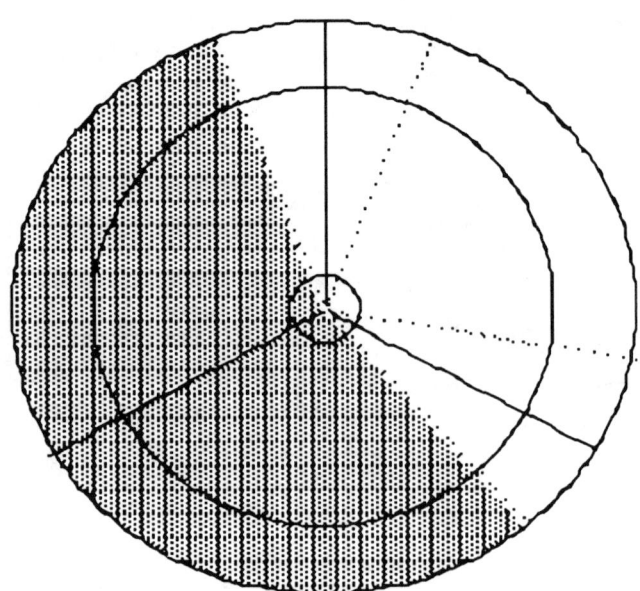

Fig. 4 Defense area divided into three sectors.

the two sectors. As in the parallel organization, each subordinate is responsible for monitoring and intercepting the threats in his sector, his area of operations. However, the two sectors overlap thus creating the possibility of conflict and misallocation of resources. If there is any conflict, that is, there is a threat in the overlap area, subordinates have to report the situation from their perspective to the supervisor. Then they have to wait for commands from the supervisor before prosecuting this threat. After receiving a command, the subordinate who received it interprets it and then makes the final decision on the resource allocation (RS). The level of interaction is higher in the hierarchical organization than in the parallel organization.

Different algorithms can be used to process information and make decisions in the SA and RS stages. The basic algorithms are: (1) quick estimation with attendant risk of errors and (2) accurate acquisition (probing) but with time delay. Which one should be chosen depends on a particular situation, i.e., on the level of uncertainty, on the time available, and so on. The characteristics of the two algorithms reflect the tradeoffs between time and accuracy. The choice of different algorithms indicates the strategy used during the execution of the task.

Model Driven Experimental Investigation

The procedure for carrying out a model driven experimental investigation in the context of the application described in the previous section consists of four stages. The first stage is the *modeling and analysis* one in which the mathematical model is constructed and then exercised to produce predictions of performance. For a given task, an organizational structure is designed. The procedure for carrying out the task is then developed. The evaluation methodology is used to generate three measures: accuracy (J^*), cognitive workload (G^i), and required number of communications (N_{rc}).

The second stage is the actual execution of the experiment and the data collection process. The output of this stage is the set of results obtained from the model-driven experiment: actual accuracy (J), response time (T_f), and actual number of communications among DMs (N_c). Because in a distributed decisionmaking organization, the members must interact to perform a task, knowledge on how the members are coordinated is essential to the design. N_c provides insights on the behavior of interacting decisionmakers for a given task, especially on their strategies for coping with increases in the workload.

The third stage is the calculation of the measures of performance (MOPs) both from the data obtained by exercising the model and from the data obtained from the experiment. These results are plotted in the performance-workload (P-W) space thus creating in each case a locus that characterizes the performance of the organization under evaluation. In general, the P-W space is an $N + 2$ dimensional space with the first two dimensions corresponding to accuracy and time and the remaining N to Workload - one dimension for each DM. Every point in this J-T-G locus is a possible operating point of the organization. The comparison of the MOPs obtained from predicted values and from the experimental results is the basis for performance evaluation.

Because the human operator is an essential element in the decisionmaking organization, cognitive activity during the execution of the task critically affects performance. Bounded rationality can be modeled as a constraint in terms of the maximum processing rate, denoted by F_{max}. In general, when the time available to do a task is reduced, the processing rate will increase so that the task (the amount of work that needs to be done) will be completed. Analytical and experimental results have shown that when the processing rate reaches a maximum value for an individual, a further decrease in available time will cause degradation in his performance because less work than that required by the task will be done. It has also been shown by Louvet et al.[5] that this bound is normally distributed across individuals doing the same task. Therefore, it is necessary to evaluate the information processing rate F required by the design to ensure that it remains less than the bound F_{max}.

By comparing the theoretical and experimental results, the evaluation of the organizational performance is completed. All the relationships characterizing organizational performance can be generated and performance can be predicted. For a given design, measures of performance (MOPs) can be computed so that a designer can evaluate and modify the design, if necessary, so that requirements are met before actual implementation.

In the last stage, experiment results combined with the task attributes can be used to compute two ratios: the time ratio T_f/T_a, which is the ratio of the response time to the available time to do the task, and the communications ratio $n_c = N_c/N_{rc}$, which is the ratio of the actual number of communications to the required number of communications as specified by the protocol design. Bounded rationality characteristics can be observed through the critical time ratio at which the performance degrades significantly. For a given task, time spent to complete it is some proportion of the available time. When the available time becomes shorter and shorter, the response time has to be faster and faster in order to complete the task. The shorter the response time, the higher the processing rate. However, when the available time is so short that the processing rate reaches the maximum capability of the DM, further decrease of the available time will not result in a faster response time. Consequently, the performance will degrade because the work required to do the task cannot be done. The time ratio at this point indicates the level at which bounded rationality limits system performance. The communications ratio indicates the interaction level for an organization: it reflects the strategies used to perform the task under different time stress levels. It may be hypothesized that the DMs will reduce the amount of interaction (needed for coordination) when time pressure is very high.

Although the response time and the available time depend on specific tasks, the ratio between the two reaches asymptotically a constant as the workload rate is increased.[5] Similarly, the communication ratio describes the fraction of communications used and it is a function of the available time - not of the task specifics. On the other hand, N_c and N_{rc} are task specific. These two dimensionless groups, derived through the use of dimensional analysis[2] show every indication of being useful for future experiment design because they appear to be independent of the specifics of the task. Further experimental evidence on a broader range of tasks is needed to confirm this observation.

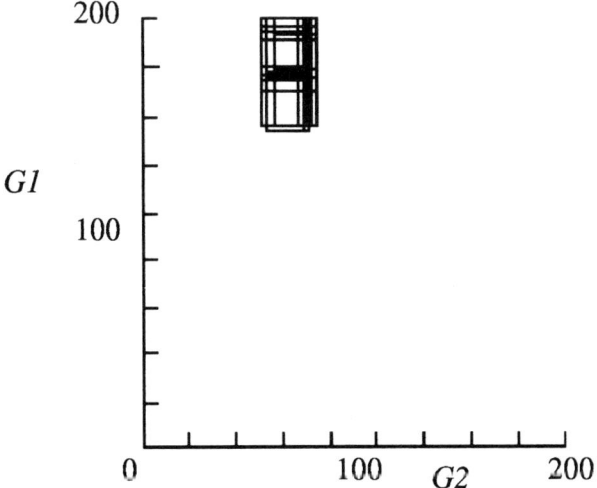

Fig. 5 Workload for two DMs in hierarchical organization: subordinate ($G1$); supervisor ($G2$).

Development of Hypotheses

On the basis of the mathematical model and the results obtained from simulations and analysis, the measures of accuracy J^* and workload G^i were computed for all possible strategies. Then, the performance-workload (J-T-G) locus was constructed to predict the organizational performance. Note that it is a five-dimensional locus since there are three DMs. For the detailed description of the computation, see Jin.[6]

In Figs. 5 and 6, the performance-worlkoad locus is projected on the two dimensional plane defined by the workload axes for decisionmakers DM1 and DM2; $G1$ and $G2$ are the workload of DM1 and DM2, respectively. For the hierarchical organization (Fig. 5), DM1 is a subordinate, while DM2 is the supervisor. The locus is the whole area contained by the boundaries of the plotted figure. The internal lines have to do with the way the algorithm computes workload; to each organizational strategy corresponds a point in the $G1$-$G2$ locus.

It can be seen from Figs. 5 and 6 that the subordinate (DM1) in the hierarchical organization has the highest workload, $G1$. Furthermore, while in the parallel organization the workload locus is symmetric -- the two DMs shown have the same range for task workload (Fig. 6) -- this is not the case for the hierarchical organization (Fig. 5). It is argued that in the hierarchical organization with a protocol requiring close interaction among DMs, when one DM's needed task processing rate exceeds his maximum processing rate, the resulting individual degradation in performance will affect organizational performance.

When the available time T_a decreases, the processing rate F increases while the task workload is kept constant.[6] If T_a decreases continuously until the processing rate reaches the maximum value $F_{\max}$, further decrease of T_a will force a reduction of workload, which is accomplished by the DM selecting a

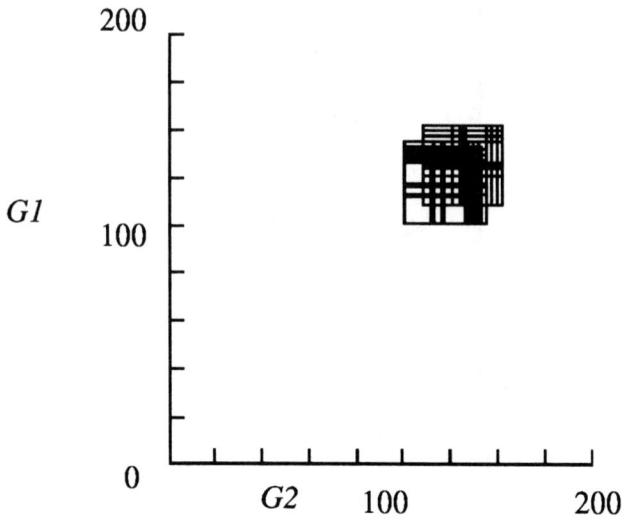

Fig. 6 Workload for two DMs in parallel organization.

strategy requiring less workload. Then, for the same time but less workload, the workload rate is reduced. Further reduction of available time forces the rate to increase again until F_{max} is reached once more. This method of coping with time pressure works until the maximum rate F_{max} is attained with the strategy with the least required workload. Then a further small decrease of T_a will result in a large degradation of performance, since no strategy is available to do the task completely. The DM will fail to complete the task and may make random errors on the portion of the task that he completes.

Let T^* denote the available time when G_{min} is chosen. Then, the maximum processing rate can be expressed as

$$F_{max} = \frac{G_{min}}{T^*} \qquad (1)$$

For a DM in the hierarchical organization, the minimum G is denoted by G_{hmin}. For a DM in the parallel organization, the minimum workload is denoted by G_{pmin}. Since F_{max} has been assumed constant for an individual DM,[5] Eq. (1) results in

$$\frac{G_{hmin}}{T^*_h} = F_{hmax} \quad \text{and} \quad \frac{G_{pmin}}{T^*_p} = F_{pmax} \qquad (2)$$

where T^*_h and T^*_p are the available times driving the DM to the maximum processing rate.

If the information theoretical model for workload[7] were exact and captured all aspects of the cognitive tasks, then for a decisionmaker

$$F_{hmax} = F_{pmax}$$

In this case, the two parts of Eq. (2) can be combined to yield

$$\frac{G_{hmin}}{G_{pmin}} = \frac{T_h^*}{T_p^*} \qquad (3)$$

Eq. (3) indicates that if the values of the minimum workload for two structures are known, the ratio of available time at which performance degrades rapidly can be predicted exactly.

However, there is another complication. While in the parallel organization the workload locus is symmetric -- the two DMs shown have the same range for task workload (Fig. 6) -- this is not the case for the hierarchical organization (Fig. 5). The question then arises as to which minimum workload should be considered, $G1_{min}$ or $G2_{min}$? It is argued now that in the hierarchical organization with a protocol requiring close interaction among DMs, when one DM's needed task processing rate exceeds his processing rate, the resulting individual degradation in performance will affect organizational performance. Consequently, the G_{hmin} in Eq. (3) is chosen as

$$G_{hmin} = \max_i \{ G_{hmin}^i \}$$

If the parallel organization's $G1$ $G2$ locus were asymmetric, then

$$G_{pmin} = \max_i \{ G_{pmin}^i \}$$

Finally, because the theoretical model for the cognitive workload is an approximate one, the exact relation represented by Eq. (3) may be expressed as an approximate relation:

$$\frac{G_{hmin}}{G_{pmin}} \sim \frac{T_h^*}{T_p^*} \qquad (4)$$

From relation (4), it is observed that if G_{hmin} is larger than G_{pmin}, then, T_h^* will be larger than T_p^*, which indicates that when the available time T_a decreases continuously, the rapid degradation of performance will occur in the hierarchical organization first. A hypothesis is established as follows.

Hypothesis 1. When the available time is decreasing, the organization with the highest minimum workload for a given set of strategies will exhibit a performance degradation at a larger value of available time than the organizations which have lower minimum workload.

The second hypothesis is derived by considering the possible strategies for doing the task. There are four pure strategies for each DM except for the supervisor in the hierarchical organization who has two pure strategies. A pure strategy for an organization occurs when all DMs in the organization use a pure strategy. The number of pure strategies for the organizations is the number of combinations of all pure strategies used by DMs, which is computed by

$$K = \prod_{i=1}^{3} k^i \quad (5)$$

where k^i is the number of pure strategies of the ith DM. Therefore, there are 64 pure strategies for the hierarchcal organization and 32 pure strategies for the parallel organization. Table 1 shows the pure strategies which lead to the maximum or the minimum workload. Strategy $D1$ corresponds to the simplest operation: estimating the threat from limited information and selecting a response. Strategy $D4$ involves probing to obtain more information.[6]

In Table 1, strategy $D1$ has three values of the accuracy measure associated with it depending on the tempo of operations. The first value is the accuracy measure for slow operation; the second for moderate operation, and the third for fast operation. Figure 7 shows a set of performance - workload loci.

From Table 1, it can be seen that the minimum workload in both hierarchical and parallel organizations is associated with $D4$ which is the strategy of probing. Note that the probing strategy results in the highest performance. To develop a hypothesis from this observation, let us consider the following.

Because of bounded rationality, DMs will change to strategies with less workload when the available time decreases. Given that in this experiment the minimum workload strategy yields the highest performance, there is no other strategy available for further reduction of the workload to accommodate a shorter available time when a DM reaches the maximum processing rate F_{max}, when using the minimum workload strategy. Then, the ways to cope with the situation are either to reduce the number of communications or to reduce the number of threats being processed. Since the objective of the naval air battle is to process completely all threats, it is hypothesized that a decisionmaker will omit some required communication in favor of processing threats in his own sector. Although this strategy may improve individual performance, it will cause a large degradation in organizational performance. Consequently, the onset of degradation of organizational performance should occur at the same time that the number of communications begins to be reduced.

This can be interpreted as selfish, local behavior. Each DM, under pressure, will attempt to respond to the threats in his sector at the expense of organizational performance. Essentially, this means that under pressure, individual DMs will tend to decouple by reducing coordination. If this were not the case, degradation of performance would occur gradually with an attendant reduction in communications and the latter will be more gradual than performance degradation. For this argument, the following hypothesis is formulated.

Hypothesis 2: Since the minimum workload strategy yields highest performance, under increased time pressure decisionmakers will reduce

Table 1 Maximum and minimum workload for pure organizational strategies

			Hierarchical		
DM1	DM2	DM3	G1/G3	G2	J
D4	D4	D4	135.62	53.65	1.0
D1	D1	D1	182.52	72.06	0.82, 0.63, 0.47

			Parallel	
DM1	DM2	DM3	All G	J
D4	D4	D4	101.70	1.0
D1	D1	D1	135.12	0.76, 0.63, 0.58

communications (coordination) with an attendant reduction in organizational performance.

These two hypotheses were tested by the experiment[6] and the results presented in Jin and Levis.[8]

Results

Hypothesis 1 predicts that the organization with the highest minimum workload will show performance degradation prior to those organizations which have lower minimum workload. Ten of the fifteen teams used in the experiment verified the hypothesis directly. Five teams did not behave in accordance with the hypothesis. The explanation of this discrepancy is as follows: the workload is allocated differently among the DMs in the hierarchical organization whereas the workload is the same for all DMs in the parallel organization. During the experiment, each subject played a role in each one of the two organizations. In the hierarchical organization, subordinates have the higher workload and the supervisor has the lower workload. We allowed the subjects to select themselves who would be the supervisor and who would be the subordinate in the hierarchical organization. They made the determination during the training sessions. Since individual DMs have different skills and capability to process information and make decisions, consider a DM in a team who is a very slow player. His effect on organizational performance will depend on which role he is playing. If a slow player plays the supervisory role in the hierarchical organization, his effect on organizational performance will be very limited since the supervisor has the least workload. However, when this same decisionmaker plays in the parallel organization, he may reach his maximum processing rate at an earlier stage of decreasing available time. In this case, his performance will significantly affect organizational performance. Therefore, organizational performance of the parallel organization starts to degrade when the processing rate of this DM reaches its maximum value. There were two teams, team 8 and team 17, which had a very slow player playing the supervisor's role in the hierarchical organizations. These two teams had the most significant discrepancy from the hypothesis.

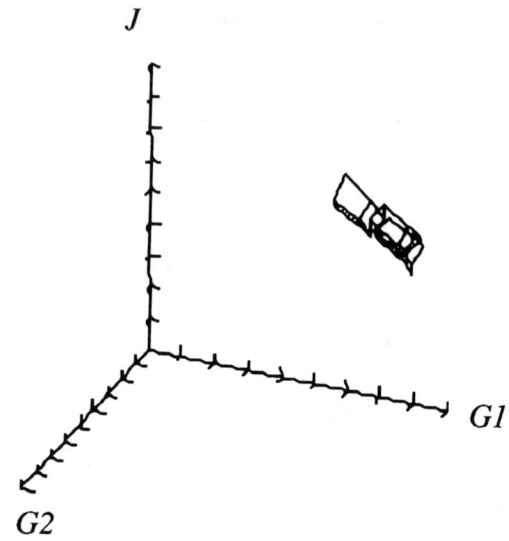

a) Performance-workload locus for hierarchical organization: Moderate speed

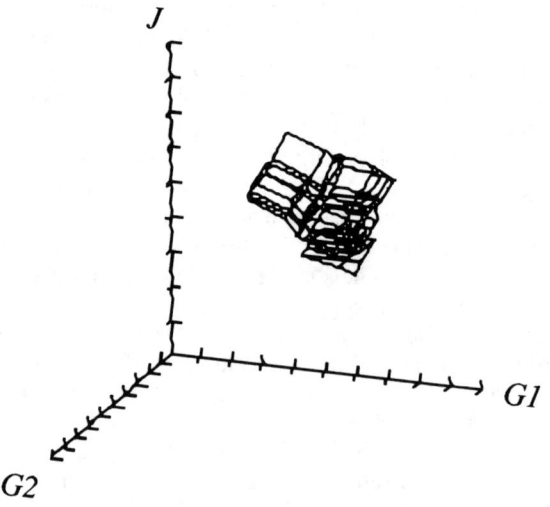

b) Performance-workload locus for parallel organization: Moderate speed

Fig. 7 Performance-workload loci.

If Hypothesis 2 is correct, the rapid reduction in the number of communications and rapid reduction of organizational performance *will occur at the same time*. Because of the difference in structures and protocols in the hierarchical organization and the parallel organization, the required number of communications (N_{rc}) is very different. Therefore, it is necessary to take N_{rc} into consideration. The communications ratio n_c is constructed by normalizing the number of communications by the task-required number of communications.

For the hierarchical organization, the time t^* at which performance as measured by accuracy J begins to degrade rapidly (see Ref. 5 for a detailed description of the process) and the time t^* at which the communications ratio decreases rapidly were found to be

$$\text{mean } t^* (J) = \text{mean } t^* (n_c) = 0.89$$

Therefore, Hypothesis 2 is confirmed. However, for the parallel organization,

$$\text{mean } t^* (J) = 0.89 \text{ and mean } t^*(n_c) = 0.94$$

Therefore, Hypothesis 2 is disproved. Since $t^*(n_c)$ is larger than $t^*(J)$, it follows that when the number of communication drops significantly, performance does not yet degrade rapidly. The explanation is the following. When time pressure is very high, the DMs attempt to reduce the number of communications in order to complete their actions against the threats in their own area of operations. However, the required number of communications in the parallel organization affects only a small portion of the threats that need to be processed. Therefore, a partial reduction in communications does not affect organizational performance significantly.

Conclusion

A multiperson, model-driven experiment has been designed on the basis of a mathematical model of distributed tactical decisionmaking. Team performance, measured in terms of response time and accuracy, was compared for a parallel and a hierarchical organizational structure. The results show that interaction among decisionmakers compensates for differences in individual performance characteristics. Individual differences have more influence on performance in the organization in which DMs have more autonomy in making decisions than in the organization in which individual decisions are coupled with the decisions of other organization members. When available time decreases, time pressure is introduced in the organization and DMs have to adjust their processing rate. The experimental results confirm a hypothesis which predicts that with decreasing available time, a significant degradation of performance occurs first in the organization which has the highest minimum feasible workload. The overall approach to the analysis and design of the experiment was based on model-driven experimentation ideas such as dimensional analysis used in the physical sciences and engineering but not previously extended to experiments with teams of human decisionmakers.

Acknowledgments

The authors would like to acknowledge support provided by the Basic Research Group of the Joint Directors of Laboratories through the Office of Naval Research under N00014-85-K-0782.

References

[1] Hunsacker, J. D., and Rightmire, B.G., *Engineering Applications of Fluid Mechanics*, McGraw-Hill, New York, 1947.

[2] Jin, V. Y., and Levis, A.H., "Command and Control Experiment Design Using Dimensional Analysis," *Proceedings of the 1988 Symposium on C2 Research*, Science Applications International Corp., McLean, VA, 1988.

[3] Boettcher, K. L., and Levis, A.H., "Modeling the Interacting Decisionmaker with Bounded Rationality," *IEEE Transactions on Systems, Man, and Cybernetics*, SMC-12, No. 3, 1982.

[4] Remy, P., and Levis, A.H., "On the Generation of Organizational Architectures Using Petri Nets," *Advances in Petri Nets*, edited by G. Rozenberg, Springer-Verlag, Berlin, 1988.

[5] Louvet, A. C., Casey, J.T., and Levis, A.H., "Experimental Investigation of Bounded Rationality Constraint," *Science of Command and Control: Coping with Uncertainty*, edited by S. E. Johnson and A. H. Levis, AFCEA International Press, Washington, DC, 1988.

[6] Jin, V. Y., "Effect of Organizational Structure on Performance of Decision Making Teams," PhD Thesis, Laboratory for Information and Decision Systems, MIT, LIDS-TH-1976, Cambridge, MA, 1990.

[7] Boettcher, K. L., and Levis, A.H.,. "Modeling and Analysis of Teams of Interacting Decisionmakers with Bounded Rationality," *Automatica*, Vol. 19, No. 6, 1983.

[8] Jin, V. Y., and Levis, A.H., "Effects of Organizational Structure on Performance: Experimental Results," *Proceedings of the 1990 Symposium on C2 Research*, Science Applications International Corp., McLean, VA, 1990.

Problem Solving Systems:
A New Concept for a Science of C2

Carl R. Jones*
Naval Postgraduate School, Monterey, California 93943

Introduction

On December 7th, 1941, the Japanese Navy's successful raid on the U.S. Pacific fleet brought the entry of the U.S. into W.W.II. It also brought a long history of dialogue concerning the behavior and ineffectiveness of the indications and warning (I&W) system used by the United States at the time. Cohen and Gooch[1] in their book study this I&W system systematically. They focus on the overall behavior of the system instead of focusing exclusively on the behavior of individual participants. They use an analytical technique called the matrix of failure to represent the set of decisionmakers involved, what information each had over time, and the command, control, and communications relationships among the decisionmakers. The reader is referred to Ref. 1 for a detailed description of the structure and processes in use in that situation. Here the general characteristics of the illustration will be used to discern the characteristics of interest in a problem solving system. The first of these general characteristics is an identified problem to be solved, i.e., when and where are the Japanese going to attack. Furthermore, this identified problem was not a novel problem for a military organization. Second, there is a set of human decisionmakers with their inherent cognitive limitations interrelated by a command structure, from the President to the operating units (e.g., the U.S.S. Oklahoma). Third, each of the decisionmakers have some shared and some not shared information available to them, with no single decisionmaker having a complete fused understanding of the situation. Fourth, the circumstances in 1941 were fraught with lack of information (uncertainty), alternative hypotheses for possible Japanese courses of action (ambiguity), and a sense making process about Japanese strategic intentions (equivocality). Fifth, the available communications media were not highly reliable. And, sixth, of course timeliness was very important. These six general characteristics

---
Copyright © 1993 by the American Institute of Aeronautics and Astronautics, Inc. All rights reserved.
*Professor, Department of Administrative Sciences and C3 Academic Group.

of a problem solving system, together with analytical constructs will be used to understand the systemic behavior of military organizational decision systems illustrated by the I&W system of 1941. Such organizational decision systems are called problem solving systems.

A problem solving system is focused on solving a particular non-novel problem encountered by the organization. It consists of a set of decisionmakers interrelated by a command and control structure and processes, with each decisionmaker having a possibly different information set. The functions that must be performed in a problem solving system are the collection of data and information (sensing), the recognition that a current organizational problem matches the problem solving system's problem (recognizing), the execution of the solution process for the problem (solving), the implementation of that solution (implementation), and the planning, directing, controlling, and coordinating of the previous functions (commanding and controlling).

To better understand the command and control function, consider the official Joint Chiefs of Staff (JCS) definition of command and control:

> [Command and Control is] . . . The exercise of authority and direction by a properly designated commander over assigned forces in the accomplishment of the mission. Command and control functions are performed through an arrangement of personnel, equipment, communications, facilities, and procedures which are employed by a commander in planning, directing, coordinating, and controlling forces and operations in the accomplishment of a mission.[2]

This definition contains five different factors: the physical means of command and control, the legal authority for command and control, the procedures of command and control, the functions of command and control, and a commander. In this paper, the focus is on the commander, the procedures, and the functions of command and control.

Although the word commander is singular, by observation, there is an hierarchy of commanders interrelated in a chain of command. This hierarchical structure is the result of the limitations of human information processing and cognition of any individual commander (See Ref. 2 for a discussion of these limitations). The required problem solving system capacity is obtained by hierarchically arranging humans with limited capacities in information processing and cognition.

The procedures of command and control are the formalized heuristics and algorithms employed to execute the functions of a problem solving system. Procedures are the processes used in a problem solving system to obtain a solution.

The functions of command and control are planning, directing, coordinating, and controlling. Implicit in these functions is a control cycle paradigm with both short and long run characteristics.

Another aspect of command and control is the relationship of command and control to the principles of war. The principles of war (adapted from Ref. 3) contains two principles that are directly focused on command and control. They are called unity of command and simplicity principles. The unity of command principle has two aspects: the unity of effort and the idea that the best command and control structure is a single commander. The unity of effort aspect can be embodied in a single word – cohesiveness. Combat effectiveness increases as the cohesiveness increases since the elements and weapons of a combat organization are more effectively used. There is systemic synergism. This is sometimes stated as command and control is a force multiplier. The other aspect, a single commander as the best command and control structure, can be understood in terms of individual human information processing and cognitive limitations resulting in an hierarchical command structure. The principle of war called simplicity tells us to use clear, uncomplicated concise language to gain understanding. That is, the structure and procedures of command and control should be clear, uncomplicated, and concise.

From this discussion, we can conclude that the concept of command and control is the use of a simple hierarchical command structure, with clear, uncomplicated, and concise procedures to accomplish the functions of planning, directing, coordinating, and controlling using a control cycle paradigm, so as to have unity of effort in a combat organization. This wisdom of the ages in military affairs is a guideline, an ideal state. A science of command and control should contain the understandings that permit the prediction of the combat effectiveness of alternative structures and processes for command and control.

These same ideas are interwoven into a new framework called Coordination theory.[4] Coordination theory draws from the literatures of computer science, economics, operations research, and organization theory. This chapter will mainly use the ideas of coordination theory in developing the concept of a problem solving system. An important aspect to note at this stage is the self-affinity of the problem solving system concept. Self-affinity is the invariance against changes in size (scale) if more than one scaling factor is involved.[5] The construct of a problem solving system is applicable to understanding an infantryman targeting and shooting, a battalion's defense against electronic warfare, or an army's air defense.

Problem Solving System Construct

A problem solving system can be formally defined as a set of actors performing interdependent activities engaged in solving a problem. This is the second characteristic discussed in the Introduction. Some of the actors will be human decisionmakers and some will be hardware/software systems–machines. The machines act according to programs. Each of the human actors use human problem solving processes. For a human or a problem solving system, a problem exists when there is a goal but no knowledge of how this goal is to be reached.[6] This dominant view of problem solving is due to Newell and Simon (see Ref. 7 for an exposition). Basically problem solving is a search through a state space, a problem space. A problem has the attributes of an *initial state*, one or more *goal states* to be reached, a set of *operators* that can transform one state into another, and *constraints* that an acceptable solution must meet. Problem solving methods are procedures for selecting an appropriate sequence of operators that will succeed in transforming the initial state into a goal state within a particular domain. The most representative general method is called "means-ends analysis." This means-ends problem solving procedure consists of the following four steps.

1) Compare the current state to the goal state and identify differences.
2) Select an operator relevant to reducing the difference.
3) Apply the operator if possible. If it cannot be applied, establish a subgoal to transform the current state into one in which the operator can be applied. (Means-ends analysis is then invoked recursively to achieve the subgoal.)
4) Iterate the procedure until all differences have been eliminated (that is, the goal state has been reached) or until some failure criterion is exceeded or stopping rule is satisfied.

When experts solve problems, they observably used specialized procedures applied to specialized knowledge domains. These specialized procedures are the heuristics of decisionmaking (for additional information on human decisionmaking, see Ref. 8 and Ref. 9). The behavior of a problem solving system is self-affline to these individual human problem solving ideas. That is, the human problem solving ideas can be scaled up to the problem solving system level. Thus, the processes (procedures) of a problem solving system are the organizational realization of the above human problem solving process. The design of a problem solving system to focus on one problem permits the use of specialized procedures for the specialized knowledge domain of that problem. The problem solving systems problem is the first characteristic discussed in the Introduction.

In a combat organization, problem solving systems focus on such major concerns as offensive, defense, and logistics. For example, in Navy warfare there are problems of antisubmarine warfare (ASW), antiair warfare (AAW), antisurface warfare (ASUW), space and electronic warfare (SEW), and strike. Each of these problems are associated with a problem solving system that has a structure and processes for sensing, recognizing, solving, and implementing the solution to the problem, along with an associated command and control (C2) system. Of course, there are analogous problems and problem solving systems in airland warfare. In both types of warfare, there are multiple combat problem solving systems, each with its own command and control. These multiple problem solving systems are interwoven in a metaproblem solving system: a "system of systems architecture." Within the metaproblem solving system, there are individual problem solving systems that are nearly decomposable from the others. That is, a problem solving system that operates nearly independently from the others, for example, the Navy's AAW combat system during an engagement. This paper focuses on understanding a single nearly decomposable problem solving system composed of a combat system and its associated command and control.

A nearly decomposable problem solving system performs the basically sequential functions of data collection (sensing), problem recognition (recognition), solution computation (solving), solution implementation (implementation), and the associated command and control. This is illustrated in Fig. 1. The abstract template for this is a control cycle, which is most commonly observed in the command and control literature as a C2 Loop. Here those abstract concepts are provided a problem solving focus. The data collection function can be performed both organically or nonorganically. For example, there are organic radars and nonorganic collectors such as a specialized reconnaissance aircraft. The data that is collected, possibly from multiple sources, is processed in predefined ways into information for the problem recognition function. In the data fusion literature, the sensing function is known as level one processing (see Ref. 10 for details of multisensor data fusion). Within the recognition function, the information is assessed to provide an understanding of the current set of problems facing the combat organization (situation assessment) and which of these problems can be solved by the problem solving system in focus. In the data fusion literature, the recognition function is known as level two processing. If there is more than one problem in the current situation that must be solved by the problem solving system in focus, and the problem solving system's capacity for multiple problem solving is exceeded, then a threat assessment occurs (data fusion level three). The greatest threat problem is then passed to the solving function to obtain a solution. The solution process may be algorithmic or heuristic. It is algorithmic if the solution

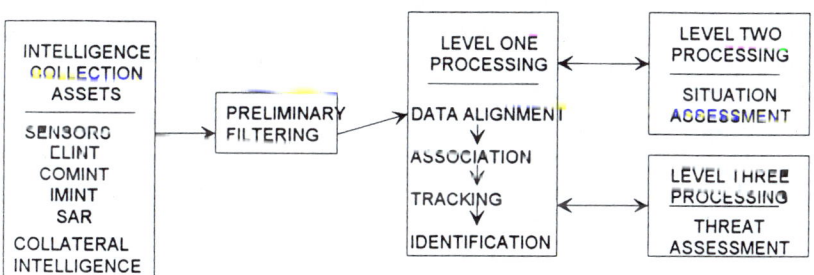

Fig. 1 Product-oriented model of the data fusion process (adapted from Ref. 10, page 16).

process is completely understood and a solution is obtained in a timely manner. Otherwise, it is known as an heuristic process, where a solution will be obtained in a timely manner but it is not an optimal solution. Since the solution methods embodied in observable problem solving systems are rarely algorithms, heuristics that yield nearly optimal solutions are called neoalgorithms. Once a solution has been obtained, it is implemented by, for example, generating firepower at a specific location, by maneuver of forces, or by deception.

Clearly, a problem solving system must operate in the presence of uncertainty, ambiguity, and equivocality. Uncertainty refers to the lack of complete information. Ambiguity refers to the existence of alternative hypotheses. For example, is the object a friend, foe, or neutral? Equivocality refers to the need to make sense out of the situation. For example, just what are the potential opponent's intentions? A problem solving system must be able to operate in the presence of all of these conditions. Stated in terms of the cybernetics literature, a problem solving system must satisfy the law of requisite variety and be a viable system. Variety is a measure of complexity based on the number of variables and parameters as well as the number of relationships between and among them. Viability refers to being in balance, effective, and in a specified environment for a stated period of time (see Ref. 11 and Ref. 12 for details of the viable systems model).

So far the construct of a problem solving system has been discussed in terms of its structure, processes, and viability. The discussion now turns to the determinants of systemic characteristics. The organization theory literature (see Ref. 1 and Ref. 13 and references therein) suggests two major determinants: 1) the state of knowledge of the relationship between action and outcome and 2) the degree of consensus about the goals or criteria for judging performance. The first of these determinants focuses on the understanding of functions of the problem solving system including the effectiveness in solving the problem. The first determinant contains the impact of uncertainty and ambiguity on the relationship of action to outcome. The second of the determinants focuses on problem recognition among the participants in the problem solving system. The participants recognition of the problem depends on 1) the degree of consensus about the collected data/information uncertainty and completeness, 2) the formulation of the current problem, and 3) the collinearity of the current problem and the problem solving system's focal problem. It contains the impact of uncertainty, ambiguity, and equivocality on the solving of the right problem. Alternatively, it focuses on a commonly perceived understanding that the right problem is being solved. The two determinants of the systemic behavior of a problem solving system can be represented as the two dimensions in Fig. 2. The vertical axis is the state of knowledge about the relationship of action to outcome. The horizontal axis is the degree of the common perception of solving the right problem. A problem solving system is represented by each point in the figure. To this extent, each problem

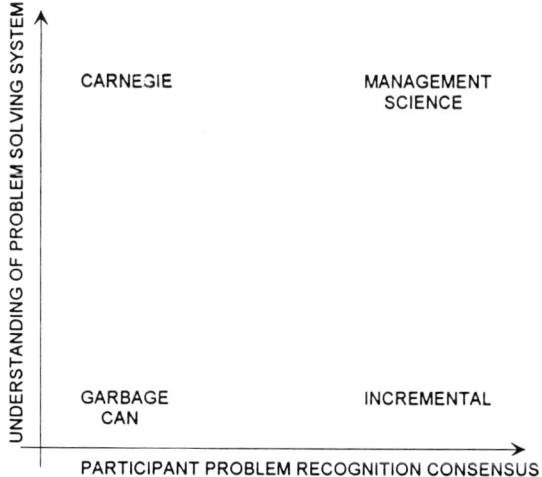

Fig. 2 Determinants of problem solving system behavior.

solving system is unique. The easiest way to understand this multiplicity of problem solving systems is to focus on four "pure" types. The four pure problem solving systems are the ones at the limit and hence, the ones labeled in the "corners" of the figure. The four types will be discussed in turn.

Management Science

When there is high problem consensus and a high state of knowledge about the relationship between action and outcome, then the problem solving system is called a management science problem solving system. The actors know what they want to do and how to do it. The identification and solution processes are well-known heuristics or algorithms. The processes are routine and standardized using a known heuristic or algorithm. Data fusion systems at level one approximate this pure case.

Carnegie

When there is a high state of knowledge about the relationship between action and outcome and a low state of consensus about the rightness of the problem being solved, the problem solving system is called a Carnegie problem solving system (after the university where the process was identified). In this case, there is a lack of consensus about whether the current situation contains the problem that the problem solving system knows how to solve. However, the problem solving and implementing processes are well understood. Systemically, the formation of coalitions of participants around the issue of what is the right problem is observed, if timeliness permits. That is, a coalition forms of individuals who identify and define the same "right" problem. Then the coalitions bargain between and among themselves. This bargaining process is problemistic oriented, i.e., it focuses on what is the problem rather than on what is a solution. Generally, the problemistic search is a local search near the current operating point of the combat organization. If the Carnegie model process results in a consensus, then there is a switch toward the management science problem solving system. If there is no resolution, then the problem solving system "hunts" for a solution, i.e., exhibits oscillations. If a solution, action, is required by outside events (for example, an action by the enemy) then a problem is chosen, solved, and a solution implemented. This is a poorly understood part of the process. Tomorrow may bring a different event, a different problem choice and consequently a different solution, and a different action. Thus, the behavior over time may be inconsistent. An illustration of this type of behavior can be seen in the planning and executing of the air campaign in support of the Normandy invasion in World War II. The interesting book by Rostow[14] contains a description of this. In any situation where the rules of engagement and mission goals are equivocal, it is expected that a Carnegie problem solving system will be observed.

Incremental

This is the situation where the problem is "right" but the means to solve it are not well known. Thus the systemic behavior is characterized by extensive search behavior. This search behavior is a series of "small" decisions that combine to produce a "large" decision. There is the possibility that the small decision will be interrupted by events (labeled interrupts). Interrupts can cause the system to recycle, permitting new ideas to be included via a learning effect. There are three phases to the process. The recognition phase provides for the matching of the extant problem to the problem solving system's problem. The diagnosis phase provides for the formulating of the search process. The development phase is where the solution shaping occurs. It is the characterizing phase of the incremental problem solving system process. There are two modes of operation in the development phase. In mode one, search for a solution occurs within the current repertoire of solutions, a local search. In mode two, a newly designed custom solution is developed. This mode is activated when it is recognized that local search has failed. This recognition occurs after search among the tried and true solutions fails to reveal a solution. It can also occur during the diagnostic phase. Solution design is a trial and error process. That is, it is a gradual, slow, groping, and incremental process. Selection occurs at the end of this phase. In mode two, there is rarely more than one custom designed alternative, so selection is acceptance rather than choice. The final section and evaluation can itself be done in three modes. In the first mode, a single decisionmaker chooses based on experience. In the second mode, analysis is used and the evaluation is systematic. This mode seems to be rarely observed. The third mode occurs when there is a group of decisionmakers. In this case, coalition formation and bargaining are observed. The last activity is the authorization of the decision, the formal acceptance of the decision. The extensive adaptation of several different types of problem solving systems for use in the recent Gulf War are examples of incremental problem solving systems in action.

Garbage Can

First, it is important to recognize that the words garbage can are technical words and not intended to be pejorative. This is the stereotypical process where there is little consensus on the right problem to be solved and little understanding of the relationship between action and outcome. This type of problem solving is also called organized anarchy. There are four main characteristics in a garbage can problem solving process: problematic preferences, poorly understood technology, turnover of participants, and choice opportunities. Problematic preferences refers to the situation where goals, problems, alternatives, and solutions are ill defined. Ambiguity and vagueness are exhibited at each step of the process. Without concrete objects, it is difficult to have stable, if any, preferences over the objects. Poorly understood technology means that there is ambiguity and vagueness here as well. Finally, the participants in the problem solving process change, turnover, due to such acts as employment changes and personal interest in the problem. Thus, there is no stability in the set of interrelated decisionmakers so participation is fluid and limited.

The garbage can problem solving process is not a sequence of steps from problem identification to solution. Rather, it is observably a random process that relates problems, solutions, participants, and choice opportunities. Choice opportunities are those events where problems and solutions must be linked. The most common choice events are budgets, new unexpected opponent capabilities, and de novo missions in new scenarios. The result of this is a systemic process where some problems get matched with solutions and get solved, some problems disappear, some solutions disappear, and some solutions get attached to problems where they are not solutions but yet do get implemented. At first consideration, this seems not only wild but dysfunctional. Certainly it can be. However, it is not always so. Consider a situation where a garbage can process is observed. The participants do not understand the problems or what solutions are available since information is limited for each participant. This is also true for the organization as a whole. Problems, solutions, and participants are randomly connected and at certain times, matched in a choice event. The result is a time path of decisions. The garbage can random process can be understood as a genetic algorithm that finds solutions to problems with a higher probability of success than any other heuristic. An analogy is attempting to climb a mountain where the topography has folds and subpeaks. This mountain is the geometric representation of a maximization problem. As one myopically proceeds up the mountain, it is possible that a subpeak will be reached and it will appear to be the real peak. What the garbage can process does to avoid this is to jump every so often somewhere else on the mountain where the myopic search continues. The result is a probabilistic global search instead of a local search. Thus, the garbage can process is a statistical process for global search when the nature of the problem and the solution possibilities are poorly understood.

From the perspective of the participants in the process, it usually does not appear as random as it does to the outside observer. In fact, the global search only works if each of the participants acts purposefully in the process. The random nature of the problem solving system systemic behavior is based on the participants behaving decidedly purposefully. An analogy is the efficient market hypothesis in economics. In a market, each of the participants acts to use all the available information to gain an economic advantage. As a result of this participant behavior, all the information is exploited. However, since all of the information is exploited, no individual participant can gain economically relative to another. Purposive action by individuals to seek economic gain results in a systemic behavior where no one gains! Such a process is frequently described by a random process. The search for the best Research, Development, Test and Evaluation (RDT&E) investment portfolio for a particular warfare area can exhibit this type of behavior. (The opponents RDT&E efforts, their and your ability to innovate to operational weapons, and the CRI structure and processes is not well known, so it is not clear what is the right problem to be solved. And finally, the nature of technology is not well understood.) So ideas, prototypes, and alternatives are aspects of a garbage can process that, on average, is performing the "best" possible search.

Dynamics

A problem solving system also exhibits dynamic behavior. That is, the processes occur over time with information flows having lags while decisions take time. This can create decisions that are "out of phase" with other decisions. If the problem solving system has a stable equilibrium, then eventually the equilibrium will be reached, i.e., the solution is obtained and implemented. Normatively, there is a need for dynamic behavior with response times that are associated with implemented solutions that win in combat. These and other issues of the dynamics of a problem solving system are important areas of research. It is beyond the scope of this particular paper.

As noted earlier, the JCS definition of command and control is generalizable to the area of study called Coordination theory.[3] A variation of this theory is used to postulate the characteristics of a problem solving system with its embedded command and control. For ease of exposition, this will be done only for the four pure types of problem solving systems. The characteristics of interest are the degree of structure in the top level problem (goal), the degree that the problem solving system's functions can be decomposed into detailed functions that operate to obtain a solution to a problem, the nature of the interdependencies between and among the problem solving system's detailed functions, the nature of decisionmaking, the degree that common objects (e.g., information) are perceived the same by participants in the problem solving system, and the methods of performing command and control (directing, mutual adjustment, standardization, and doctrine). Table 1 contains the author's hypotheses about the characteristics for each pure problem solving system. These hypotheses are based on a reading of the literature and discussions with officers.

The first characteristic concerns the nature of the problem the problem solving system is designed to resolve. If the problem solving system's problem is well understood and the mapping of a current problem to it is well understood, then the problem solving system's problem is called well structured. If not, it is called either moderately structured or unstructured. The degree of problem structureness depends on the degree of ambiguity [A], equivocality [E], and uncertainty [U] present. Ambiguity refers to alternate views of the likelihood of a correct match of a current problem and a problem solving system problem. Equivocality focuses on the degree that the mapping of a current problem to a problem solving system's problem is well understood, i.e., that it makes sense to the participants. Uncertainty refers to the lack of data and information to instantiate the current problem and/or the mapping of the current problem to the problem solving system's problem. For the purposes of this paper, the degree of ambiguity, equivocality, and uncertainty will be measured as low or high. In the management science and incremental problem solving systems, it is hypothesized that there is low ambiguity, uncertainty, and equivocality. In the Carnegie and garbage can cases, it is hypothesized that there is high ambiguity, uncertainty, and equivocality. In these cases, equivocality is high since the problem solving system's participants have different judgments about the match of the problem solving system's problem and the current problem–the "rightness" of the problem. The participants have different senses of the current problem and/or the mapping of the current problem into the problem solving system's problem.

The second characteristic refers to the understanding of the solution procedure. If the solution procedure can be decomposed into detailed functions and performance requirements set for each of these detailed functions, and the procedure provides a solution, then the procedure is well structured and called algorithmic. If the procedure is not well structured, and the solution only "usually works," then it is called an heuristic process. Since algorithmic procedures are not usually observed, a procedure that is close to an algorithmic is called neoalgorithmic. The management science and Carnegie problem solving systems are hypothesized to be neoalgorithmic. The incremental and garbage can problem solving systems, because of the nature of the search activity, are hypothesized to be heuristic procedures.

The quantity of interdependence in the solution procedure depends on the nature of the interfaces between and among the functions with their associated requirements at the bottom level of the decomposition. These interdependencies arise from the sizing, sequencing, and synchronizing of the functions for the specialized knowledge domain of the problem being solved. The entries in the table are the author's judgment generalizing across multiple knowledge domains. Thus relative to one another, a management system is low, the garbage can is high but random, whereas the others are in between.

The nature of decisionmaking by the individual participants in the process is the next characteristic of interest. If each individual participant's problem and solution procedure is well structured, and there is a neoalgorithmic systemic process, then systemically decisionmaking is labeled neoalgorithmic. Otherwise, it is heuristic. The author hypothesizes that the neoalgorithmic case applies to the management science and Carnegie systems, whereas the incremental and garbage can systems are heuristic.

The communications characteristic refers to the relative degree of communications required to find and implement a solution. It is a relative measure. It is lowest with management science systems and highest with a garbage can system. The others are in between. There is a clear implication for the communications capacity required for the particular problem solving system to operate effectively.

The decision aiding characteristic measures the degree to which decision support systems are possible to aid the decisionmakers, individually or collectively. If the decisionmaking is neoalgorithmic, and the decision support system is based on models and decision theory, then it is hypothesized that high decision aiding is present. If no support is possible, then that is stated. Otherwise, decision aiding is judged to be low. Table 1 contains the author's judgment of a high for management science system to a low for a garbage can system.

The next characteristic is labeled the perception of common objects (e.g., shared information). The degree that the participants in the process (or some subset thereof) perceive an object as the same object is the notion of this measure. For example, if there is a complete data dictionary for the problem solving system so that all data objects are names, then there is very high perception of common objects. If there is no data dictionary, even tacitly, then the perception is very low. The management science system is hypothesized to have a very high perception of common objects while the garbage can system has very low. The others are judged to be moderate.

When the problem solving system's command and control function is implemented, there are four methods available: direct supervision, mutual adjustment, standardization, and ideology (doctrine). Direct supervision means the use of a hierarchy of commanders where each commander internalizes the needed coordination and executes it via orders. Mutual adjustment refers to the resolution of issues between individuals at the same hierarchical level in the command structure. The creation of standard operating procedures, standard participants through education and training, and the specification of output and/or inputs are the means by which standardization provides coordination. Last is the development of a common framework for all the participants to use in solving problems. In military organizations, this is called doctrine, whereas the academic literature calls it ideology (terminology associated with study of religious organizations). Management science systems are hypothesized to exhibit high direct supervision, low mutual adjustment, high standardization, and high doctrine. An incremental problem solving system is hypothesized to have moderate direct supervision, high mutual adjustment, moderate standardization, and moderate doctrine. The bargaining process inherent in a Carnegie system leads to the hypothesis that there is high mutual adjustment and very low direct supervision. Otherwise, mutual adjustment and standardization are low. In the case of a garbage can system, direct supervision, mutual adjustment, and standardization are hypothesized very low and doctrine low. The above hypotheses are shown in Table 1. The author hopes and expects that they will change based on observation and discussion.

Postscript

The thrust of this paper is an exposition of a construct to understand the systemic behavior of individual, nearly decomposable problem solving systems; that is, individual combat systems and their associated command and control. The prescribed length of this paper permits only a general overview of the construct. Among the interesting aspects of problem solving systems is the dynamic behavior of the system; the allocation of functions to physical assets, human and machine; the detailing of specific algorithms and heuristics for specialized problem areas; and the problem of the command and control of a single physical asset that hosts multiple functions. These issues need to be pursued in the field, the laboratory, and theoretically.

In addition to the work on a single problem solving system, there is the issue of understanding the behavior of a "system of problem solving systems," a meta-architecture problem solving system. When nearly decomposable problem solving systems are interwoven into a tapestry of meta-architecture, what is the systemic behavior of that meta-architecture? Do the determinants of behavior remain the same for this meta-problem solving system? Notice that a function within a decomposable problem solving system can itself be a problem solving system, e.g., recognition. These are questions for future research. It is interesting to note the dearth of articles on the system of systems phenomenon.

Understanding both individual problem solving systems and a meta-architecture of them is a very challenging task that is also very important. The rapidly changing technology of C4I reduces the cost of performing command and control. Thus for the same expenditure, more activities can be coordinated with an associated increase in the unity of effort in the combat organization. Or the same degree of unity of effort can be achieved at lower cost. Whatever choice is made, including combinations, careful attention must be paid to the choice of what to coordinate and by implication, what not to coordinate. The coming decade will also include an increase in information war considerations. What is the best meta-architecture for the age of information war with rapid change in technology, and very high variety in the possible situations that will be faced? That is the challenge that the construct of a problem solving system can help resolve.

References

[1] Cohen, E. A., and Gooch, J., *Military Misfortunes*, The Anatomy of Failure in War, The Free Press, N.Y., 1990.

[2] Joint Chiefs of Staff, *Department of Defense Dictionary of Military and Associated Terms*, (Washington D.C. GPO), June 1986, p. 77.

[3] FM100-5, "Principles of War," *Operations*, Appendix A, Washington D.C. GPO, 1986, pp. 173-177.

[4] Malone, T. W. and Crowston, K., "What is Coordinative Theory and How Can It Help Design Cooperative Work Systems?," *Groupware: Software for Computer-Supported Cooperative Work*, Marca, D. and Bock, G., ed., IEEE Computer Society Press, Los Alamitos, California, 1992.

[5] Schroeder, M. R., *Fractals, Chaos, Power Laws: Minutes from the Infinite Paradox*, W. H. Freeman, N.Y., 1991, p. iv.

[6] Duncker, K., "On Problem Solving," *Psychological Monographs*, Vol. 58, Whole No. 270, 1945, p. 5.

[7] Newell, A., and Simon, H. A., *Human Problem Solving*, Prentice Hall, Englewood Cliffs, N.J., 1972.

[8] Holland, J. H., Holyoak, K. J., Nisbett, R. E., and Thagard, P. R., *Induction: Processes of Inference, Learning, and Discovery*, MIT Press, Cambridge, Massachusetts, 1986.

[9] Waltz, E., and Llinas, J., *Multisensor Data Fusion*, Arteck House, Boston, Massachusetts, 1990.

[10] Klir, G. J., *Architecture of System Problem Solving*, Plemm Press, N.Y., 1985.

[11] Espejo, R., and Harden, R. J., *The Viable System Model: Interpretations and Applications of Stafford Beer's VSM*, J. Wiley, N.Y., 1989.

[12] Daft, R., *Organization Theory*, 3rd ed., West Publishing company, N.Y., 1989.

[13] Rostow, W. W., *Pre-Invasion Bombin Strategy: General Eisenhower's Decision of March 25, 1944*, University of Texas Press, Dustin, 1981.

Colored Petri Net Model of Command and Control Nodes

Alexander H. Levis*
George Mason University, Fairfax, Virginia 22030

I. Introduction

COMMAND, control, and communications (C3) systems support the command and control process. That process is becoming increasingly distributed as larger and more diverse geographically dispersed forces are used to carry out complex missions. As a result, more demands are being placed on the C3 systems themselves and on the way they are used to support specific missions. The latter is (or should be) described by the command and control architecture in an explicit and unambiguous manner. The description of such an architecture must be based on a set of models that capture the diversity of roles a particular asset can play and on a procedure for describing the interactions between the various nodes. The emerging theory of distributed intelligence systems (DIS) provides a conceptual and mathematical framework for modeling and evaluating command and control architectures.

Distributed intelligence systems may be defined as those in which the capacity for reasoning is dispersed across their component subsystems. This definition, however, begs the issue by relating intelligence to reasoning, a term that requires definition in its own right. A more appropriate pair of definitions which may not be formal, but capture the concept of distributed intelligence used in this paper, can be found in Minsky.[1] In a *distributed* system, each function is spread over a number of nodes so that each node's activity contributes a little to each of several different functions.

The notion that a function is decomposed and its components (or subfunctions) assigned to different nodes is an old one. What this definition of "distributed" makes explicit is that each node contributes to the execution of *several* different functions. The types of systems being considered carry out a number of functions, sometimes in sequence and sometimes concurrently. Thus, the problem is not solved by doing a simple allocation of a decomposed function

---

Copyright © 1993 by A. H. Levis. Published by the American Institute of Aeronautics and Astronautics, Inc., with permission.
*Professor of Electrical, Computer, and Systems Engineering, Department of Systems Engineering.

to the available resources, human and machine ones. One must allocate several decomposed functions in such a manner that the resulting workload does not exceed the capacity of each node.

The modeling of (cognitive) workload in distributed intelligence systems must take into account not only the workload associated with each subfunction carried out by a node, but also the workload associated with coordinating the execution of the subfunctions and with coordinating the interactions between intelligent nodes. An information theoretic framework has been used to model cognitive workload.[2]

The concept that will be introduced as the basis for modeling the distributed aspects of the system is that of a role; a role is capable of executing a subfunction. The role represents the lowest level of functional decomposition for a particular application; a role must be executed in its entirety by a single node. Two types of interactions among roles will be defined: 1) those that are among roles within the same node, called internal interactions, and 2) those that are between roles in different nodes, called external interactions. The latter are the ones that determine the organizational structure or architecture of the DIS, and by extension, of the command and control architecture.

The origins of the model of the role can be traced back to the four-stage model of the interacting decisionmaker with bounded rationality introduced by Boettcher and Levis.[2] The formal specification of the allowable interactions between interacting decisionmakers was made by Remy and Levis.[3] This specification led in turn to an algorithm, the lattice algorithm,[3] which generates all feasible fixed structure architectures that meet a number of structural and user constraints. Andreadakis and Levis[4] introduced an alternative model that was not based on the decisionmaker model, but on the function carried out by a resource, whether that represented a human or a machine. Although this was a five-stage model, it was very similar to the four-stage one in terms of the allowable interactions. That model formed the basis for a different algorithm for organization design, the data flow structure (DFS) algorithm.[4] Andreadakis' approach to organization design (determine first the required data flow structure for the system and then assign functions to resources) led to the formulation of the notion of role by Demaël and Levis.[5] In a parallel effort, Monguillet and Levis[6] formulated a model of variable structure architectures. A synthesis of the idea of the role and of a variable structure led to the generalization of the Lattice algorithm to a special class of variable structure architectures: those who adapt their structure to the task or the input they have to process.[5]

The various models mentioned in the previous paragraph have been used to address a number of problems in the design, analysis, and evaluation of distributed decisionmaking organizations supported by decision aids and decision support systems, i.e., of command and control organizations. The analytical and experimental work of the last ten years, when combined with theoretical developments in the representation of discrete event systems through Colored Petri Nets as defined by Jensen,[7] has led to the reassessment of the various models and their variants and to the conclusion that a slightly more general model could subsume all previous ones without invalidating any of the cognitive modeling or the design algorithms.

Such a model is presented in this paper. The formalism of ordinary Petri Nets is used to introduce the model in the next section, whereas in the third section, interactions between roles are defined. In the final section, the most general form of the model is folded into a Colored Petri Net representation.

II. Model of Role

A distributed intelligence system, such as a command center, is designed to carry out a mission. Following accepted terminology, a mission is decomposed into functions and functions are decomposed into tasks. This hierarchical decomposition, mission→function→task, shown in Fig. 1, has some interesting properties.[8]

One property of interest is that it is a nested functional decomposition. What may be considered a function from a certain vantage point becomes a mission from another. This is shown in Fig. 2.

Consequently, this functional decomposition can be carried to any level of detail. In practical terms, it is carried out to the point that the lowest level tasks (the leaves of the tree) must be executed by a single resource. A role is used to model the execution of such a task by a single resource. The basic model of the role can be represented in block diagram form as shown in Fig. 3. It consists of three processing stages and two interaction stages for a total of five stages.

A role receives inputs or data x from the external environment (sensors) or from other nodes of a system. The incoming data are processed in the first block marked situation assessment (SA) to obtain the assessed situation z. This variable may be sent to other nodes, as shown by the outgoing arrow. If the role receives data about the assessed situation form other nodes, these data z' are fused together with its own assessment z in the information fusion (IF) stage to obtain the revised assessed situation z''. The assessed situation is processed further in the task processing (TP) stage to determine the strategy v to be used to select a response. However, if hierarchies exist in the organization, the particular role may receive command inputs v' from superordinate nodes (or higher echelon decisionmakers) that may restrict the strategies available for selecting a response. This is depicted by the use of the command interpretation (CI) stage. The output of that stage is the variable w which contains both the revised situation assessment data and the response selection strategy. Finally, the output or response of the role y is generated by the response selection (RS) stage.

Note that if there are no situation assessment and command inputs from other nodes, the five-stage model reduces to the common two-stage situation assessment and response selection model of the single, noninteracting decisionmaker.[2] If there is no information fusion, then the IF stage disappears and the SA and TP stages can be merged into a single SA stage. Finally, if there are no command inputs, then the CI stage disappears and the TP and RS stages can be merged into a single RS stage. The five-stage model and its special cases correspond very closely to the flow types defined by Ref. 4 with the main difference being the explicit meaning given to each stage. Therefore, by introducing the middle processing stage, task processing, a new model is obtained that contains the fea-

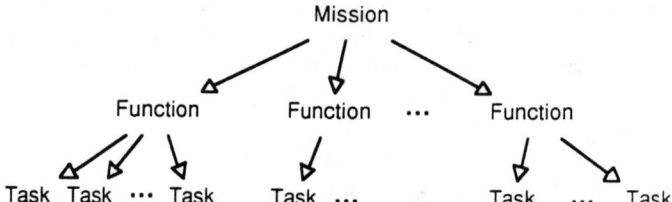

Fig. 1   Functional decomposition.

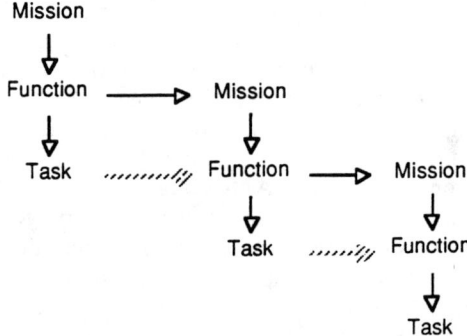

Fig. 2   Nested decomposition of functions.

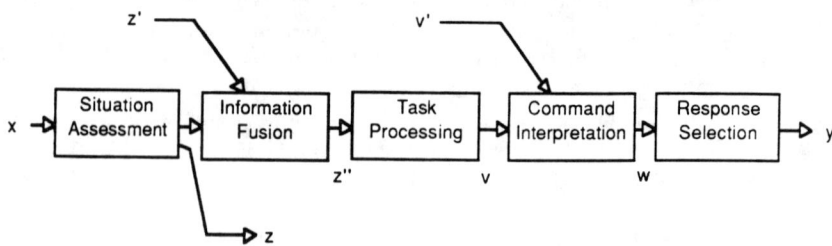

Fig. 3   Five-stage model of a role.

tures of both the four-stage interacting decisionmaker model of Boettcher and Levis[2] and the Andreadakis and Levis[4] model.

If ordinary Petri Net notation is used, the model of Fig. 3 takes the form shown in Fig. 4. The rounded box enclosing the five stages has no formal meaning; it is used to indicate the internal structure of a role and show explicitly the inputs and outputs of the role.

In accordance with Petri Net conventions, transitions are denoted by solid bars, and they represent processes or events. Signals or conditions are depicted by places that are shown as circles. Directed arcs indicate the relationships between the two types of nodes. Since Petri Nets are bipartite directed graphs, arcs can exist only from a transition to a place or a place to a transition. A detailed description of Petri Net modeling of decisionmaking organizations is given in Levis.[9]

III. Interactions Between Roles

The model of the role shown in Fig. 4 is the one used to define the interactions that can exist between two roles in different nodes. Consider two roles, role i in node k and role j in node m. Figure 5 shows the possible flow of data from role i to role j. A symmetric set of flows are defined from role j to role i.

Consider role i. The first question to be answered is whether it receives data from the external environment, from the sensors. This is denoted by the coefficient e_i. If e_i is equal to 1, then role i receives data from sensors; if it is equal to zero, it does not. The coefficient G_{ij} indicates whether the output of role i is an input to the situation assessment stage of role j. This type of interconnection is needed to represent the tandem (series) connection of roles. The coefficient F_{ij} represents the sharing of the assessed situation z by role i with role j. This is one type of information sharing. The second type, the sharing of results, is represented by H_{ij}. In this case, role i communicates to role j the output of the response selection stage. The link H_{ij} is used in place of G_{ij} when there is an input from the external environment to role j. Whether to communicate the assessed situation or to communicate the decision a role has made is an interesting design question that has been addressed by many researchers. The basic tradeoff is the amount of data that needs to be transmitted; the situation assessment usually requires more bits than the decision. On the

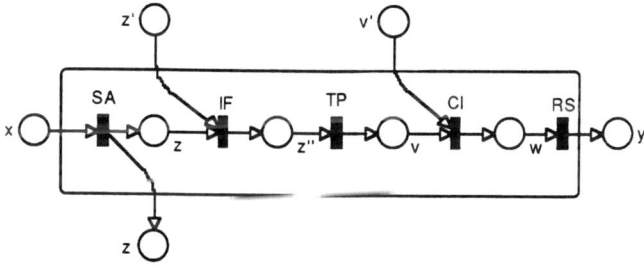

Fig. 4 Petri Net representation of the five-stage role model.

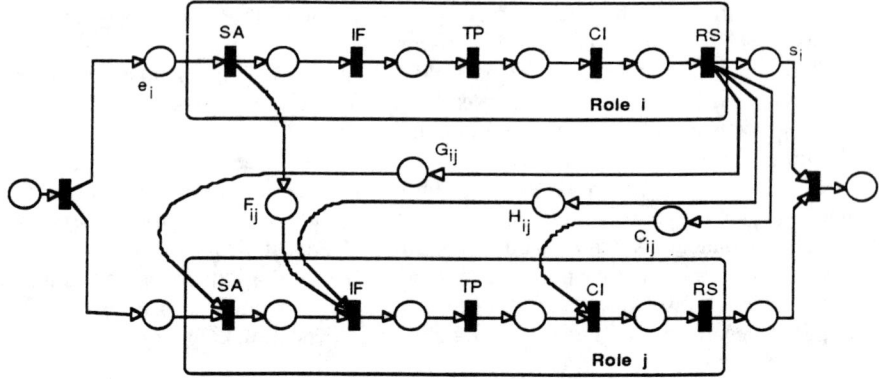

Fig. 5 Admissible data flows from role i to role j.

other hand, under some rather restrictive conditions, it is possible to reconstruct the assessed situation when the decision is known. The final type of interaction is the issuing of a command from role i to role j, as shown by C_{ij} from the response selection stage of i to the command interpretation stage of j. Finally, the coefficient s_i denotes whether role i produces an output to the environment.

When fixed command and control architectures are considered, then each node contains only one role and these six coefficients are constant and take values in $\{0, 1\}$. If there are n roles/nodes in a fixed structure organizational form, then all of the interactions can be represented by a set of six arrays

$$\Sigma = \{e, F, G, H, C, s\}$$

where e and s are $n \times 1$ and F, G, H, and C are of dimension $n \times n$. From the definition of the arrays (see Fig. 5) it is apparent that the diagonal elements of F, G, H, and C are identically zero. Therefore, any possible set of interconnections among n roles can be represented by assigning the value of 0 or 1 to q elements where

$$q = 2n + \left(4n^2 - 4n\right) = 4n^2 - 2n$$

The number of admissible organizational forms is $2q$. Note that all of these structures may not necessarily represent feasible organizations, i.e., organizations that satisfy a number of structural constraints. The basic structural constraints are as follows:

1) The Petri Net that corresponds to Σ should be connected; a directed path should exist from the single source place to every node of the net and from every node of the net to the single sink place.

The single source and single sink places are modeling artifices used to coordinate the source model and to collect all the outputs of the net into a single node.

2) The net Σ should have no loops; it should be acyclic. Note that this constraint applies only to the basic information flow. Loops can be added to represent resource constraints or coordination conditions.

3) There can be at most one link from the RS stage of node i to another node j, i.e.,

$$G_{ij} + H_{ij} + C_{ij} \leq 1$$

4) Information fusion can take place only at the IF and CI stages. Consequently, the SA stage can receive either inputs from the source model or from a single other node.

5) This last constraint is not necessary; its inclusion eliminates some awkward interactions between nodes.

$$G_{ij} + F_{ij} \leq 1$$

Additional constraints may be introduced to reflect the specifics of a particular application. The lattice algorithm can be used in conjunction with this model to generate alternative fixed structure architectures.

IV. Intelligent Nodes

One way of representing an intelligent node without resorting to the use of Colored Petri Nets is to use a special type of transition, namely, a switch. A switch is a transition with multiple output places and some decision rule which directs the generation of a token after firing in one and only one of its output places.

Since a switch is really a transition, the firing rules for a switch are identical to the firing rules for a transition: a switch will fire if all of its input places contain at least one token. Unlike regular transitions, however, all of the output places of a switch will not receive a token. Only one of them will. This place will be chosen by the internal decision rule associated with the switch. An example of a switch with two input places and three output places is shown in Fig. 6. The output places of the switch are called the branches of the switch. In Fig. 6a, the switch is enabled. The rule embedded in $s1$ is used to determine in which output place a token will be generated. In this case, the token is generated in $p11$. Note that it is necessary to associate attributes with the tokens so that the rule embedded in the switch can differentiate among inputs.

Consider an intelligent node that can instantiate any one of several roles depending on the input it receives. A switch can be introduced between the source-generating tasks (inputs) and the roles with a rule that directs the particular task to the appropriate role. This is shown schematically in Fig. 7 for the case when there are four roles.

Consider the changes that are needed in the modeling of the source. Let the source generate n distinct tasks x; then the set of tasks X can be partitioned into four disjoint subsets X_1 to X_4. Then the rule embedded in the switch takes the form: if the input $x_i \in X_j, j = 1$ to 4, then role j is activated. This particular generalization is straightforward. A task is generated by the source, it is directed by the switch to the appropriate role, and an output is generated. The complications arise when interactions between intelligent nodes are considered.

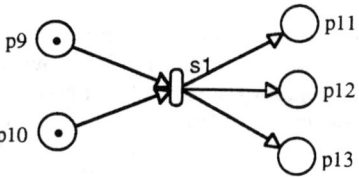

a) Switch $s1$ is enabled

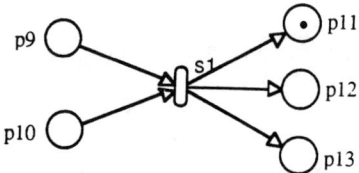

b) Only output place $p11$ contains token

Fig. 6 Example of a three-branch switch.

As shown in Ref. 6, the rules governing the switches are not independent. If the interactions between nodes depend on the task (i.e., type of token that is being processed) then a node i that interacts with node j must know what role node j has assumed so that it can select a compatible one. If this does not happen, then deadlocks can occur. For example, node i chooses a role that does not include the transmission of situation assessment information to node j. On the other hand, node j chooses a role that requires situation assessment information from node i. The IF transition of node j will not be enabled and that node will be deadlocked. To address this problem, a table must be created that contains all of the admissible combinations (the intercorrelations) of switch settings. The difficulty is that this information is not included as part of the Petri Net formalism.

The answer to this problem is the introduction of Colored Petri Nets and the addition of several more structural constraints to the list of five that was presented in the previous section. In Colored Petri Nets, the form of High Level Nets introduced by Jensen,[7] tokens are distinguishable. A set of attributes

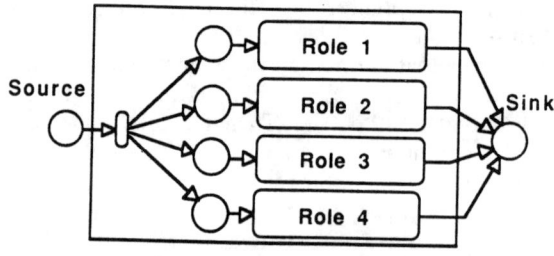

Fig. 7 Model of intelligent node.

is associated with each token, where each attribute can take a number of values. The color of the token denotes a particular choice of attribute values. All of the possible combinations of attribute values (all of the colors) constitute the universal color set for a particular problem. There is a color set associated with each place in the Petri Net: the color set specifies the colors of the tokens that may reside in that place. This color set is a subset of the universal color set. Arcs are inscribed with expressions of the form

[Boolean] %expression

where Boolean is a Boolean expression. When this expression evaluates as true, then the arc inscription evaluates to a set of colors according to the normal expression on the right.

A transition is enabled, if there exists at least one binding for the variables in the inscriptions of the input arcs such that each input place contains at least as many color tokens as specified by the arc inscription. In addition, when the transition contains a guard function, the condition indicated by the guard function must also be satisfied. Figure 8 demonstrates these definitions.

The universal color set contains three colors: red, white, and blue. All three places have as their color set the universal set; they can hold all types of tokens. A variable x has been defined that can take values in (be bound to) the same universal set S. The transition has a guard function that restricts the variable x from taking the value w if the transition is to be enabled. The inscriptions translate as follows: if the variable x takes the value r and there is at least one r token in the input place on the right, and if there are two blue tokens in the place on the left, then the transition will fire. Two blue tokens will be removed from the left input place, one red token from the right input place, and two red tokens will be generated in the output place. If x takes the value b, then the right place must have at least one blue token. However, the arc inscription on the left will not evaluate as true and, therefore, there is no enablement condition for that binding of the variable. Finally, because of the guard function, there

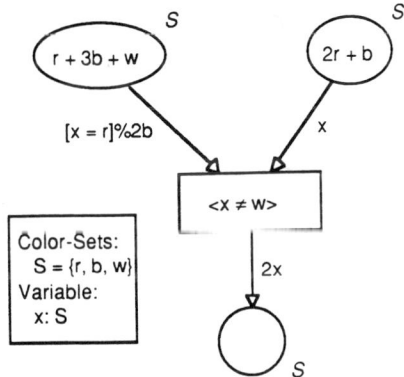

Fig. 8 Colored Petri Net example.

is no point checking what will happen to the arc inscriptions when x takes the value w.

Consider now the example in Fig. 7. There are n distinct inputs (token colors) in the set X and four color sets, each set corresponding to one of the partitions X_i of X. The arcs from the input place to each one of the roles are inscribed as follows:

$$[x_j \in X_i] \% x_j \qquad \text{for} \qquad i = 1 \text{ to } 4 \qquad \text{and} \qquad j = 1 \text{ to } n$$

Note that instead of an arc inscription, a guard function of the form $\langle x_i \in X_i \rangle$ could have been placed in a transition preceding the ith role. The places on the right in Fig. 9 are the input places of the individual roles; a particular role will have an input and will be activated only if a token or tokens belonging to the right color set appear in its input place.

The Colored Petri Net (CPN) model of each role is then attached to the model of Fig. 9 to represent the intelligent node that can instantiate a number of nodes. Note that the CPN model makes it very simple to model concurrent operation of several roles; if tokens $x2$, $x6$, and $x17$ are present in the source place, then the transitions corresponding to roles 1, 2, and 3 will be concurrently enabled and can fire. The output place of the intelligent node is identical in structure to that shown in Fig. 7. The color set for the output place contains all of the output colors of each role.

One condition that must be maintained is that of the roles within a node not directly interacting with each other. There can be resource constraints that may force the execution of the roles to be serialized, but the roles cannot exchange information about the task they are processing. Consequently, the model of the interactions between roles presented in Section III still holds, but with an additional important condition: *the roles must belong to different nodes*. However, the introduction of the color dependent conditions for the enablement of transitions brings back the issue that led to the discarding of the switch transition as an acceptable solution—the coordination between conditions. This is addressed by the introduction of one more condition (Ref. 5) to the five discussed in Section III.

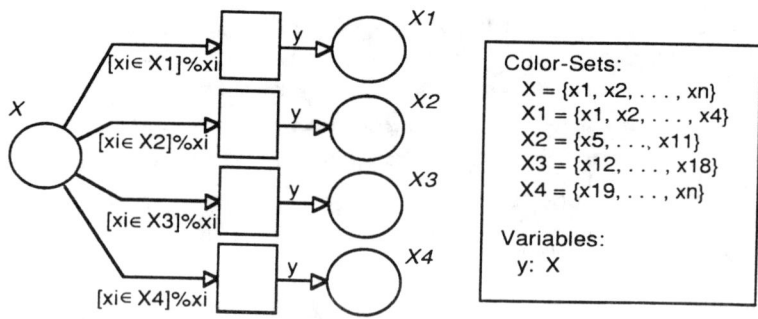

Fig. 9 Colored Petri Net model of role inputs.

To formulate the new condition, several concepts must be introduced. First, it is recognized that the variable structure architecture Π that can be constructed with this CPN model of the node is nothing but a superposition, or folding, of a number of fixed structure architectures Σ of the type defined earlier: to each input x_i corresponds some fixed architecture Σ. For the example of Fig. 9,

$$\Sigma_j = \Pi\left(x_i \in X_j\right)$$

i.e., role j is instantiated when the input is an element of the subset X_j. A color set X_j is said to be accessible at a transition t if and only if there is a directed path from the source to the transition t in the fixed structure obtained when $x_i \in X_j$. The reason for introducing this concept is that the variable interaction between two stages of roles in different nodes must be based only on information that is accessed jointly by the roles that interact. For example, the SA stage of role 1 in node 3 must determine, based on some information it has accessed, whether it must send information to role 2 in node 5. Similarly, role 2 in node 5 must infer from some of the information it has already received whether or not it must wait for a message from role 1 in node 3 before initiating information fusion at the IF stage. These conditions can be expressed by inscribing the arcs in the manner described in Fig. 8. In that example, the new condition is manifested by the assumption that the left arc knows that the binding on the right is $x = r$, even though no red tokens are required on the left for the transition to be enabled. Thus, constraint 6 takes the following form. In an architecture that contains intelligent nodes, the following condition must hold:

1) If there is an interaction between the SA stage of a role in a node and the IF stage of a role in another node, then in every $\Pi(x)$, for any $x \in X_j$ that activates those roles, there must exist directed paths from the source to these two transitions.

2) If there is an interaction between the RS stage of a role in a node and the IF stage of a role in another node, then in every $\Pi(x_i)$, for any $x \in X_j$ that activates those roles, there must exist directed paths from the source to these two transitions.

3) If there is an interaction between the RS stage of a role in a node and the CI stage of a role in another node, then in every $\Pi(x_i)$, for any $x \in X_j$ that activates those roles, there must exist directed paths from the source to these two transitions.

For a formal description of these conditions in the context of a special class of Colored Petri Nets, see Ref. 5. The software used to model the intelligent nodes is *Design/CPN*^{TM} by Meta Software Corporation.[10] The software has embedded in it syntax checks that ascertain that the information needed by condition 6 is accessible to the arcs representing interactions between nodes. They are the conditions needed for the model to be able to execute, i.e., for a simulation to take place.

V. Conclusion

In this paper, a generalization of the model of the interacting intelligent node has been presented. This model, based on the Colored Petri Net formalism, subsumes the model used in Ref. 5, and is supported by commercially available software. It is forming the basis for ongoing research on variable structure command and control architectures and, more generally, of distributed intelligence systems.

References

[1] Minsky, M., *The Society of Mind,* Simon and Schuster, New York, 1987.

[2] Boettcher, K. L., and Levis, A. H., "On Modeling the Interacting Decisionmaker with Bounded Rationality," *IEEE Transactions on Systems, Man, and Cybernetics,* SMC-12, May/June 1982.

[3] Remy, P. A., and Levis, A. H., "On the Generation of Organizational Architectures Using Petri Nets," *Advances in Petri Nets 1988,* edited by G. Rozenberg, Springer-Verlag, Berlin, 1988.

[4] Andreadakis, S. K., and Levis, A. H., "Design Methodology for Command and Control Organizations," *Proceedings of the 1987 Symposium on C^2 Research,* National Defense University, Fort McNair, Washington, DC, June 1987.

[5] Demaël, J. J. , and Levis, A. H., "On Generating Variable Structure Architectures for Distributed Intelligence Systems," *Proceedings of the 11th IFAC World Congress,* Pergamon Press, Oxford, UK, 1990.

[6] Monguillet, J. M., and Levis, A. H., "Modeling and Evaluation of Variable Structure Command and Control Organizations," *Toward a Science of Command, Control, and Communications,* edited by C. Jones, Progress in Astronautics and Aeronautics, AIAA, Washington, DC, 1993 (elsewhere in this volume).

[7] Jensen, K., *Coloured Petri Nets,* Springer-Verlag, Berlin, 1992.

[8] Bonwit, W. R., Kidd, T. W., and Levis, A. H., "Battle Force Level C3I Assessment Methodology," *Proceedings of the 1990 Symposium on C^2 Research,* Monterey, CA, June 1990.

[9] Levis, A. H., "Quantitative Models of Organizational Information Structures," *Concise Encyclopedia of Information Processing in Systems and Organizations,* edited by A. P. Sage, Pergamon Books Ltd., Oxford, UK, 1988, pp. 368–375.

[10] Anon., *Design/CPN,* Meta Software Corp., Cambridge, MA, 1990.

Modeling and Evaluation of Variable Structure Command and Control Organizations

Jean-Marc Monguillet* and Alexander H. Levis†
Massachusetts Institute of Technology, Cambridge, Massachusetts 02139

I. Introduction

The need to meet ever increasing performance levels and to satisfy conflicting requirements has led to the investigation of organizations whose structures are variable. Variable structure organizations could be a possible design solution when no fixed structure organization can meet such requirements as robustness or survivability. The modeling of variability in the structure of organizations constitutes another step toward the representation of more realistic decisionmaking organizations.

The mathematical formulation of the modeling and analysis problem is based on the theory of Predicate Transition Nets, which is an extension of the Petri Net theory using the language of first order predicate logic.[1] The information processing and decisionmaking organizations that have been modeled and analyzed in earlier work[2,3] have been depicted as systems performing tasks in order to achieve a mission. These organizations are now viewed from a new perspective. The types of interactions which can exist between the decisionmakers are first considered without taking into account the identity of the decisionmakers themselves. The latter are represented by individual tokens (instead of subnets of a Petri Net) moving from one interaction to the other, and as such, are treated in the same manner as any other resources needed for the processing of a task. Interactions, resources, and tasks are modeled independently, and this new way of describing decisionmaking organizations allows the development of a modeling methodology with a modular architecture. By modular is meant that the representation of the basic components of the information processing (interactions, resources, and tasks) is done in separate

---

Copyright © 1993 by the American Institute of Aeronautics and Astronautics, Inc. All rights reserved.
  *Laboratory of Information and Decision Systems; currently, Ministry of Industry, Paris, France.
  †Laboratory of Information and Decision Systems; currently, Professor of Electrical, Computer, and Systems Engineering, School of Information Technology and Engineering, George Mason University, Fairfax, Virginia.

modules, and that modifications in one module can be made without affecting the others.

In the next section, variable structure organizations are defined, and in the third one the modeling methodology is described. A case study is presented in the fourth section; it illustrates the whole procedure through the design of a set of three candidate structures for a given mission, one of which is variable. Measures of effectiveness are used to select the most effective candidate for a specific mission.

II. Variable Structure Organizations

A variable structure decisionmaking organization (VDMO) is one in which the topology of interactions between the elements or components can vary. Analogously, a decisionmaking organization (DMO) that has a constant pattern of interactions among its components, i.e., a fixed structure, is called an FDMO.

The relationships that tie the components together are defined at three different levels: physical arrangements, links between components, and protocols ruling the arrangements of these links. The architecture of the organization allows the topology of interactions to vary. The way it does vary is implemented in the protocols themselves. The rules setting the interactions can be of any kind. We distinguish three types of variability, each corresponding to characteristic properties that a VDMO may exhibit; an actual VDMO may very well have these properties (to some extent) together and simultaneously.

1) *Type 1 variability:* The VDMO adapts its structure of interactions to the input it processes. Some patterns of interactions may be more suitable for the processing of a given input than others.

2) *Type 2 variability:* The VDMO adapts its structure of interactions to the environment. The performance of a DMO depends strongly on the characteristics of the environment as perceived by the organization. For example, an air defense organization may be optimized for some types of threats and their probabilities of occurence. Now, if the adversary's doctrine changes, or the deployment of his assets changes, then the probability distribution of the occurence of the threats is modified. The organization (with the interactions set as before the changes in the environment) may not meet the mission requirements any more.

3) *Type 3 variability:* The VDMO adapts its structure of interactions to the system's parameters. The performance of a system changes when assets are destroyed or become unavailable because of countermeasures such as jamming of communications.

These three different types of variability can be related to the properties of flexibility, reconfigurability, and survivability. A DMO is survivable when it can achieve prescribed levels of performance under some wide range of changes in the environment, in the characteristics of the organization, or in the mission itself. The extent to which a DMO is survivable depends on the extent to which it is flexible, and reconfigurable. Flexibility means that the DMO may adapt to the tasks it has to process, to their relative frequency, or to its mission(s). Reconfigurability means that it can adapt to changes in its resources. Both properties overlap, and their quantitative evaluation clearly falls outside the scope of this paper.

The organizations under consideration are restricted to the class of teams of boundedly rational decisionmakers (DMs).[4] Each DM is well trained and memoryless. The Petri Net formalism has been found to be very convenient for describing the concurrent and asynchronous characteristics of the processing of information in a decisionmaking organization. The internal processing which takes place in any decisionmaker has been modeled by a subnet with four transitions and three internal places. A simplified version of this so-called four stage model is shown in Fig. 1. This model allows the differentiation among the outputs and the inputs of the decisionmaker, and the description of the types of interactions which can exist between two decisionmakers.

The decisionmaker receives an input signal x from the environment, from a preprocessor, from a decision aid, or from the rest of the organization. He can receive one input to the situation assessment stage (SA) at any time. He then processes this input x with a specific algorithm which matches x to a situation the decisionmaker already knows. He obtains an assessed situation z which he may share with other DMs. He may also receive at this point other signals from the rest of the organization. He combines the information with his own assessment in the information fusion (IF) stage, which leads to the final assessment of the situation, labeled z'. The next step is the possible consideration of commands from other DMs which would result in a restriction of his set of alternatives for generating the response to the input. This is the command interpretation stage (CI). The outcome of the CI stage is a command v which is used in the response relection (RS) stage to produce the output y (the response of the decisionmaker) which is sent to the environment or to other DMs.

As shown in Fig. 1, the decisionmakers can only receive inputs at the SA, IF, and CI stages, and send outputs from the SA and RS stages.[5] The interactions which are the most significant are shown in Fig. 2. For the sake of clarity, however, this figure only accounts for the interactions as directed links from DM_i to DM_j. Symmetrical links from DM_j to DM_i exist as well.

Two kinds of places can be distinguished: internal places or memory places, where the decisionmaker stores his own information: between SA and IF, IF and CI, or CI and RS. The places between the DMs and the sensors, the preprocessors, or the actuators, as well as those between two DMs are called interactional places. Knowledge of the set of interactional places is equivalent to that of the whole structure of the net.

A decisionmaker may not have all of his four stages present. Depending on the interactions he has with the rest of the organization and with the environment he may exhibit different internal structures: SA alone; SA, IF, CI and RS (IF and CI can be simple algorithms that copy the signal); or IF, CI, and RS.

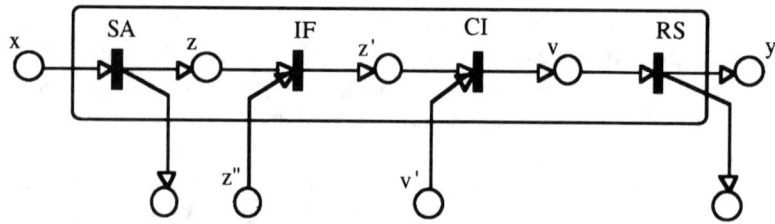

Fig. 1 Four stage Petri Net model of a DM.

Depending on what the designer of the organization requires, different constraints on the allowable interactions can be expressed, which limit or expand the set of possible organizations.

In the Petri Net representation, the transitions stand for the algorithms, the connectors for the precedence relations between these algorithms, and the tokens for their input and output. The places act like buffers, hosting the tokens until all of the input places of a transition t are non-empty, in which case the algorithm embodied in t can run and remove the tokens. The time taken by the algorithm to run is the transition processing time $\mu(t)$. The tokens in this model are all indistinguishable. A token in a place p means simply that an item of information is available there for the output transition(s) of p.

Attributes can be used to describe what the tokens represent. For instance, if the decisionmaker has to identify an incoming threat and to respond to it, then a token on the input place of his SA stage may be just a blip on the DM's radar screen. The token that the SA algorithm produces is in turn formatted information which includes the DM's measurement, or assessment, of the position, speed, nature, behavior, or size of the threat. The DM can receive from elsewhere in the organization other formatted information, not necessarily of the

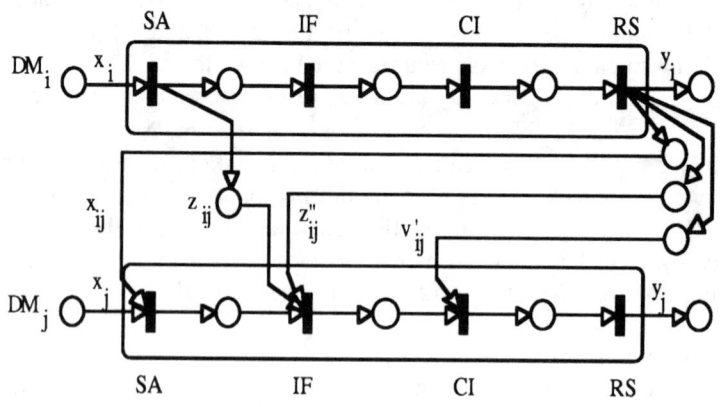

Fig. 2 Allowable interactions from DM_i to DM_j.

same format, provided that it matches what his IF algorithm expects as inputs formats. The different tokens in the different places have then different formats and different attributes. But as long as the protocols ruling their processing do not vary from one set of attributes to the other, they are indistinguishable tokens.

What is needed is a tool which would allow the differentiation of tokens, and which would have the capability to implement logic able to determine explicitly what interaction and what DMs have to be active for the processing of a given input. Individual tokens, predicates, and operators can meet these requirements. The application of the Predicate Transition Nets to that purpose is developed in the next section.

III. Modeling Methodology for Variable Decisionmaking Organizations

In this section, a step-by-step procedure for the modeling of VDMOs using Predicate Transition Nets is developed. An example of a three member organization with type 1 variability illustrates the methodology. The methodology has a modular architecture (Fig. 3). There are five modules, 1) interface with the environment, 2) scarce resources, 3) interactions, 4) switching module, and 5) algorithm implementation. Each of the first three modules can be executed independently and in arbitrary order. The first three modules address the subproblems a) of modeling of the inputs that the DMO receives and the responses that it gives, b) of representing the scarce resources that the DMO needs, and c) of modeling the possible interactions which can exist between the components.

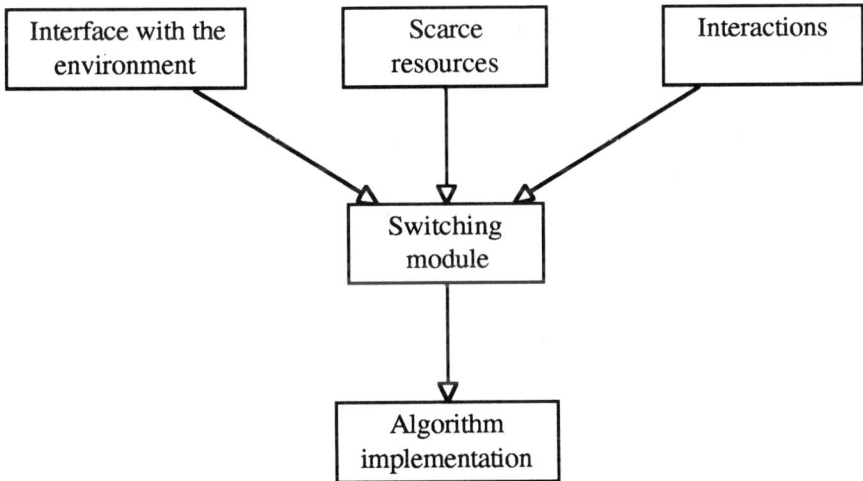

Fig. 3   Architecture of the modeling methodology.

When the first three modules have been completed, the switching module is executed. The switching module is the part of the model where the logic, which controls the variability of the organization, is implemented. For each incoming input, this is the part of the model that decides what particular resources and what particular set of interactions will be adopted. The way this choice is made will determine what type of variability the VDMO exhibits.

What is obtained at this point is a Predicate Transition Net where only the nontrivial operators are indicated in the corresponding transitions. The fifth and last part of the methodology consists of the rigorous labeling of the nodes, connectors, and tokens of the net. It also gives precise meaning to what the individual tokens stand for (i.e., the list of their attributes), depending on the places that host them, and on what algorithm, or what set of algorithms, a particular transition models. The processing time of the different algorithms is also specified. The steps of that methodology are independent enough to allow changes in any subproblem, without threatening the functioning of the whole model. The modular architecture is also very convenient for the implementation of extensions of the model, which simply become new modules or new well-defined subproblems.

The following subsection focuses on the modeling of type 1 variable DMOs. An example of a three member organization with type 1 variability serves to illustrate the methodology. Examples of VDMOs exhibiting type 2 or type 3 variability are included in Ref. 6.

Interface with the Environment

The goal of this subproblem is to achieve a representation of the input and output alphabets. In the modeling of decisionmaking organizations, the discrete representation of information sets is done in the form of lists of attributes, an instance of which is called a token. In the ordinary Petri Net representation of a DMO, the values of the attributes were of no importance; no matter what these values were, the treatment of the token was the same: the interactions between the components were the same.

In the case of type 1 VDMOs, the alphabet X of inputs is partitioned in r classes, namely X_i, for $i = 1, ..., r$. All inputs x belonging to the same class are processed with the same resources used with the same pattern of interactions. A given input x cannot belong to more than one class, which implies that it can only be processed with one specific set of resources, and one specific kind of interactions. The identity of a token is the class X_i to which it belongs; it is denoted by the index number i. The variable "class of inputs" is denoted by x and has the following set of allowable identities:

$$x = \{1, ..., r\}$$

Since the environment is not modeled, the tokens which model the outputs of the organization need not have an identity. They are instances of the 0-ary variable $¢$.

Example: Step 1. The example consists of a three member organization with four possible interactions between the decisionmakers. The DMO consists of two field units, FU1 and FU2, and one headquarters, HQ. The possible interactions are the following:

Int 1 - FU1 and HQ: HQ fuses its assessment with FU1's, and issues a command to him;
Int 2 - FU2 and HQ: HQ fuses its assessment with FU2's, and issues a command to him.
Int 3 - FU1 alone: SA and RS stages.
Int 4 - FU2 alone: SA and RS stages.

The alphabet of inputs X is therefore partitioned in four classes X_i, $i = 1, ..., 4$. The variable x representing the class of the inputs has a set of identities $\{1, 2, 3, 4\}$. The outputs are not partitioned. The model of the organization which is obtained at this point is shown in Fig. 4.

Scarce Resources

Resource is a generic name which designates elements needed for the processing of a task. A resource is scarce when it cannot be allocated freely to the processing of any incoming input because of insufficient or limited supply. The scarcity of resources results in an upper bound to the performance of the organization. Scarce resources are modeled in a convenient way in the Petri Net formalism. They are represented by places with multiple input transitions and multiple output transitions, and nonzero initial marking. Examples of scarce resources can be common databases with limited access, communication links with limited capacity, mainframes with shared processing time, or weapons platforms capable of handling a limited number of threats at a time.

In this modeling methodology, the decisionmakers are treated as scarce resources: they are assigned to an incoming input; once they have been assigned to a certain number of inputs, the other inputs have to wait in line to be processed. The pool of decisionmakers that implements the organization is partitioned in classes of DMs who have the same function within the organization, i.e., who possess the same kind of algorithms. Two decisionmakers who belong to the same class are then interchangeable. The DMs of a class are represented by individual tokens of a variable and placed in the corresponding resource place. If there is only one class of DMs, then the DMs are represented by indistinguishable tokens. The other resources that the

Fig. 4 Example: Step 1.

organization may need are partitioned and associated with variables and places in the same way.

Connectors are labeled with a formal sum of variables, which indicates the kinds of tokens they can carry. The input and output connectors of a given resource place R where the corresponding variable is x are labeled by elements of $L^+(x)$, the set of all applications from x to the non-negative integers.

Example: Step 2. In this example, two DMs are interchangeable as far as their interactions with the rest of the organization are concerned; these are the field units FU1 and FU2. There is a single HQ that performs a unique function in the DMO. The three DMs are then represented by the following variables:

1) Resource place FU: associated with the variable $s = \{1, 2\}$. The individual token 1 models the decisionmaker FU1. Token 2 stands for FU2.

2) Resource place HQ: since there is only one HQ, the place carries an indistinguishable token $¢$, shown as a dot in the place HQ.

The modeling of the DMO at this point is shown in Fig. 5.

Interactions

The allowable interactions between components are represented without considering the identity of the resources they involve. What is of interest, at this point in the modeling, is only the topology of interactions that can be found in the DMO. The typical model obtained at this point is shown in Fig. 6; it is a list of the possible patterns of interactions depicted in their most aggregated form. Had these interactions been considered alone as DMOs with fixed structure, the input and output places would have been the source and sink places.

The possible interactions can be partitioned in four generic types, as illustrated in Fig. 6.

1)*Type a:* the pattern of interactions is that of an organization with a fixed structure which processes the inputs without resources. It is represented by an ordinary Petri Net which can be aggregated in a supernode Int 1.

2)*Type b:* the pattern of interactions has the same characteristics as in type a, but the net which models that pattern exhibits some properties of symmetry. A

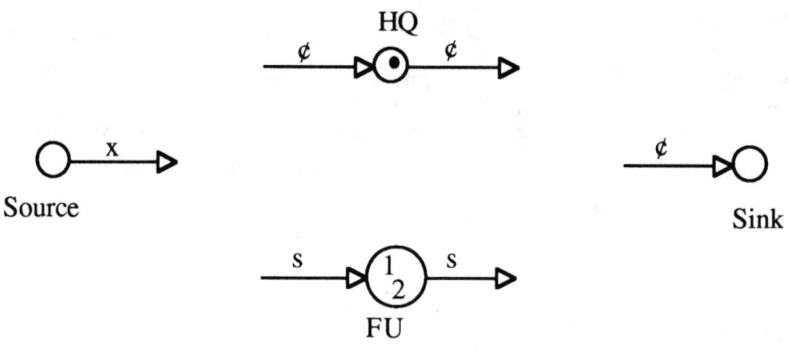

Fig. 5 Example: Step 2.

more convenient representation is obtained by folding the net. The Predicate Transition Net which is obtained is aggregated in turn in a supernode Int 2.

3) *Type c:* the pattern of interactions is the same as type a, but the DMO with that pattern requires a resource R_1 for the processing of the inputs. This resource is used from the beginning of the processing until its completion. The ordinary Petri Net which models that pattern is therefore aggregated in a supernode and the resource place R_1 is both an input and an output places of that macrotransition Int 3 (however, the underlying Petri Net is still pure).

4) *Type d:* the pattern of interactions is similar to type c, except that resource R_2 is not used during the processing of the inputs. In the particular case of Fig. 6d, it is only needed at the beginning. The ordinary Petri Net representing that pattern is then aggregated in two supernodes, (Int 4, 1) and (Int 4, 2). The former stands for the part of the processing that uses resource R_2, whereas the latter accounts for the remaining processing.

Any other combination of type a - d can be encountered as well. In particular, the number and diversity of resources required and the lack of symmetry of the pattern of interactions may make aggregation in supernodes inappropriate. In that case, the net which would appear in Fig. 6 would show in detail all the stages of the decisionmaking process.

No matter where the resource places are connected, the subnet which is subsumed in a macro-transition represents a decisionmaking organization where the internal processing of the input is modeled by the four stage representation that was described in Fig. 1. That net stands, therefore, for an organization with fixed structure, which is to say, that it may contain some switches, but the setting of these switches does not affect the structure of the interactions between the decisionmakers (whose identities are not defined). If each switch is aggregated in a macrotransition, then the ordinary Petri Nets which are obtained are all marked graphs, i.e., a place can have only one input transition and only one output transition.

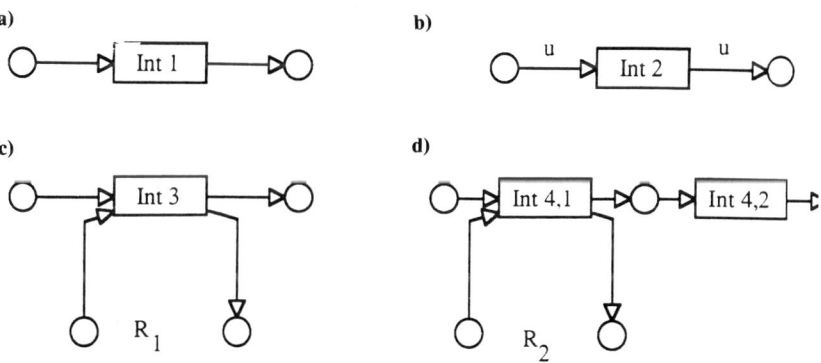

Fig. 6   Allowable interactions.

Example: Step 3. In the example, only two patterns of interactions are actually distinct: one where the HQ interacts with a FU, and one where the FU processes the task alone. The first part of the modeling consist of representing these patterns in detail (Fig. 7). Then an aggregated model comparable to Fig. 6 can eventually be produced.

For the first pattern of interactions, two resources are required, namely, HQ and FU. The resource HQ is not used in the decision process until a response is chosen and can be free before that. However, the resource FU is needed from the beginning of the processing to the end. Finally, this pattern of interactions is such that no aggregation in supernodes is possible. For the second pattern, the only resource used is FU, and it is needed during the whole processing of the input.

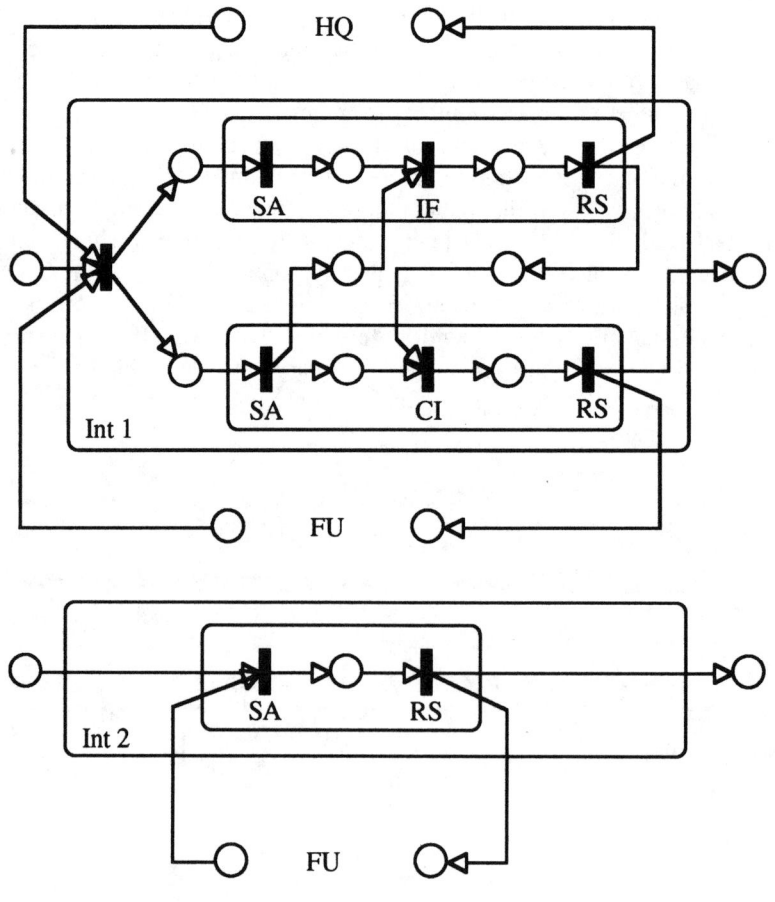

Fig. 7   Example: Step 3.

Switching Module

The objective of this module is the representation of the decision rule which determines, for any incoming input, what the actual configuration of the organization will be. The switching module is the part where the type of variability of the organization will be modeled. It supposes that the first three subproblems have been already completed.

A switch is implemented as an output node of the source and the resource places. This switch consists of a set of transitions with operators, whose arguments are the individual tokens in the source and resource places. Recall that a DMO with type 1 variability is being modeled, and that it has been assumed that each class of inputs has associated only one possible pattern of interactions. Thus, if the number of classes of inputs is r, there are at most r branches in that switch.

A decisionmaking organization needs an interaction and some resources to process an incoming input. The type 1 variable DMO that has been considered so far adopts, for each class of inputs, a specific interaction and set of resources. The formal notation for the inputs, resources, interactions, and their relations is the following.

1) *Inputs:* a) An input is an individual token of variable x. b) The source place SO is associated with variable x. c) The set of allowable identities for x is $x = \{1, ..., r\}$. d) An input of variable x belongs to the class X_i, where $x = i$.

2) *Resources:* a) The resource places are R_k for $k = 1, ..., K$. b) The resource place R_k is associated with the variable s_k. c) The set of allowable identities for s_k is $s_k = \{1, ..., S_k\}$.

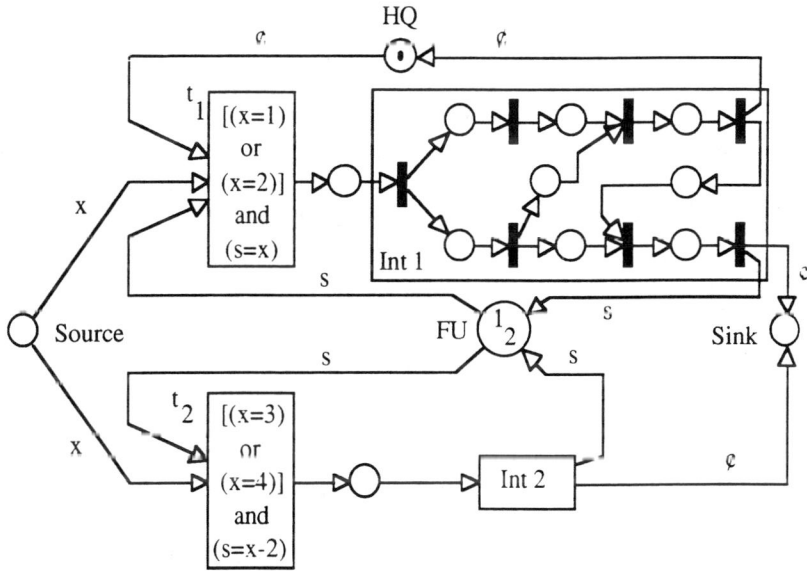

Fig. 8   Example: Step 4.

3) *Interactions:* a) The patterns of interactions are Int (γ), for $\gamma = 1, ..., \Gamma$. b) There are J transitions t_j in the switch. c) t_j is associated with the operator Op_j. d) t_j is associated with the pattern of interactions $\phi(j)$, i.e., Int [$\phi(j)$].

4) *Relations:* a) The input x requires a pattern of interactions $\gamma(x)$, i.e., Int [$\gamma(x)$]. b) The input x requires some resources from R_k, which are

$$\text{res}(k, x) = \{s_{k,n}(x) \mid n = 1, ..., N(x)\}$$

c) $\gamma(x)$ and res(k, x) for any k are functions of x. d) $\phi(j)$ is a function of j; ϕ is attached to the switch.

An incoming input, modeled as an instance of an individual token x, belongs to the class X_i. The organization is type 1 variable, and it adapts the pattern of its interactions to the class of the incoming input. The processing of the input x requires a specific pattern of interactions, namely Int [$\gamma(x)$]. Since the same interactions can be adopted for different classes of inputs, the function γ is not bijective, and the number Γ of interactions is necessarily smaller than the number r of classes of inputs. The processing of this individual token x also needs some resources of type R_k, given by the set of individual tokens res(k, x). The transition of the switch which corresponds to the pattern of interactions Int [$\gamma(x)$] is the transition t_j such that $\phi(j) = \gamma(x)$; there is only one j such that this relation is verified, which is denoted as $\phi^{-1}[\gamma(x)]$.

If all the conditions stated above are fulfilled, then the input x is processed i.e., for $\phi(j) = \gamma(x)$, the transition t_j is enabled and fires. The operator Op_j associated with t_j expresses in logical terms the above conditions, and can be written as follows:

$$(\exists x \in SO) \wedge [\gamma(x) = \phi(j)] \wedge [\exists \text{ res}(k, x), R_k \supseteq \text{res}(k, x)] \quad (1)$$

Since the transition t_j corresponds to Int [$\phi(j)$], and since this pattern of interactions may be needed for more than one class of input, the actual operator associated with t_j is the logical OR ($\vee$) of the operators (1) for the inputs x such that $\gamma(x) = \phi(j)$, i.e., for all the inputs x in the set $\gamma^{-1}[\phi(j)]] = \{x \mid \gamma(x) = \phi(j)\}$. The operator Op_j associated with t_j is finally the following:

$$Op_j = \bigvee_{x \in \gamma^{-1}[\phi(j)]} [(\exists x \in SO) \wedge (\exists \text{ res}(k, x), R_k \supseteq \text{res}(k, x))] \quad (2)$$

The operators $(Op_j)_{j=1,...,J}$ which are attached to the transitions t_j (the branches of the switch) are such that the following conflict resolution rule is verified: for any input x in the place SO, there is exactly one transition in the set $\{t_j\}$ which is enabled, the one with the number $j = \phi^{-1}[\gamma(x)]$. There is, therefore, no conflict and as soon as the required resources res(k, x) are available, t_j can fire.

The connectors from the place R_k to transition t_j are labeled by the set $L_{conn}(R_k, t_j)$ whose elements are the symbolic sums of the individual tokens in res(k, x). If the set res(k, x) is non empty, the connector from R_k to t_j has the following label:

$$L_{conn}(R_k, t_j) = \left\{ \lambda \in L^+(s_k) \,\middle|\, \lambda = \sum_{n=1}^{N(x)} s_{k,n}(x) \text{ and } \gamma(x) = \phi(j) \right\} \quad (3)$$

Example: Step 4. In the example, the switching module contains two transitions t_1 and t_2. Therefore, the inputs are $x = \{1, 2, 3, 4\}$; the resources are R_1 = HQ, associated to the 0-ary variable ϕ., and R_2 = FU, associated to the variable s, with $s = \{1, 2\}$. The interactions are Int 1, corresponding to transition t_1, and Int 2, corresponding to transition t_2. For any input x, the pattern of interactions Int [$\gamma(x)$] is $\gamma(1) = 1$, $\gamma(2) = 1$, $\gamma(3) = 2$, and $\gamma(4) = 2$.

For any input x, the required resources are

$$\text{res}(1, 1) = \text{res}(1, 2) = \{1\phi\}$$
$$\text{res}(1, 3) = \text{res}(1, 4) = \emptyset$$
$$\text{res}(2, 1) = \{1\}$$
$$\text{res}(2, 2) = \{2\}$$
$$\text{res}(2, 3) = \{1\}$$
$$\text{res}(2, 4) = \{2\}$$

The operators Op_1 and Op_2 can then be written (without mentioning the quantifiers) as follows:

$$Op_1: [(x = 1) \wedge (s = 1)] \vee [(x = 2) \wedge (s = 2)]$$
$$Op_2: [(x = 3) \wedge (s = 1)] \vee [(x = 4) \wedge (s = 2)]$$

The operators can actually be aggregated into a more convenient form

$$Op_1: [[(x = 1) \vee (x = 2)] \wedge (s = x)]$$
$$Op_2: [[(x = 3) \vee (x = 4)] \wedge (s = x - 2)]$$

In the net obtained up to this point the patterns of interactions, the resources, the source, the sink, and the transitions of the switch are connected together, and the transitions show the operators assigned to them. However, the patterns of interactions are still in their most aggregated form, and the connectors are not all labeled (Fig. 8). This net is not yet fully defined. The purpose of the next module will be to make this net functional by completing its annotation.

Algorithm Implementation

This fifth module of the methodology deals with the labeling of the connectors, with the definition of the attributes of the tokens that can reside at different places, and with the algorithms that the various transitions represent. The rules of firing must also be established.

Labeling of Connectors

The connectors from the source to the transitions of the switch are labeled x, i.e., with the variable designating the class of the inputs. Those from the input nodes of the sink to the sink itself are labeled $¢$. The labels of the output connectors of the resource place R_k have already been given in Eq. (3). The input connectors of R_k are labeled accordingly.

Each pattern of interactions Int (γ) is adopted whenever the incoming class of input x is such that $\gamma(x) = \gamma$. When x describes the set of classes of inputs x, the number of times Int (γ) is activated is equal to the number of times $\gamma(x) = \gamma$. The connectors which are involved in the representation of the organization with a pattern of interaction Int (γ) can then be labeled with a variable ρ_j whose set of allowable identities is:

$$r_j = \{1, 2, ..., \gamma_0\}$$

These labeling rules are the most general that can be presented, and can be applied to any case.

Firing rules:

The firing rules are actually problem dependent and can be revised at any time. However, they are generally the following:

1) The transitions that constitute the switch are enabled and fire consecutively, i.e., with one input at a time.

2) The transitions that are part of the subnets representing the possible interactions with ordinary Petri Nets are enabled and fire in the same consecutive manner. In other words, if a given place in one of these subnets contains more than one token, its only output transition ("only" because the subnet is an event graph) is enabled by more than one token. But it will fire them only one by one.

3) The transitions that are part of the subnets representing the possible interactions with Predicate Transition Nets, i.e., when the original Petri Net has been folded, can allow simultaneous firing; depending on the circumstances, two tokens in the same place can enable the same transition at the same time and leave the same place simultaneously.

Depending on the identity of the individual token of variable ρ_j which enables it, a particular algorithm, or a particular switch, is activated and processes the input that the token represents. Depending also on the organization that the net models, this transition can very well consist of only one algorithm, which is always activated and executed when the transition is enabled and fires, regardless of the identity of the individual token which has triggered that process. The rule that selects the algorithm which will process the token that enabled the transition is problem dependent, and as such, defined for each particular case.

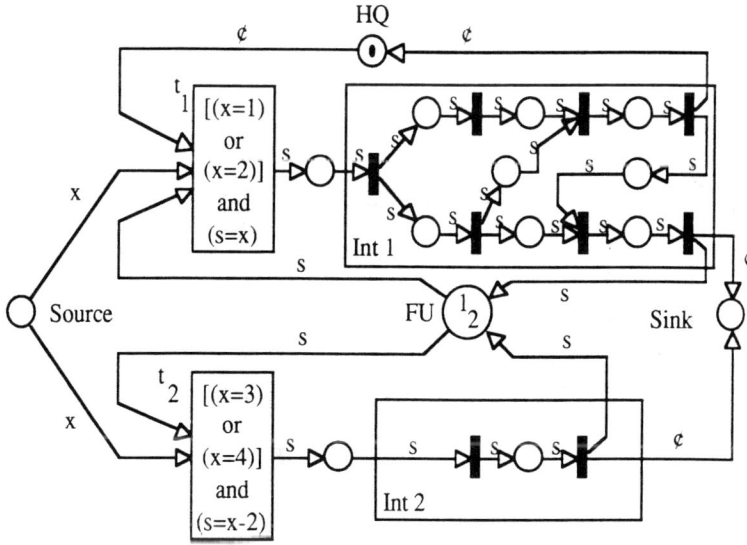

Fig. 9   Example: Step 5.

Example: Step 5. The final representation of the example is given in Fig. 9. Since the organization is fairly simple, a simplified and self-explanatory labeling has been adopted.

IV. Effectiveness of a Type 1 Variable Decisionmaking Organization

In the previous section, a methodology for the modeling of VDMOs was presented, and it was assumed that the inputs were partitioned in classes, corresponding to specific patterns of interactions, before being processed by the organization. An example of a three member variable structure organization for which this assumption is relaxed is considered in this section.

Organization and Its Model

We consider an organization composed of three decisionmaking units, the headquarters (HQ) and two field units (FU1 and FU2). Its mission is the defense of a given area against aerial threats, aircraft or missiles. Each incoming threat is identified by HQ, and its location determined by both field units. HQ communicates then the identity of the threat to the FUs who decide to fire or not to fire, depending on that information.

Decisionmaking Organizations with Fixed Structure

Different settings for the interactions between the DMs are possible. In the first case (FDMO1), the HQ and the FUs receive simultaneously the input and

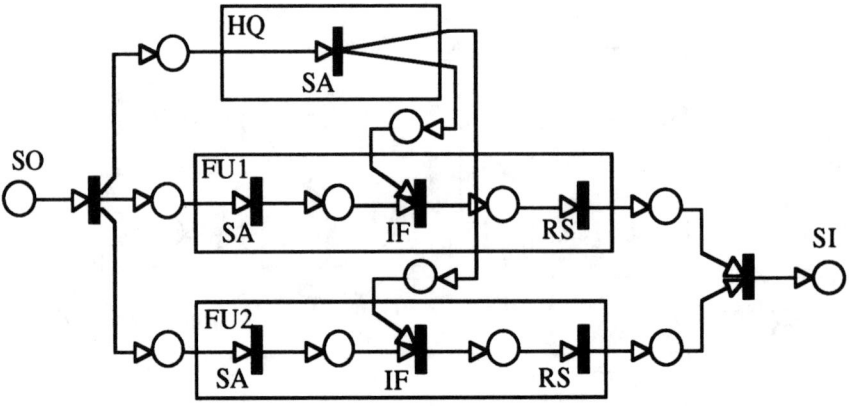

Fig. 10   Candidate 1: FDMO1.

HQ sends its information on the identity of the threat to each of the FUs at the same time. They each fuse their assessement of the situation with that information and give a response to the threat in a simultaneous way. In the second case (FDMO2), only FU1 receives information from HQ, which he fuses with his own assessment of the situation and sends to FU2. FU2 in turn fuses this information with his own assessment and produces the final response of the organization (Figs. 10 and 11).

Type 1 Variable Decisionmaking Organizations

In general terms, it is legitimate to suspect that FDMO1 would take less time to respond than FDMO2, since in the first case the two Field Units have parallel activities, but in the second case they have to interact. However, for the same reason, the response of FDMO2 may be more accurate than the one of FDMO1.

An organization in which the three decisionmakers would concurrently and simultaneously assess the situation and in which headquarters would decide the type of interactions to be adopted between the FUs for their final processing is likely to perform better; i.e., lower processing delay and higher accuracy. The organization which would be obtained that way would be type 1 variable, and the headquarters in that case would play the role of a preprocessor. The inputs arrive and are indistinguishable; then the HQ attaches to each of them an attribute, or class, which determines the type of interactions that are best suited for their processing. There are, therefore, three candidates for that air defense mission, two organizations with a fixed structure (FDMO1 and FDMO2), and a variable structure organization (VDMO).

Predicate Transition Net Model

The variable organization is modeled with a Predicate Transition Net (PrTN) using the methodology developed in the previous section. The situation

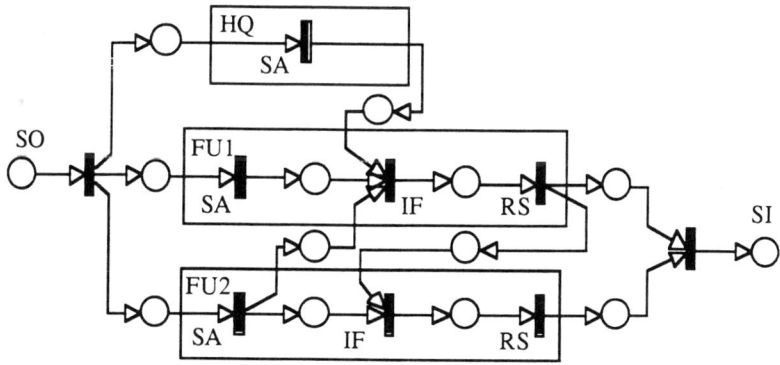

Fig. 11   Candidate 2: FDMO2.

assessment stage of the HQ acts as a source of information and associates an attribute u to the incoming token. What results is an hybrid representation, using the formalisms of both ordinary Petri Nets and Predicate Transition Nets. The VDMO is shown in Fig. 12. The variable controlling the variability is called u, whose set of allowable values is $\{0,1\}$. The situation assessment stages of the field units are modeled with the conventional representation. After an input has been processed in these stages, the FUs are modeled with individual tokens of a variable x. The set of allowable values for x is $\{1, 2\}$, with token 1 (resp. 2) standing for FU1 (resp. FU2).

Inputs

The three decisionmakers are geographically dispersed. They communicate with the help of wired links or radio. The threats are characterized by their radial distance, i.e., they are modeled as occurences on a line. Their position on this line is measured by a variable x, $x \in [0, 3]$. They appear one at a time, and they are independent. The line is divided in three sectors, namely, [0,1],]1,2[, and [2,3]. Since the field units are placed close to the extreme sectors, they perform the same algorithm that determines the position of the target on the line but with different accuracy, depending on the sector in which the target appears. For instance, FU1 is accurate when a threat appears in [0,1], less accurate when it appears in]1,2[, and even less accurate when in [2,3]. The accuracy of FU2 is analogous to FU1's accuracy.

The inputs are instances of elements x of an alphabet X. A given instance is modeled by the pair $x = (z, \text{name})$, where z is a real in [0,3] and name is a string in $\{00, 10, 01, 11\}$. The name of the input represents the identity of the threats. They can be thought as being types of aircraft or types of behavior. The threats whose names are 00, 01, or 10 represent foes and have to be destroyed. Only 11 represents friends.

The position of the threat on the line is denoted by z. This is the actual position, but the field units, who are in charge of determining it, only achieve

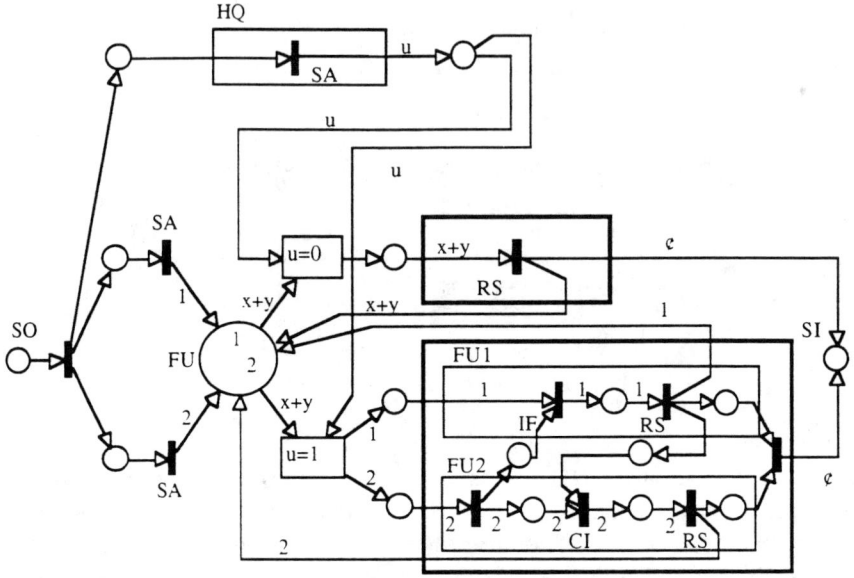

Fig. 12 Candidate 3: VDMO.

their own measure [z] of z. In other words, each of them has an interval (of uncertainty) for the value of z. The accuracy of their measure decreases with remoteness. To keep the computations simple, the position z in [0, 3] is discretized so that only 30 different positions are allowed, namely, 1, 2, ..., 30. Any input which appears actually in [0.1*(i - 1), 0.1*(i)] is called z_i, where i is an integer between 1 and 30. For completeness, the last interval is [2.9, 3.0]. Consequently, the alphabet X consists of elements $x = (z_i, name_j)$, with

$$z_i \in \{1, 2, ..., 30\}$$
$$Name_j \in \{00, 01, 10, 11\}, \quad \text{for } j = 1, ..., 4$$

Strategies of the Decisionmakers and Cost Matrix

For any incoming input x_i, the field units determine the position of the threat, and the headquarters identifies its name.

Situation Assessement

Each FU has the same set of two algorithms in the SA stage, called SA1(FU) and SA2(FU). SA1(FU) is more accurate than SA2(FU), and, as a result, takes more time to produce a response. Each algorithm yields a measure of the position of an input x_j with precision δ represented by an integer. A precision of 1 means that there is no uncertainty in the knowledge of z_i, and that the measure of its position $[z_i]$ is equal to z_i. The interval of uncertainty is reduced

VARIABLE C2 ORGANIZATIONS 211

to $\{z_i\}$. A precision of 3 means that the measure $[z_i]$ can be at any one of three different positions: $\{z_i - 1, z_i, z_i + 1\}$.

The algorithms used in the situation assessment of the field units are characterized by the precision they can achieve. In this model, precision is taken as a function of the sector to which the threat belongs: the precision δ is supposed to be a linear function of the remoteness, at least in this range of positions of the threat.

Algorithm SA1(FU) for FU1
$$1 \leq i \leq 10 \Rightarrow \delta = 1$$
$$11 \leq i \leq 20 \Rightarrow \delta = 3$$
$$21 \leq i \leq 30 \Rightarrow \delta = 5$$

Algorithm SA2(FU) for FU1
$$1 \leq i \leq 10 \Rightarrow \delta = 3$$
$$11 \leq i \leq 20 \Rightarrow \delta = 5$$
$$21 \leq i \leq 30 \Rightarrow \delta = 10$$

The precision of measurements for FU2 are deduced from the above by setting

$$i' \to (30 - i)$$

The values of δ are quantized so that they are the same wherever the threat appears in a given sector. Their dependence on the distance has been set to account for a rapid decrease in accuracy when the distance increases. The delay of the second algorithm has been set arbitrarily at one unit of time. At this point, we assume that if one obtains a measurement with precision δ but spends T units of time in that operation, then one will require more than $2T$ units of time to obtain a precision $\delta/2$. Since the first algorithm is twice as accurate as the second one, the processing delay of the first one is set to three units of time.

The headquarters possesses a set of two algorithms in its SA stage. The first one, SA1(HQ), identifies the name of the threat by reading the two characters of the string. In that case, the threat is completely identified. The second algorithm, SA2(HQ), only reads the first character of the string and is less accurate than the first one. The same argument as above leads to a processing delay of two units of time for SA2(HQ) and four units of time for SA1(HQ).

Internal Strategies

The set of alternative algorithms that the decisionmakers possess leads to the definition of their *internal strategies*. The variables u_1, u_2, and u_3 are first defined to have their set of values equal to [1, 2], and to correspond to the settings of the switch of the situation assessement stage of FU1, FU2, and HQ, respectively. The variable u_1 for instance is set to

$$u_1 = 1 \quad \text{if FU1 processes its input with the algorithm SA1}$$
$$u_1 = 2 \quad \text{if FU1 processes its input with the algorithm SA2}$$

The variables u_2 and u_3 are determined accordingly. Now the internal strategy of FU1, D(FU1), is the probability distribution of the variable u_1, as indicated in the following:

$$D(FU1) = p(u_1) = \{p(u_1 = 1), p(u_1 = 2)\}$$
$$D(FU2) = p(u_2) = \{p(u_2 = 1), p(u_2 = 2)\}$$
$$D(HQ) = p(u_3) = \{p(u_3 = 1), p(u_3 = 2)\}$$

A decisionmaker uses a *pure strategy* when he always processes the incoming input with the same algorithm. Otherwise, he uses a *mixed strategy*. In the present case, each DM possesses two pure internal strategies.

Information Fusion Stages
The time delay of the information fusion stages is a function of the number of inputs to be fused. If two inputs have to be fused, the processing delay is one unit of time. If three inputs have to be fused, the delay will be two units of time. All other algorithms have associated a delay of one.

When the two field units fuse their measurements of the position of the threats, precision is increased, if these measurements are consistent. If two measurements of the same input with precision δ_1 and δ_2 are fused into a measurement with precision $\delta = \text{Fus}(\delta_1, \delta_2)$, then the results are as given in Table 1.

Response Selection Stage and Cost Matrix
The decisionmaker in each field unit can either allocate a missile to the target or do nothing. If he sends a missile to the position where he has measured the threat to be located, then he can either hit the target or miss it, depending on the accuracy of his measure. The FU's response is denoted by y, the target coordinates: y can take the values x, if the missile is sent exactly where the target is; $\neg x$, if a missile is sent to a wrong position; and † if no missile is sent as seen in Table 2.

Table 1   Precision of fused information

Fus(1, -)	= 1
Fus(3, 5)	= 2
Fus(3, 5)	= 2
Fus(3, 10)	= 3
Fus(5, 5)	= 3
Fus(5, 10)	= 4
Fus(10, 10)	= 10

The ideal response for a friend (name 11) is to do nothing, whereas the ideal one for a foe is to destroy it. There is, furthermore, a penalty for an overconsumption of missiles. The cost associated with any discrepancy between the ideal and the actual responses is indicated in the cost matrix given in Table 2.

In Table 2, the left column corresponds to the ideal response of the organization. The top row labeled x_1 indicates the response of FU1, whereas the one labeled x_2 represents the response of FU2. The costs are adjusted to reflect subjectively the ranking of the actual responses of the organization. For example, the ideal response for a friend input is for the field units to take no action, i.e., x_1 and x_2 to be inactive (†). If one missile is targeted to the wrong coordinates, in other words if $x_1 = $ †, and $x_2 = \neg x$, (or the reverse), then the cost of wasting one missile is estimated to be one. The cost of targeting accurately a friend is three. These values can be modified to account for any other set of beliefs.

The probability distribution of the occurences of the inputs is assumed to be uniform, unless otherwise specified. The probability for the input x of the alphabet X of having its name equal to a given name$_j$ is then 1/4, whereas the probability that this input has a position equal to a specific z_i is 1/30. We have then

$p(x = (z_i, \text{name}_j)) = 1/120$,
for all z_i in $\{1,..., 30\}$ and all name$_j$ in $\{00, 01, 10, 11\}$

Measures of Performance

Measures of performance (MOPs) are quantities which describe the system properties. The MOPs are functions of the system parameters and of the organizational strategy adopted by the organization. The two MOPs considered here are *accuracy* and *timeliness*.

Accuracy, denoted by J, is a measure of the degree to which the actual response of the organization to a given input matches the ideal response for that same input. If we denote by X the alphabet of inputs x_i: $X = \{x_1, x_2, ..., x_n\}$, Y the alphabet of outputs y_j: $Y = \{y_1, y_2,..., y_q\}$, $p(x_i)$ the probability of occurence of the input x_i, with $\sum p(x_i) = 1$, $y_d(x_i)$ the ideal (or desired) response to x_i, $y_{aj}(x_i)$, $j = 1,...,q$, the response that the DMO actually produces, and

Table 2 Cost matrix

x_1	x			¬x			†		
x_2	x	¬x	†	x	¬x	†	x	¬x	†
Foe: Y = D	1	1	0	0	6	6	0	6	6
Friend: Y = ND	3	3	1	3	2	1	3	1	0

$C(y_d, y_a)$ the cost of the discrepancy between the ideal and the actual responses, then a measure of accuracy of the DMO is

$$J = \sum_{i=1}^{n} p(x_i) \sum_{j=1}^{p} C[y_d(x_i), y_{aj}(x_i)] \, p[y_{aj}(x_i) \mid x_i] \qquad (4)$$

Timeliness, denoted as T, is the ability to respond to the input with a time delay T_d that is within the allotted time $[T_{min}, T_{max}]$, called the window of opportunity. If we denote by $T_d(x_i)$ the average processing delay of x_i, and 1_Ω the characteristic function on the set Ω, then a measure of timeliness of the DMO is the expected value of the processing delay

$$T = \sum_{i=1}^{n} p(x_i) \, T_d(x_i) \qquad (5)$$

The performance of the organization is a function of the strategy of the organization as a whole, or organizational strategy, which is given by the triplet

$$S = \{D(FU1), D(FU2), D(HQ)\}$$

Since the three switches which are present in the organization are in the situation assessment stages, the internal strategies are not formulated with probabilities conditioned by the inputs. There are, therefore, eight pure organizational strategies, which are the triplets of the pure internal strategies. These pure strategies S_i, $i = 1,..., 8$, can be defined by the algorithms the DMs are using, as follows (the order is FU1, FU2, HQ):

$$S_1 = (SA1, SA1, SA1)$$
$$S_2 = (SA1, SA2, SA1)$$
$$S_3 = (SA2, SA1, SA1)$$
$$S_4 = (SA2, SA2, SA1)$$
$$S_5 = (SA1, SA1, SA2)$$
$$S_6 = (SA1, SA2, SA2)$$
$$S_7 = (SA2, SA1, SA2)$$
$$S_8 = (SA2, SA2, SA2)$$

The application of Eqs. (4) and (5) gives immediately the values for accuracy J and timeliness T for FDMO1 and FDMO2, for each pure strategy S_i. The results are shown in Table 3, with T in units of time.

The type 1 VDMO being considered adapts the interactions between the field units to the inputs that they have to process. We consider the case where the inputs are distinguished on the basis of the sectors in which they have appeared. HQ is assumed to be able to determine the sectors of occurence of the threat, which the FUs either cannot do, or can do but have to wait for the HQ's

Table 3 Accuracy and timeliness for the pure strategies

Pure Strategies	Decision Makers			FDMO1		FDMO2		VDMO	
	FU1	FU2	HQ	T	J	T	J	T	J
S_1	SA1	SA1	SA1	5.00	3.59	9.00	2.48	6.67	3.39
S_2	SA1	SA2	SA1	6.00	2.82	10.00	1.96	7.67	2.63
S_3	SA2	SA1	SA1	6.00	2.82	10.00	1.96	7.67	2.63
S_4	SA2	SA2	SA1	6.00	2.18	10.00	1.42	7.67	2.00
S_5	SA1	SA1	SA2	7.00	3.05	11.00	2.42	8.67	2.81
S_6	SA1	SA2	SA2	7.00	2.08	11.00	1.13	8.67	1.73
S_7	SA2	SA1	SA2	7.00	2.08	11.00	1.13	8.67	1.73
S_8	SA2	SA2	SA2	7.00	1.31	11.00	0.56	8.67	1.06

command. HQ, therefore, sets the interactions between the FUs to be as in FDMO1 when the threat occurs in the extreme sectors [0, 1] and [2, 3], and as in FDMO2 when the threat is in]1, 2[. In the former case, there is no real need for the field units to interact since at least one of them has an accurate measurement of the position of the threat. In the latter case, however, the precision of the measurement is increased because the FUs fuse their information, and, in doing so, reduce the interval of uncertainty of their respective measurements.

When compared to FDMO1, VDMO is likely to have an improved accuracy of response when the threat appears in]1,2[. When compared to FDMO2, VDMO will have a lower response time when the threat appears in the extreme sectors. The results for accuracy and timeliness for the VDMO are shown in Table 3, for the eight pure strategies.

A behavioral organizational strategy is constructed by considering the probability distributions of choosing a particular algorithm at each switch. In the present case, such a strategy is completely defined by the triplet $\{p(u1), p(u2), p(u3)\}$. The resulting strategy space for the organization is the set $[0, 1]^3$. The system loci for the two organizations with a fixed structure, i.e., FDMO1 and FDMO2, are depicted in Fig. 13. They are disjoint, and no matter what

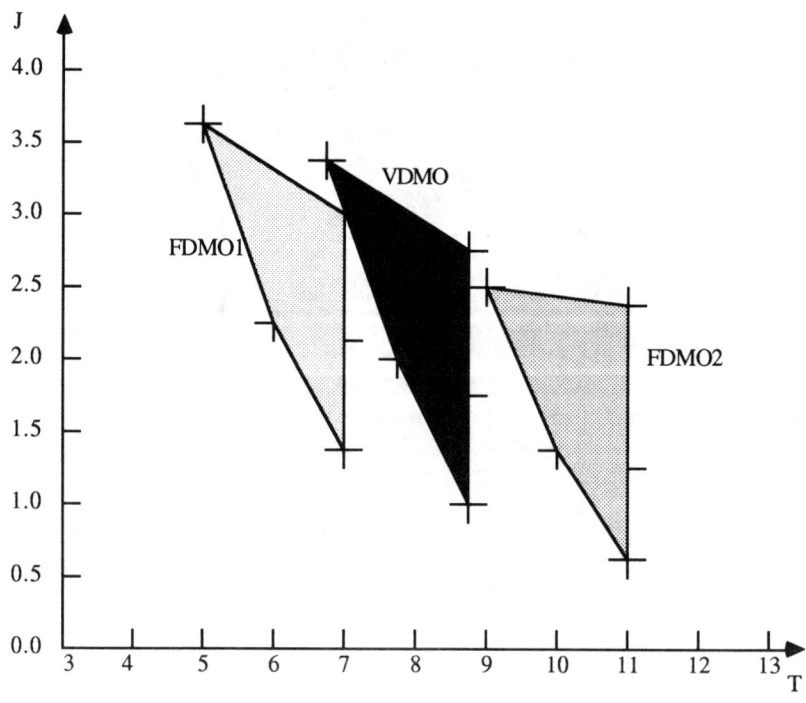

Fig. 13   System loci for VDMO.

organizational strategy is used in any of the two organizations, FDMO2 needs more time to respond. As indicated in Fig. 13, the whole locus for FDMO1 is to the left of the line T = 7 units of time, whereas the one for FDMO2 is to the right of the line T = 9 units of time.

The same methodology for evaluating the MOPs applies to the organization with a variable structure, the VDMO. The system locus of VDMO is also shown in Fig. 13. As expected, the variable structure organization is, on the average, faster to respond than the fixed structure organization in which the field units have to interact (FDMO2), precisely because they do not always interact in VDMO. VDMO is also, on the average, more accurate than FDMO1, since the FUs in the VDMO interact as needed to improve their measurements of the position of the target.

The computation of the performance of an organization for any behavioral strategy and the representation of its system locus are not sufficient to allow the designer to select the best organization among a set of candidates. The mission the organization has to fulfill must be taken into account. This mission may be described in terms of a pair (T^0, J^0) of constraints on performance. A convenient measure of effectiveness of a DMO is $E(T^0, J^0)$, the percentage of strategies for which the performance of the DMO (T, J) meets the requirements of the mission $(T \leq T^0, J \leq J^0)$. This effectiveness measure $E(T^0, J^0)$ takes a value between 0 and 1, with 0 corresponding to no strategy at all satisfying the mission, and 1 meaning that all admissible strategies lead to satisficing performance.

After the measure of effectiveness for each of the three design candidates has been computed, for any given mission defined by its requirements (T^0, J^0), the organization that has the highest effectiveness for a specific mission can be selected. More than one organization can, of course, achieve the same effectiveness. Then each organization has associated a range of mission requirements (T^0, J^0) in the MOP space, such that for any mission requirements (T^0, J^0) within that subset, that organization will have higher effectiveness than all of the other candidates. This defines a partitioning of the requirements space (T, J) in areas corresponding to each organization, or set of organizations, if the maximum effectiveness is obtained for several designs for the same mission requirements. For this example, FDMO1 has a higher effectiveness than either of the other two organizations in the region of stringent constraints on timeliness but not on accuracy. Conversely, FDMO2 is the most effective when the mission requires high accuracy.

The computation of the measure of effectiveness E for each design candidate has been done for discrete values of T^0 and J^0. Values (33) for the timeliness requirement T^0, ranging from 4.00 to 12.00, and values (36) for the accuracy requirement J^0, ranging from 0.50 to 4.00, have been used. This resulting grid of 33 × 36 values for the effectiveness of each candidate was then used to determine the ranges of mission requirements for which each candidate is the most effective. The precision of the determination of these ranges is of course a function of the size of the grid. This explains, for instance, the occasional piecewise linear border between zones.

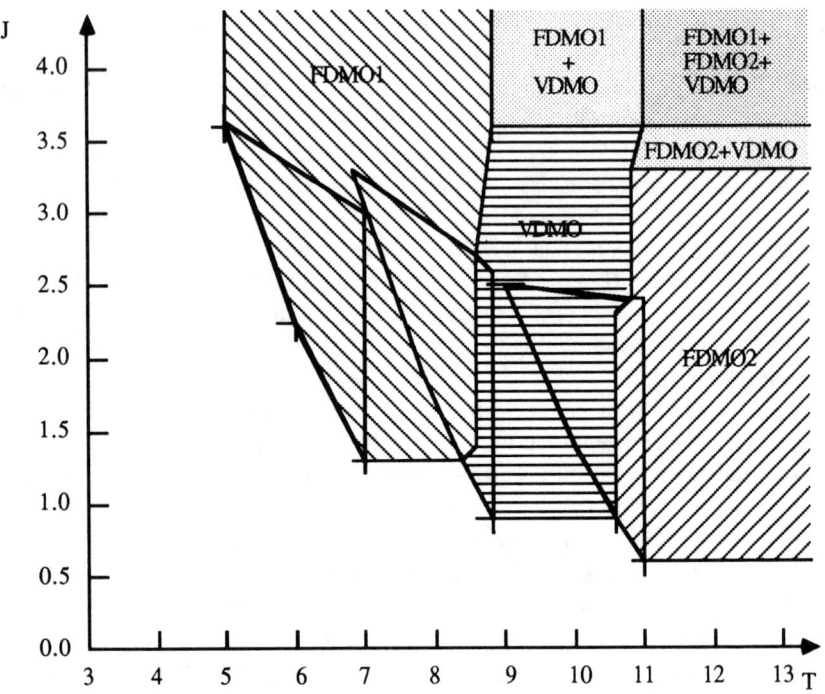

Fig. 14 Partitioning of the requirements space for fixed and variable structure DMOs

Such a partitioning is represented in Fig. 14. There are seven distinct areas. The first area, with no shading pattern, corresponds to the set of mission requirements for which all of the organizations have an effectiveness equal to 0, i.e., there is no organizational strategy that can meet the mission requirements. The area labeled FDMO1 is the one in which FDMO1 is the most effective; its nonzero measure of effectiveness is higher or equal to the measure of effectiveness of FDMO2. The areas labeled VDMO and FDMO2 are the ones for which VDMO and FDMO2 are most effective. In the fifth area, which is labeled FDMO1+VDMO, both organizations have an effectiveness of 1, which means that for both any organizational strategy will meet completely the requirements of the mission. There is no rationale in that case to select one organization over the other. There is no region corresponding to FDMO1+FDMO2. In the region (FDMO1+FDMO2+VDMO), all three designs meet totally the requirements.

V. Conclusions

In the previous sections, the need for variable structure organizations has been described and the concept of variability discussed. A methodology for modeling variable structure decisionmaking organizations that is based on Predicate

Transition Nets has been presented. The approach was then used to model a variable structure organization, and then analyze it with the tools that have been developed earlier for fixed structure organizations. It has been shown that one cannot decide whether a VDMO performs better than an organization with a fixed structure, unless the specific mission requirements are taken into consideration. Then ranges of mission requirements have been identified for which specific organizational designs are most effective. If the requirements are such that the best design is the one with variable patterns of interactions, then the VDMO should be considered. If they are not, then there is no need to introduce variability, since a VDMO would not perform any better. A fixed organizational structure would require a simpler C3 system to support it.

If the requirements are met both by a variable structure organization and an organization with a fixed structure, then other criteria may be used at this point, such as, for instance, the robustness of a design, which would favor a fixed structure DMO since it is less sensitive to noise or jamming. These criteria have not been addressed in this paper, but would constitute the next step toward the modeling of more realistic decisionmaking organizations.

Acknowledgments

The authors would like to acknowledge support provided by the Office of Naval Research under contract no. N00014-84-K-0519 (NR 649-003).

References

[1] Genrich, H. J. and Lautenbach, K., "System Modeling with High-Level Petri Nets," *Theoretical Computer Science*, No.13, 1981, pp. 109-136.

[2] Levis, A. H., "Information Processing and Decision-Making Organizations: a Mathematical Description," *Large Scale Systems*, No.7, 1984, pp. 155-163.

[3] Levis, A. H., "Human Organizations as Distributed Intelligence Systems," *Proceedings of the IFAC Symposium on Distributed Intelligence Systems*, Pergamon Press, Oxford, UK, 1988.

[4] Boettcher, K. L. and Levis, A. H., "Modeling the Interacting Decisionmaker with Bounded Rationality," *IEEE Transactions on Systems, Man and Cybernetics*, Vol. SMC-12, No.3, 1982.

[5] Remy, P., and Levis, A. H., "On the Generation of Organizational Architectures Using Petri Nets," *Advances in Petri Nets*, edited by G. Rozenberg, Springer-Verlag, Berlin, 1988.

[6] Monguillet, J. M., "Modeling and Evaluation of Variable Structure Organizations," MS Thesis, Lab. for Information and Decision Systems, MIT, Report LIDS-TH-1730, Cambridge, MA, 1988.

Command and Control Reference Model

Israel Mayk*
U.S. Army Communications-Electronics Command, Fort Monmouth, New Jersey 07703

Role of the C2 Reference Model

This chapter is an introduction to the C2 reference model (C2RM).[1] The C2RM provides a framework for the evolution of a coordinated and detailed definition of a command and control (C2) discipline. The C2RM embodies an integrated multidisciplinary approach to describe intelligent C2 systems and the way in which they cope with conflicts and uncertainty. It is intended to be complete and self-consistent for the highest levels of abstractions which one often encounters in various other models, simulations, functional descriptions, paradigms and metaphors of C2. The scope of the C2RM embraces C2 using all key physical and logical interactions associated with C2 systems. It is concerned with interactions, involving not only communications (e.g., radios), but transportations (e.g., vehicles), identifications (e.g., sensors), and inflictions (e.g., weapons), which may take place between resources of the same, friendly, hostile or neutral C2 systems. High levels of abstractions of user requirements for C2 across the broad spectrum of military and civil domains have led to the development of the C2RM. The role of the reference model in general is depicted in Fig. 1. It applies to all phases of system acquisition from the laboratory to the field and from conceptualization to realization. The idea of establishing a reference model for coordinating standards and technical specifications has been adopted by the International Standards Organization (ISO) for open system interconnection (OSI) communications protocols[2] and for open distributed processing (ODP) and by the Object Management Group (OMG) for open applications object-oriented analysis, design and programming.[3] The C2RM is in the process of being coordinated by the U.S. Department of Defense (DoD) and developed for C2 applications and interoperability.

This paper is declared a work of the U.S. Government and is not subject to copyright protection in the United States.
*Chairman, JDL TPC3 BRG C2RM Subgroup, and Research scientist, C2 and Systems Integration Directorate

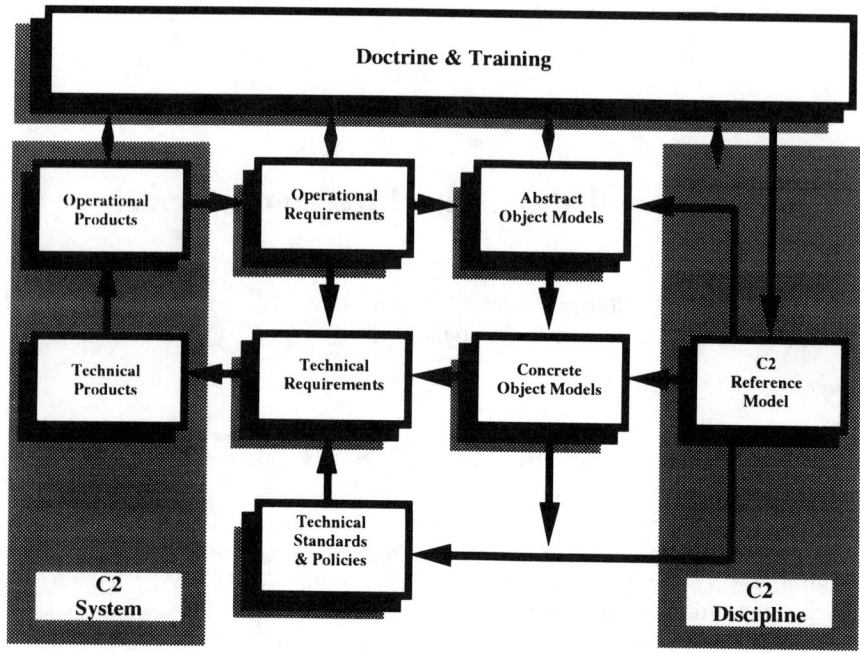

Fig. 1 Role of the C2RM.

As a technical starting point, the C2RM is based upon generic and analog extensions to the ISO open system interconnection (OSI) reference model (RM). ISO started work on its OSI RM in 1977. It was accepted as an international standard in 1983. More recently, the OSI/Network Management (NM) Forum was established to develop an architecture and protocol specifications for the mangement of communications networks.[4] Network management for communications at a higher level of abstraction is very similar to C2. OSI/NM Forum also uses the ISO OSI RM for its underpinnings. The C2RM builds on the success of the ISO OSI RM. However, since it requires reinterpretations and generalizations which go far beyond the scope of the ISO OSI RM, it is expected to take much longer in gaining widespread visibility, acceptance, and utility. An understanding of this chapter is only one small step in that direction. Any constraint or feature which may be present within the ISO OSI RM may not be automatically or unqualifyingly present, relevant, and applicable to the C2RM. Nevertheless, the major theme of layering features of services in layers which are common across resources is preserved to facilitate understanding, reuse of design, implementation, and interoperability to the maximum degree possible and commensurate with the state-of-the-art. Services of a given layer are relative to the requirements of the layer above. Fully compliant resources are "open" in the sense that interoperability may be enhanced by adding more resources to the system or upgrading existing subresources to satisfy shortfalls without a penalty of incompatibility and without a high price for overhead associated with resources which are relatively "closed." Therefore, the C2RM should serve to raise the level of modularity and standards compliance which may be possible within any C2 system.

Breadth of C2

Informally, the fundamental notions of C2 have existed since man began to understand himself and resolved or attempted to resolve potential conflicts which lurked in his path. Formally, however, fragments of these notions evolved in narrow contexts, scattered and imbedded in a variety of distinct but broad disciplines such as management science, behavioral science, operations research, physical sciences, cybernetics, automatic control, communications, and computer science as well as more specialized disciplines such as artificial intelligence, robotics, distributed processing and signal processing. A coherent merger of the theories supported by each of these disciplines as they apply to C2 systems through common definitions constitutes the theory of C2.

The many aspects of C2 are often hidden in war stories, doctrine and technological products which compete for the opportunity to be fielded. In effect, however, C2 may be found individually in any resource involved in a conflict and collectively in all resources teamed as a coherent fighting force. C2 is multidimensional corresponding to many perspectives which must be pulled together. Command is typically associated with exercising "authority" whereas control is associated with exercising "direction" over assigned forces in the accomplishment of missions. The definitions for command and control (C2) usually imply the combination of command and control in an additive fashion. Definitions of command and control system, however, typically include the facilities, equipment, communications, procedures, and personnel essential to the commander for planning and controlling operations of assigned forces pursuant to the assigned missions. Note that in many paradigms of C2 (Ref. 5) the force and the commander are generally excluded from the C2 system, yet both the commander and the C2 system are included in the table of organization and equipment (TOE) of force units. The above definitions are insufficient to resolve the many related issues. Are commanders part of C2 systems or are they just users of them? If commanders are not part of C2 systems, what is the name of the system that includes the commander and the C2 systems? What is the name of the integrated system that includes not only the commander but his forces as well? What is the difference between C2 and C3 and between C2 system and C3 System? From an ISO OSI RM perspective, a C2 system is a network of systems, and the environment provides the media for communications. From the C2RM perspective, a C2 system is a network of resources, and the environment provides the media for interactions.

Often C2 is augmented by an additional "C." The third C in C3 stands for communications of all contextually meaningful information objects necessary to accomplish the mission of the C2 system. At the physical end of its domain, communications may be carried out through any media in the environment capable of *signalling*, i.e., forced vibrational behavior propagating conventionally through electronic, photonic, sonic, or audio phenomena and nonconventionally through chemical, biological, or nuclear phenomena. At its highest form, it is carried out through the attribute of leadership. Leadership is an aggregated quality of motivation and decisiveness which are principally conveyed through communications. The meaning of communications, therefore, goes far beyond the mere technical aspects of transmitting and receiving bits of information. Ultimately, communications relevant to C2 address and involve the use of all possible means of interactions needed to be ready for, to cope with, and to resolve a given conflict or crisis.

Decision-Theoretic C2 Process

The common denominator of all C2 paradigms can be shown to be decision theoretic in nature. As shown in Fig. 2a, to each system-level (global) *observation* (**O**) there corresponds a system-level *action* (**A**). A system-level *decision rule* (**D**) is invoked to select a system-level action for a given system-level observation. Sequences of global observation-action pairs characterize the *dynamics*, i.e., the rules of behavior and evolution, of the system-level *decision process*. This structure and associated overall cyclical process (as conveyed by the arrows), is called the *C2 process*. Functionally, each of the **O**, **D**, and **A** subsystems of **F** or **G** represents a complex, collective and compound process which must be logically interconnected as shown. When tightly coupled, the physical perspective is identical to the functional perspective shown in Fig. 2a. When loosely coupled, the physical perspective becomes more complex as typified in Fig. 2b. As a result of weight, size, power, and survivability considerations, physical resources must be established to effect the functional subsystems in a distributed and dispersed manner. Their logical interconnections

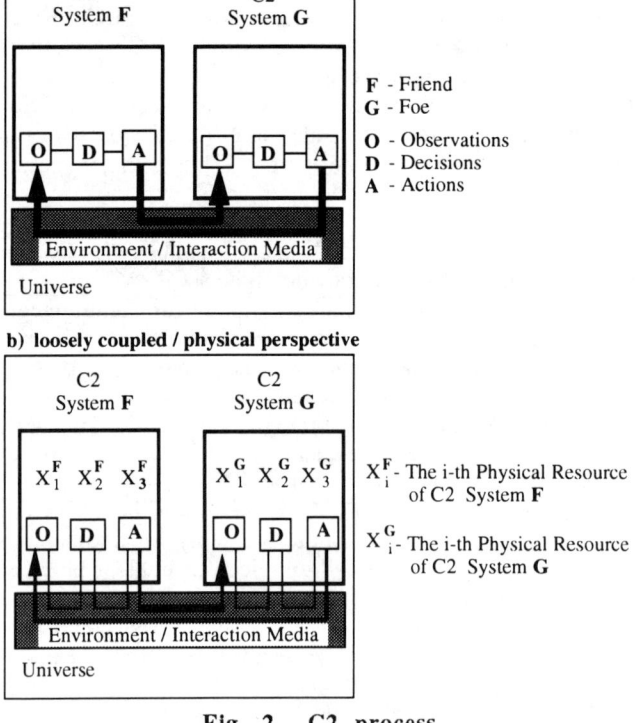

Fig. 2   C2 process.

become highly uncertain due to the environment through which they must communicate.

Thus, a C2 *system* F/G is a collection of *resources* $X^{F/G}_i$, capable of *interactions* $Y^{F/G}_i$ and their underlying dynamics subject to *conflicts* $C^{F/G}_i$. Interactions are responsible for environmental impacts which trigger potential observations. A primitive resource is a collection of interrelated *entities* which *form* a meta-object able to *perform* a useful set of *processes* $W_j(X^{F/G}_i)$ for C2 system F/G. Physical processes of F or G generate and receive *physical objects* in the course of interacting with other resources of either F or G. Physical processes may be grouped and characterized by *logical processes*. Logical processes generate and receive *logical objects*. Logical objects represent selected variables of the state of meta-objects, objects, or processes. A logical object consists of a *record* from which (and in conjunction with other records) conclusions, recommendations, and judgments may be derived and made a part of the record in support of any decision process.

Each C2 system must continuously and recursively go through the global C2 cycle regardless of how the subsystems are partitioned. Each element in the paradigm includes and must be fully supported by implementation of supply, equipment, tools, objects, information, knowledge, and experience-based reasoning. It is multisided and hierarchically nested. In nested architectures, lower level resources may become assets of higher level resources. It is possible to add feed-forward connections which use the latest observations for adjustment of decisions to provide a priori predictive actions and thereby speed up the main C2 cycles. It is also possible and often desired or required to incorporate feed-backward connections for fast corrective actions which use previous decisions as desired states to react a posteriori to unpredictable changes in the environment and other systems.

The environment E is depicted as a separate factor. Given the universe as the totality of all physical objects, potentially adversarial C2 systems, F,G,H,. . ., and the environment E are disjoint sets of physical objects which together form the universe. Note that independent C2 systems structure and dynamics are generically identical. In addition, physical elements depicting land, air, sea, space, and weather are considered as part of the environment. Since E also includes *environmental assets* (natural resources) which C2 systems require for habitat and consumption, a primary concern for any C2 system is its ability to survive and thrive within E.

Layering Resources for C2

The description of resources of C2 systems provide a vehicle to consolidate perspectives from different fields into a common architecture. In a very general way, each specializing discipline attempts to describe the dynamic behavior of what is essentially a two-body problem from its own point of view. Two-body interactions are subsequently described in the presence of a network of interacting bodies. Depending upon the discipline, the two bodies assume different names and definitions as shown in Table 1. In the C2RM these bodies are meta-objects instantiated by resources, subresources, or entities to support specific services.

Table 1 Scientific discipline-oriented resources

Scientific Discipline	Meta-object A	Meta-object B
Control	Controller	Plant
Communications	Transmitter	Receiver
AI	Planner	Agent
Queueing	Server	Customer
Distributed processing	Server	Client
C2	Commander	Controller

Each one of the meta-objects in Table 1 may interact with its dual on multiple levels. Interactions on a given level are grouped to form a layer. The layer of the architecture in which a process is embedded, therefore, becomes important since it provides the context within which to formalize definitions in a coherent and complete fashion. The notion of layering is not new and has been known and used effectively for many years in many disciplines. In particular the ISO OSI RM[1] was developed using the generic principles of layering similar to the ones given below:

1) Number of layers should be manageable.
2) Boundaries between layers should be clear with minimum interactions.
3) Each layer should have unique functions.
4) Related functions should be grouped into a single layer.
5) Past experience should be considered.
6) Functions within a layer should be tightly coupled.
7) Only layer boundaries should be standardized.
8) Layers should be distiguished by hierarchical abstractions.
9) Layers should be encapsulated and highly transparent to other layers.
10) A layer boundary should be required only for adjacent layers.
11) Layers may be sublayered.
12) Sublayers should have a common interface with adjacent layers.
13) Sublayers may be degenerated .

An ISO OSI RM system is a set of one or more computers, the associated software, peripherals, terminals, humans, and application processes which form an autonomous whole capable of communicating with other ISO OSI RM systems in the prescribed seven-layered fashion. The ISO OSI RM, however, is primarily focused upon the transmitter-receiver meta-object problem. It does not delve into the architecture of the application processes. The ISO OSI RM (communications) application layer mediates between the lower six communications layers and any supported application process such as the C2 process. Thus from an ISO OSI RM perspective, the observation, decision, and action processes of the global C2 process are supported by the layered communications process as shown in Fig. 3.

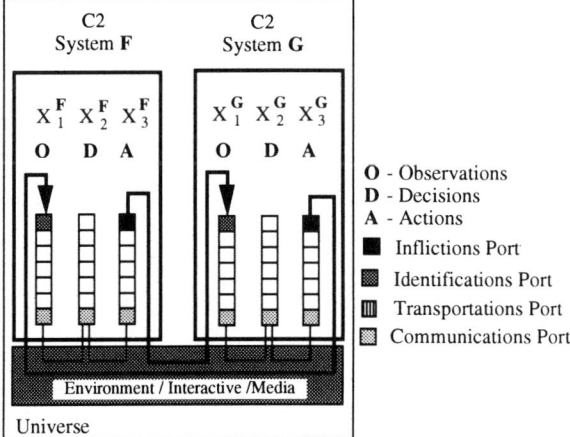

Fig. 3   C2 process through the ISO OSI RM.

The C2RM, however, is a more complete reference model from a C2 perspective since it provides layers for the global C2 process and for transportation, identification and infliction processes which at their highest level of abstraction complement the communications process in support of the C2 process. As shown in Fig. 4, an ISO OSI RM system is expanded at the ISO OSI RM application/application process layers to provide an isomorphically layered architecture for the C2RM application/interaction ports layer. Note that a local C2 process is confined within a resource or a subresource. Not every C2RM resource must have the full complement of interaction ports. As shown in Fig. 5, local C2 processes within each C2RM resource support the loosely coupled global C2 process of Fig. 2b and provide the missing structure from a C2 perspective to the loosely coupled global C2 process modeled by its ISO OSI RM counterpart of Fig. 3. The ISO OSI RM inherently applies only to loosely coupled systems as shown in Fig. 2b. The C2RM, however, applies equally well to depictions of tightly coupled C2 systems as shown in Fig. 6 where communications is assumed to exist but does not play a critical part, and all resources are aggregated to form a single C2 system.

C2 Applications Layers

As mentioned in the previous section, the ISO OSI RM application /application process layers are instantiated to address the global and local C2 process. The C2RM C2 application layer consists of seven sublayers, as illustrated in Fig. 7, such that each sublayer provides services to the sublayer directly above. Conversely, the services defined at each sublayer rely upon the services provided by the sublayer directly underneath. Thus the ISO OSI application layer becomes a sublayer of the **C2 asset** layer. The highest C2 application sublayer is called the **conflict layer**.

The conflict layer is further layered as shown in Fig.7. Each of the conflict sublayers provides services to further define and refine any potential conflict and its associated space-time boundary. *Peace* is the state of a space-time region in

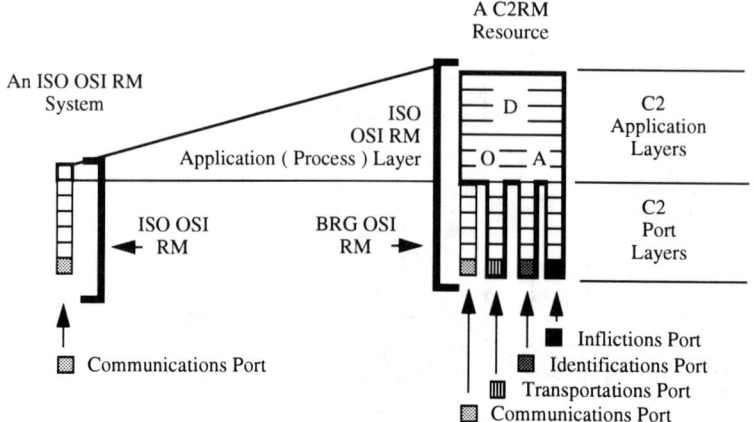

Fig. 4 Extending the application layer/application process of the ISO OSI RM to C2.

E which denotes a perceptively attainable desired degree of freedom which is void of militant conflicts between and among co-existing C2 systems. Competition for natural resources, however, as well as other apparently rational or irrational C2 system requirements may lead to the development of space-time regions of militant conflict. Each space-time region of conflict must be supported by a number of varied resources or processes which depends on the scale of the conflict. The highest region of militant *conflict* is a space-time region characterized by the state of *war* nested in the region of peace. Any *discontent* or lack of freedom may lead to conflicts of *concerns, interests, influences, maneuvers, impacts,* and *lethalities*. These motivations are nested hierarchically to characterize the space-time regions of *war, campaign, battle, combat, engagement,* and *armament*, respectively.

A commander is the key official resource of a C2 system. The commander is responsible for every C2 system unit of action, force, or power used in peace or in conflict. Conversely, no unit of action, force, or power should act without the authority of its commander. The authority of the commander may be formalized but must always retain the flexibility to take initiatives as required to meet unexpected conflicts and exploit opportunities which enhance the survivability posture of the C2 system. Thus, the system which executes commands is strongly coupled to the system which generates commands, and, as such, command must be distributed hierarchically in a C2 system.

The *mission* of the C2 system defines the *goal, aim, objective, purpose, intent, decision requirement, function,* or *desired state* of the C2 system. The mission must be derived from the conflict in which the C2 system is involved. A mission may be decomposed into a series of sub-missions. At the system level, the mission is supported by layered *services* global to all resources and associated process entities which are local to each resource. The service process entities, in turn, may be modularized and refined, stepwise, in a formally descriptive manner, to support a series of C2 *modes, activities,* and *functions*

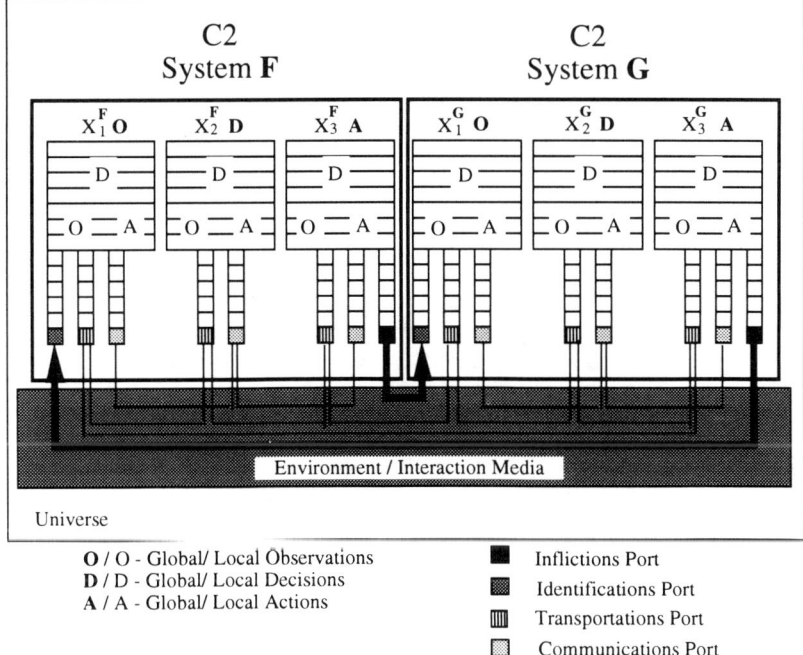

Fig. 5 C2 process through the C2RM (loosely coupled {O, D, A} subprocesses).

through a set of *utilities* and *facilities*. Note that whereas the generic structure of C2 system is common and independent of echelon, the meaning of its concept, requirements, and specifications will vary depending upon the frame of reference for instantiating the C2 system.

C2 systems consist of a hierarchy of nested levels called *echelons*. Therefore, this model may pertain to each echelon if it is autonomous. It is more practical, however, to consider that the C2 application layer is distributed to the lower echelons rather than to be nested in the lower echelons since lower echelons should provide more specificity and refinement of implementation of common goals rather than independent goals which may conflict. In reality nesting of C2 does occur. This may lead to conflicts between local and more global C2. The role of global C2, therefore, is to minimize such conflicts where possible. At each echelon of command, the highest level of service available to the user is the capability to select the resources necessary to carry out the commander's mission, and the establishment of an understanding of the commander's mission by the resources selected to carry it out. The C2 conflict layer is the highest sublayer of the C2 application layer. It provides for formulating the mission and communicating the commander's intent, guidance, and leadership.

Reflecting the views of many commanders, every resource should have the same view of the universe as the commander. This notion, which is key to promoting greater morale and motivation, implies that ideally, under perfect communications, every resource should be fully aware of the mission in which it

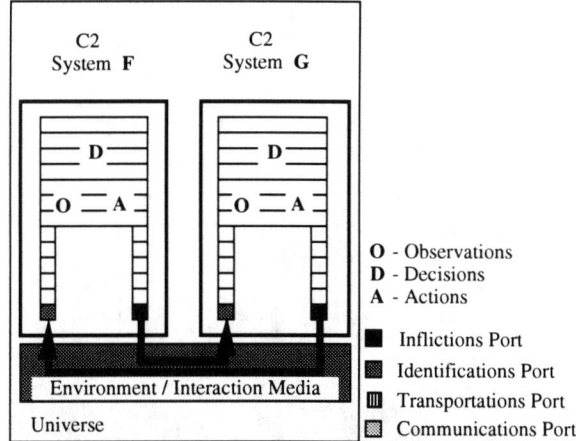

Fig. 6 C2 Process through the C2RM (tightly coupled {O, D, A} subprocesses).

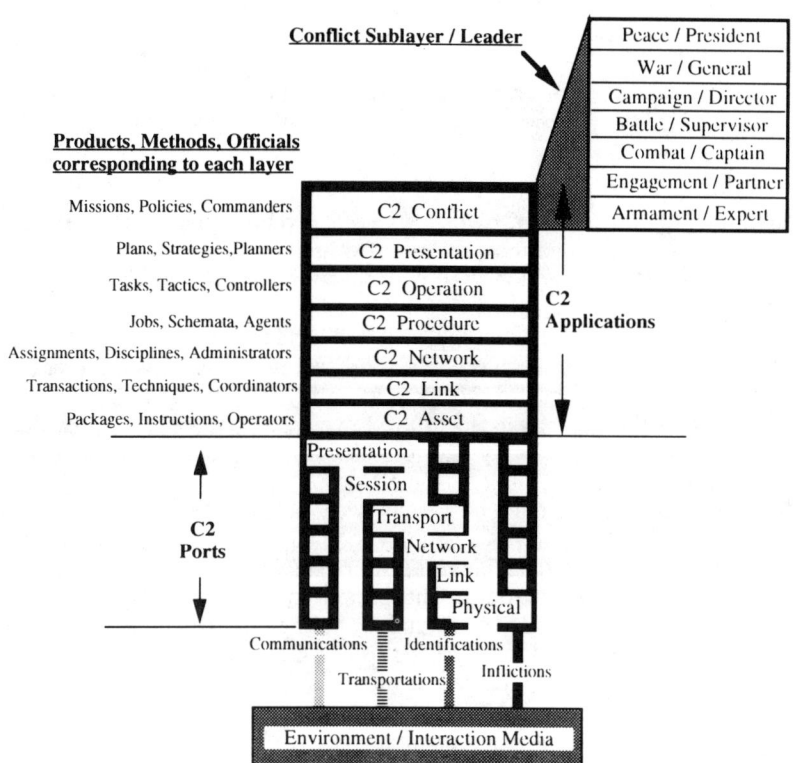

Fig. 7 C2 application and interaction port layers.

plays a part. The C2 conflict layer must, therefore, be served by planning functions responsible for the organization and deployment of the selected resources in the best possible posture. The C2 presentation layer is responsible for the preparation, coordination, approval, and dissemination of the plans in support of the mission. To ensure proper implementation of the plans, the C2 presentation layer is serviced by the C2 operation layer. The C2 operation layer initiates control functions which generate orders to the committed resources and monitors their status. Orders are given in broad descriptions of the intended outcome of submissions. The C2 operation layer is serviced by the C2 procedure layer which implements the doctrine concerning rules-of-reporting status and the rules-of-engagements of targets. C2 procedures are detailed prescriptions of various C2 techniques which may be selected for implementing orders by fully networked resources. The C2 procedure layer is served by the C2 network layer which provides for the synergistic effect of fusing status, observations, intelligence and communications data to achieve combined, coordinated weapons engagements and maneuver of resources across the theater of operation. The C2 network layer is supported by the C2 link layer which ensures the reliability of individual pair wise resource-resource interactions whether 1) friendly-friendly, one with one, e.g., sensor-sensor, sensor-weapon, weapon-weapon or transmitter-receiver or 2) friendly-enemy, one-against-one, e.g., sensor-target, weapon-target, and other such warfare interactions. Finally, the C2 link layer is dependent on the performance of the physical resources

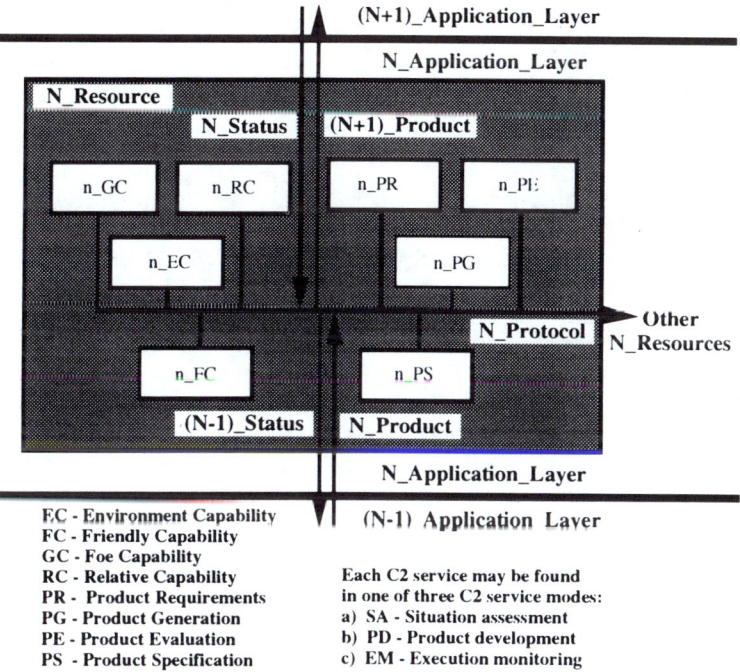

EC - Environment Capability
FC - Friendly Capability
GC - Foe Capability
RC - Relative Capability
PR - Product Requirements
PG - Product Generation
PE - Product Evaluation
PS - Product Specification

Each C2 service may be found in one of three C2 service modes:
a) SA - Situation assessment
b) PD - Product development
c) EM - Execution monitoring

Fig. 8 Services within a C2 application layer sublayer.

employing behavioral, biological, mechanical, chemical, nuclear, or electromagnetic principles, rules, and forces.

A C2 application layer sublayer process typically involves eight types of services depicted in Fig. 8. These services provide utilities for situation assessment, product development, and execution monitoring with respect to: a) n+1 product requirements; b) friendly capabilities essential to support the n+1 product; c) adversarial capabilities opposing the n+1 product; d) environment associated with the relevant space-time region of conflict; e) key factors in a,b,c,and d which are critical to continue or modify the current n product; f) generation of n product alternatives; g) evaluation of the n product alternatives; and h) specification of the n product for dissemination. Note that each of these application services may be supported by facilities to generate scenarios, snapshots of scenarios for past, present and future situations, as well as gaming with candidate products to assess potential impacts.

C2 Interaction Ports and Layers

At the highest level of abstraction, the application layer of the ISO OSI RM is also extended sideways and downward to include **C2 interaction ports** which are associated with the other types of interactions for C2, i.e., transportations, identifications, and inflictions. The layers of these interaction ports are analogous to the presentation and lower layers of the ISO OSI RM for communications. In the C2RM, the presentation and lower layers of the ISO OSI RM represent only the C2 communications ports. Thus, the C2RM embraces anologous architectures for all the key types of physical interactions and uses the C2 application layer to provide integrated command and control over all types of interactions. The four fundamental types of interactions are identifications, communications, transportations, and inflictions.

Identification

Identification is an interaction which directly results in the recognition of objects in the environment. Identification is used to determine the stages, phases, and targets required for each layer of conflict. Signals retrieved from the environment result in the recognition of objects in the environment. The identification port is responsible for correlation, fusion, aggregation, and exploitation of signal parameters processed by its own sensor assets. Identification port interactions range from signal acquisition from objects in the environment through the exploitation of associated data, information, knowledge, and individual experiences to identify targets

Communication

Communication is an interaction which directly results in an exchange of data, information, knowledge, and experience. Communication is used to command, control, and coordinate among the resources. Signals transmitted and received from other resources allow for the exchange of data objects through the environment. The communications port interactions range from signal generation to data updates.

COMMAND AND CONTROL REFERENCE MODEL 233

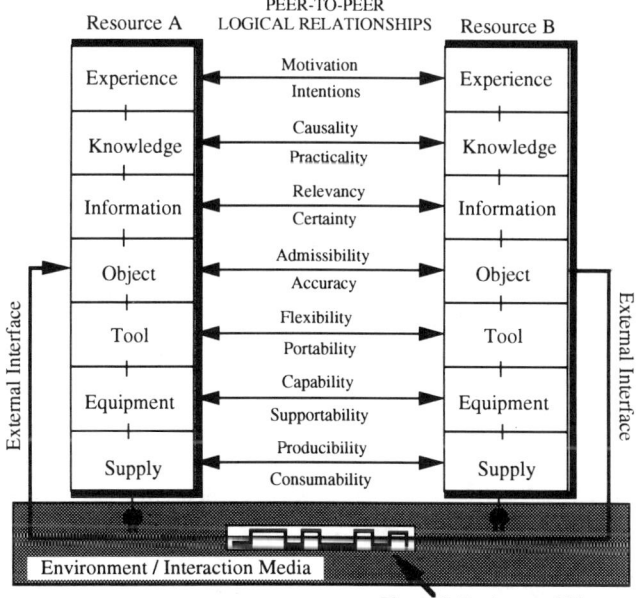

Fig. 9   C2 implementation layers.

Transportation

Transportation is an interaction which directly results in the motion of objects. Transportation is used to carry, supply, strengthen, equip, and load resources with the necessary physical assets. Vehicles, manual or motorized, provide for the translation and rotation of physical objects as well as for the movement of resources through the environment. The transportation port interactions range from the packaging of physical objects to the unpacking for expenditure, consumption, or coordinated maneuver.

Infliction

Infliction is an interaction which directly results in the destruction, damage, degradation or disruption of objects. Infliction is used to reduce the capabilities of adversarial C2 systems through a variety of means called armaments which are capable of causing destruction, damage, degradation, and disruption of target resources involved in the conflict. Infliction port interactions range from the use of armaments in single shots at isolated targets to the general use of armaments in cooordinated engagements.

C2 Implementation Layers

The key entities of implementing C2 are the physical assets which are crucial in realizing and characterizing the performance of a resource. These entities underlie a set of layers as shown in Fig. 9 which provide a resource with the

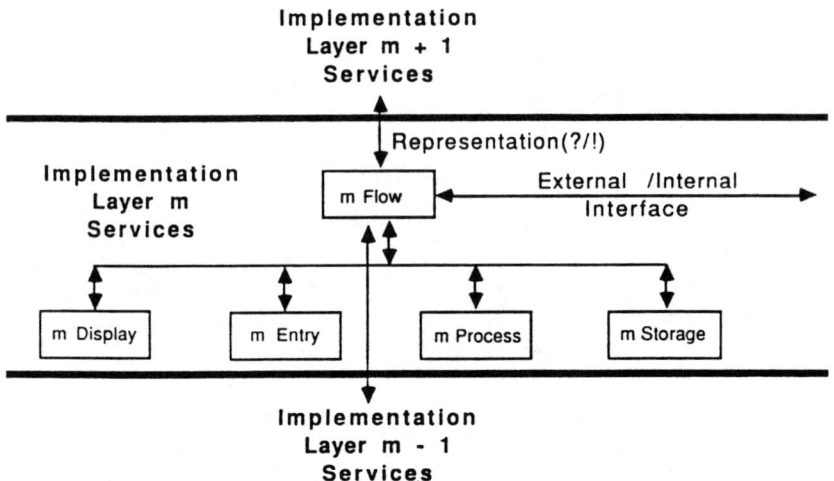

Fig. 10 Services within a C2 implementation layer.

capabilities to utilize its *supplies, equipments,* and *tools* to build and use its own base of *objects, information, knowledge,* and *experiences.* These layers are called *implementation* layers. The implementation layers are associated with each of the C2 Layers. The generic implementation services are shown in Fig. 10. Resources are capable of consuming or producing a wide variety of *energy forms* through a diversified set of supply services. Supportability of any capability is dependent on equipment services which generate *carriers* to facilitate the transition of energy involved in various types of interactions. Flexibility and portability are inherent through on-board tool services. A variety of modes or representations may be admissible. An impact in the environment, for example, may propagate and enter through the object layer of any one of the C2 interaction port. The associated carrier is processed by the tool services and converted into an input mode for more efficient processing and storage. Admissibility into the object base and the accuracy of the representation are taken into account by the object services. The object services manage *arrays* of similar objects. A new array is analyzed by information services in conjunction with other information to provide new information in the form of a candidate conclusion about the relevancy, certainty, and significance of the observed impact. The new conclusion is consolidated by the knowledge services with previously derived conclusions to generate new knowledge in the form of a candidate recommendation. Recommendations should be practical and predict potential causal consequences. The new recommendation is subsequently reviewed by experience services in conjunction with previously accumulated case histories, motivation and known intentions to judge whether or not to accept the recommendation and prioritize it with respect to previously accepted recommendations.

Two-dimensional (n x m) service access is needed to exploit the status maintained by the layered implementations of the other C2 layers. In this respect the C2RM provides a structured interface for both vertical (C2 layers) and horizontal (implementation layers) boundaries to facilitate open system interconnection within a resource. Implementation services may be structured to be activated within each layer of C2, with a request which propagates explicitly or implicitly through the seven implementation layers starting top-down with the experience layer and culminating in the supply layer. Responses to requests are structured to proceed also in real or virtual time utilizing Supply Services and finalized with the experience services. Responses may take on a variety of forms. Requests for a service may be conditionally or unconditionally 1) subject to acknowledgment, 2) granted, 3) denied or 4) delayed.

Illustrative Scenario

Consider the topological positioning of resources and their mutual physical interactions for a hypothetical but illustrative example scenario as shown in Fig. 11a. Resource C is the commander of system F. Resource S is a sensor of F which includes its own controller. Resource W is a weapon of F which also includes its own controller. Resource G belongs to system G. Systems F and G are involved in a mutual conflict. More specifically, the conflict has deteriorated down to the level of engagement (conflict layer 7.7.2). Note that C, S, and G are fixed sites whereas W is mobile as indicated by the textured arrow for transportation. As shown by the other textured arrows, S is within identification range of G as a potential target. C and S as well as C and W are within communications range. Finally, W is within range to inflict damage upon G. Having identified the types of interactions which are key to a given resource in a given scenario, each resource may be expanded in terms of the layers which correspond to each interaction type and the application sublayers which couple among the interactions across resources. In the following scenario the layers involved are identified in parentheses.

Consider a single thread analysis, as shown in Fig. 11b, for the context of one hypothetical sequence of events chosen to illustrate what could occur from the time of detection by S to the time of firing a round of ammunition by W upon G. S detects (identification layer 1/S) an activity on top of an adjacent hill. S tracks (identification layer 2/S) the activity for a short time period (identification layer 3/S-5/S) and notices a large antenna. S performs additional correlation (identification layers 3/S-6/S) to determine that G at space-time coordinate (x, y, z, t) is an observation post. At layer 7.1/S, the target information is stored and displayed as a **package** data object. At layer 7.2/S, the target information is **transacted** either for immediate transfer to the commander C, using its available communications port (communications layers 6/S-1/S) or for local assessment of its relevancy with respect to any one of the several available **assignments** generated at layer 7.3/S. As an unexpected target of opportunity, an assignment cannot be made by S, and layer 7.4/S must decide if any procedures have been authorized for handling the target as a part of any one of several on-going **jobs**. Assuming that **tasks** activated at layer 7.5/S did not anticipate this type of target at the given location, the **plan** generated at layer 7.6/S must be reviewed to see if, indeed, the recognized target is of interest to the **mission** derived at layer 7.7/S for any resource of system F. Assuming that the mission is sufficiently broad, S decides to report G as a target of potential interest to C. S reinterprets, revises or regenerates a plan

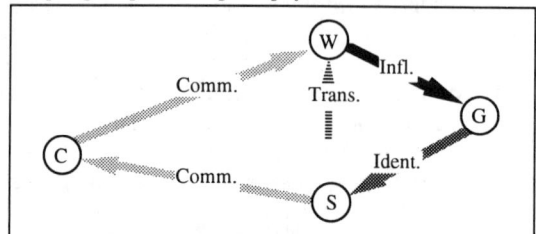

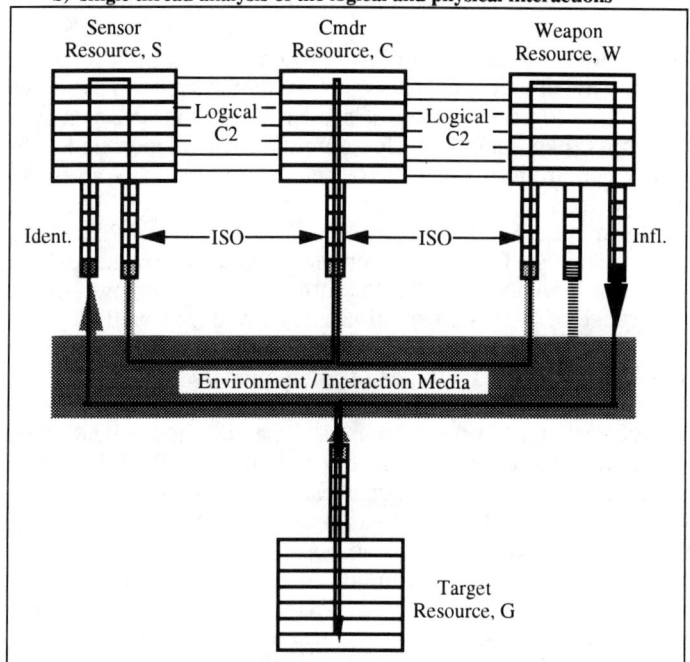

Fig. 11   C2 scenario cell for illustration.

(layer 7.6/S), task (layer 7.5/S), job (layer 7.4/S), assignment (layer 7.3/S) and transaction (layer 7.2/S) in that order to enable a communications package (layer 7.1/S) to be motivated and formatted down through the communications port (communications layers 6/S-1/S). The resulting message is transmitted to C. Resource C demodulates and decodes the message (communications layers 1/C-6/C) for presentation to layer 7.1/C.

Stored and displayed by layer 7.1/C, the package is made available to layer 7.2 /C for retransacting. Time is of the essence and the commander must decide immediately whether the target is a threat to his immediate mission or some future mission. Being caught by surprise, the package which is automatically retransacted for potential infliction is pre-empted by a review process for consistency and practicality with respect to currently active assignments (layer

7.3/C), jobs (layer 7.4/C), tasks (layer 7.5/C), plans (layer 7.6/C), and missions (layer 7.7/C) as understood by C. C establishes that the target is a threat to current and future plans and operations. C decides to select (layer 7.7/C) W to engage the G. C generates an engagement mission (layer 7.7.2/C-7.7.1/C), redraws or revises the plan (layer 7.6/C), issues a new task (layer 7.5/C), to initiate the job to fire on the given target (layer 7.4/C), and remakes an assignment (layer 7.3/C) to repackage (layer 7.2/C) the target information in a manner suitable for a transaction (layer 7.1/C) to W. The repackaged information about G is processed by layer 7.1/C to create a communications transaction which follows the ISO OSI RM services down through the communications port of C (communications layers 6/C-1/C), through the environment and up through the communications port of W (communications layers 1/W-6/W).

Resource W (layer 7.1/W) stores and displays the package and makes it available to layer 7.2/W for potential retransaction. A transaction (layer 7.2/W) allows for cross coupling of communication, transportation, and infliction packages. At this point, W must decide (layer 7.3/W) whether it is close enough to cause the required damage or whether it should move closer to improve the probability of impact. Once an assignment is contemplated for infliction (layer 7.4/W) it is presented higher up to be scheduled as part of an on-going job. The candidate job is then subject to review for priority, consistency, and urgency with respect to existing tasks (layer 7.5/W), plans (layer 7.6/W), and missions (layer 7.7/W). W decides to respond with a hasty fire mission (layer 7.7/W). The overall deployment plan is checked (layer 7.6/W) to assure that S is a safe distance away (layer 7.5/W). The job is approved (layer 7.4/W), and an assignment is made for infliction (layer 7.3/W). The target information available from storage (layer 7.1/W) is repackaged with supplemental timing constraints (layer 7.2/W). It is then reprocessed to create an inflictions package (layer 7.1/W) which is serviced by the inflictions port. Infliction port services may range from a survey of the weapon position to the loading armaments and pulling the trigger (inflictions layers 6/W-1/W). The armament is launched and hits G and destroys his antenna.

Acknowledgment

I would like to thank not only those individuals who have worked very closely with me to develop the C2RM but many others who also took time out of their valuable schedule to share their perspectives on C2 in general and on the C2RM in particular. For a complete list of contributors please see Ref. 1.

References

[1] Joint Directors of Laboratories, Technical Panel for C3, Basic Research Group, "The C2RM" Coordination Draft #6, May 11 1992.

[2] International Standards Organization, "OSI Basic Reference Model" - ISO 7498(Draft) Information Processing Systems, 1984.

[3] Object Management Group, "Standards Manual," Draft 0.1, edited by Soley, R.M, OMG TC Document 90.5.4, May 1990, OMG, Framingham, MA.

[4] OSI/Network Management Forum, "Release 1 Package: Protocol Specifications, Application Services, Object Specification Framework, Architecture, Glossary, Library of Managed Objects, Name Bindings and Attributes, Managed Objects Naming and Addressing, Conformance Requirements, and Shared Management Knowledge," Bernardsville, NJ, 1990

[5] Mayk,I.and Rubin I., "Paradigms for Understanding C3 Anyone?" <u>Science of Command and Control: Coping with Uncertainty,</u> edited by S.E. Johnson and A.H. Levis, National Defense Univ./AFCEA International Press, Washington, D.C., 1988, pp.48-61

Stochastic Modeling, Analysis, and Calibration of C3 Systems

Izhak Rubin*
IRI Corporation, Tarzana, California 91356
and
Israel Mayk[†]
U.S. Army Communications-Electronics Command, Fort Monmouth, New Jersey 07703

I. Introduction

A C2 system consists of a set of resources which are able to interact among themselves, with adversary resources and with the environment. The specification and modeling of a C2 system thus requires proper definitions of the spaces of resources and interactions to be included. These are defined to a level of quantization and aggregation which is commensurate with the underlying extent of detail relevant to the key events of interest and to the echelon (layer) which is being modeled. The functioning of a C2 system is dictated by the underlying methods, procedures, command policies and doctrines which describe the actions taken by the system under various system (friendly, adversary and environmental) states. To assess the performance of a C2 system, we need to examine its behavior under a specified set of battle scenarios and command missions.

In this Chapter, we describe a number of stochastic mathematical models and architectures which we have recently developed to provide computationally efficient tools for the modeling, evaluation, analysis, planning and decision aiding for C3 systems. These models are parametrically and generically structured, so that they accommodate a large variety of conflict conditions, mission objectives, C2 resources and their interactions, and command policies. They can, as a result, be applied, through modeling approaches compatible with the C2 reference model (C2RM)[1,2], with views specialized to the layer and organizational/functional level of interest.

In Section II, we present an architecture which describes the underlying C2 system and its dynamic evolution under specified mission

---
Copyright © 1993 by the American Institute of Aeronautics and Astronautics, Inc. All rights reserved.
*President; also, Professor, Electrical Engineering Department, University of California, Los Angeles, Los Angeles, California 90024.
†Research Scientist, C2 and Systems Integration Directorate.

and scenario conditions. It has been implemented as a stochastic C3 graph (SCG) model. The graph consists of nodes, where each node describes the state and procedures associated with a group of (friendly and/or adversary) resources within a time-space cell. A key "event" (or sequences of such events) is used to characterize the essence of the nodal process. Force units (group resources) flow, in accordance with the underlying maneuvering policy and battle conditions and results, from one node to another. Interactions between nodes are designated by corresponding marked lines connecting these nodes. Interactions between resources are classified into four main categories: infliction (firing, weapons), identification (sensing, sensors), transportation (maneuvering, vehicles) and communications (communicating transceivers). Stochastic process based analytical models are used to describe the operation, effects and performance behavior associated with the procedures and methods of each node and its corresponding interactions. They involve dynamic random process representations of the joint C2 policies for firing, sensing, maneuvering and communications.

This C2 architecture has been structured to fit an object oriented system model. The ingredients of an object oriented SCG (OOSCG) model implementation are described in Section III.[‡]

Each C2 system continuously goes through the basic cycle involving the three primitive C2 processes of: observation, decision and action. Such a generic structure is exhibited by each C2 system involved in a conflict. Many C2 process paradigms have been constructed based on these basic procedures[1,3]. As stated in Ref. 1, a C2 subsystem is a subset of C2 system resources, including their embedded dynamics. The observation process is employed (by the observation based resources) to observe, identify and track the state of the system resources and of the underlying environment. The decision process is then undertaken (by the decision makers) to evaluate the observations in relation to the stated mission (rules, doctrines, and objectives). As a result of this evaluation, actions are being taken; the latter are executed by the action subsystem. These actions lead, in turn, to changes in the states of the environment and of the system resources, which are recorded by the observation subsystem and the cycle continues.

In Secs. IV and V, we describe specific stochastic process based mathematical C2 models which can be used as the basis of the SCG nodal (as well as a serial/parallel collection of nodal) procedures and methods describing the evolution of the C2 state processes and the

---

[‡]The SCG model was developed by Dr. Izhak Rubin at IRI Corporation, Tarzana, California. It has been implemented as an objected oriented architecture (OOSCG) and applied to the development of stochastic C3 system models and analysis techniques, including those presented in this Chapter, at IRI under research contract support provided by U.S. Army CECOM, Ft. Monmouth, N.J.

associated system performance behavior. These models incorporate the elementary observation-decision-action paradigm governing the generic structure of any C2 system.

Also in Sec. IV, we present a semi-Markov attrition model which involves the employment of a general maneuvering policy at the times of occurrence of each key event (or subevents which can occur as part of the nodal process). We present a set of equations which are recursively solved to provide for the computation of system measures of effectiveness, including the mission win probability (for the nodal process or for a process describing a group of nodes being modeled), the average battle duration and the average number of losses incurred to win, given a win event. We refer the reader to other such stochastic process based models developed by us in Refs. 4-14.

In Section V, we present the elements of a multistage and multiphase stochastic C2 model which has been used by us to describe an outer-air followed by an inner-air battle scenario. Two battle stages are identified (the outer-air and the inner-air stages), each being further decomposed into battle phases. Each combat phase occurs within a predefined cell. Each such cell corresponds to an SCG node. Thus, each SCG node describes the evolution and performance of the battle during a single "phase," involving the corresponding cellular spatial and resource (friendly and adversary force) levels. Using probabilistic models, recursive equations are derived to compute the distribution of the state of the system at the end of each stage and phase.

We further discuss and demonstrate in Section V, how such analytically based simplified lower granularity stochastic models can be calibrated by a corresponding higher granularity large simulation model. The calibration process is used to establish the appropriate procedural functions and their parameters for the SCG based model. Following the calibration process, the SCG model is used on its own to provide extensive modeling, analysis, and tradeoff evaluations, serving as an efficient training, planning, and real time decision aid tool.

II. Stochastic C3 Graph Model

In this section, we describe the key elements of the stochastic C3 graph (SCG) model. Its object oriented implementation is described in the next section. The SCG model provides for the establishment of an architecture which is used to model the structure, dynamics, and performance of a C3 system operating under specified mission objectives, scenario conditions, and combat doctrines.

In accordance with the C2 reference model (C2RM),[1] a layered structure is used to describe the architectural elements and operational functions and services of a C2 system. Such a description is generically applied at each organizational level (echelon). The highest layer of the C2RM is the conflict layer. It consists of layers of conflict which include war, campaign, battle, combat, engagement and armament. For example, consider an underlying battle conflict. The battle commander issues battle level missions which are needed to win a campaign. The commander may launch a number of combats, conducted in parallel or serially to exploit and occupy key areas. In turn, the combat commander to win the battle issues combat-type missions which call for the maneuvering of the forces. In doing so, this commander initiates a series of engagements in sequence or in parallel. In turn, the engagement commander initiates engagement missions which launch a number of armaments against individual or aggregated targets. Armaments are selected by analyzing the target as a threat potential and the lethal capabilities of the underlying weapon systems.

Under the SCG model, we analyze the outcomes and effectiveness of a C3 system, for a stated conflict level and the corresponding specified missions. For example, to accomplish a campaign mission, parallel or sequential battles are specified. Each battle can consist of multiple combats, which in turn initiate multiple engagements. The dynamic process characterizing each conflict level based mission is defined through the identification of key "events," and the corresponding quantification of the joint (time × space × resources) space into cells. Under the SCG model, each cell is described by one or more nodes. The node corresponds to a certain set of resources (such as a group of friendly tanks) that are operating within a specified geographical environment over a period of time, which corresponds to one or more conflict phases and conflict event(s).

The SCG node (cell) is thus constructed in terms of its defining mission event(s), its time and space boundaries, its associated resources, and its associated C2 functions, procedures and methods. The associated mission event(s) define the objectives of the nodal conflict actions that are to take place (such as crossing a river mission). The nodal time and space boundaries are related to the underlying mission event type; they can be set to be fixed quantities (so that this event occurs within specified time and space boundaries), or they can be determined dynamically (so that, for example, the nodal mission terminates when the forces reach a certain topographical line, and/or at a time when the river crossing is completed to a specified level of success).

The resources space consists of the following key C2RM assets:

S_A = {infliction, identification, transportation, communications}

Infliction resources include weapon systems; identification systems involve sensor mechanisms and information collection systems; transportation systems provide maneuvering, reinforcement, logistics and supply services; and communications systems are essential to provide for coordination, and information distribution related activities.

The basic interactions between nodes (identified as the interactions between physical assets in accordance with the C2RM) fall into the following four-dimensional space of interactions:

S_I = {inflicting, identifying, transporting, communicating}

Thus, an infliction interaction between two (or more) nodes occurs when force units in one node fire at force units at another node. Identification interactions represent sensing (detecting, identifying, acquiring and tracking) of the states of assets of a node by another node. Transportation interactions are associated with maneuvering, reinforcing, merging/splitting of forces, terrain traversing, logistics, and supply movements. Communications interactions are essential to the coordination and distribution of data, information, knowledge, and experience among nodes. Note that the above nodal characterization applies both to friendly and adversary forces, so that nodes can contain friendly only, adversary only, or both friendly and adversary forces and resources.

The SCG model represents each interaction between nodes by lines connecting these nodes, with indicators and weights applied to these lines to identify the type of interaction involved (being one of the four interactions in S_I) and its parameters (relevant to the underlying nodal method computing the effect of this interaction).

The associated C2 methods embedded in the SCG node are the ones required to model and analyze the effects of the underlying interactions associated with the node, during the specified temporal (phase) limits and spatial boundaries. Methods are included for statistical calculations of the effects of actions including: acquisition of targets and sensing of target parameters, firing and engagements leading to attritions, maneuvering and movements, and message distributions across communications networks and electronics warfare measures. The SCG model incorporates various stochastic based mathematical models which describe such actions and their results using probabilistic mechanisms. In the following sections, we will illustrate a number of such stochastic process based models which combine statistical descriptions of sensing, firing and maneuvering actions to describe the statistical outcome of an engagement during each phase, involving the underlying set of interacting nodes.

The associated C2 procedures and methods embedded in the SCG node also involve specification of command policies. Dependent on the C2RM layer under consideration, and the organizational echelon modeled, such policies include various selected doctrines, resource allocation laws, tactical approaches, and rules of engagement. Such policies and rules can be prescribed for the duration of the evaluated conflict or can be dynamically determined based on intermediate outcomes of the conflict (embedded in the mathematical method used or through a 'man in the loop' process). Command policies are thus defined for each basic action and interaction. For the basic four interactions, such policies involve schemes for weapon engagement and fire allocation; sensor employment; maneuvering, reinforcing, supplying, terminating an engagement and starting a new one; and data distributions and coordination.

In summary, the SCG model thus consists of nodes which involve specified force levels over time and space boundaries, which interact in accordance with the basic four interactions in S_I. In accordance with the classification of the assets into their four categories in S_A, each node is further characterized by the parameters of its resources, by the associated command policies, and by the methods employed to model and analyze the effects of the interactions which take place in accordance with the commander's missions (further layered by the C2RM as plans, tasks/orders, jobs, assignments). Our models involve stochastic descriptions and probabilistic analysis, so that the state of the conflict at the start and end of any phase is computed and described through probabilistic means. This is essential for the modeling of combat systems and their performance under the employment of various C2 mechanisms and policies, since the occurrence of nonaverage events can critically dictate the desired outcome of a battle and thus be crucial in determining the effectiveness of the underlying C2 methods and systems used.

Measures of effectiveness and indices of performance are expressed through the calculated state distributions involving the probability distribution of the sizes of surviving and attritioned (lost) assets of each class for each node, the probability of mission accomplishment, the distribution of time durations for various conflict levels (e.g., battle durations), the distributions of position gains, and others.

In Fig. 1, we outline the SCG model development process. Two key graphs are defined for each scenario which is modeled and analyzed:

    a. A scheduling graph, which describes the temporal (in terms of phases and stages) and spatial evolution of the underlying combat process.

    b. Per-phase interaction graphs; each such graph describes the interactions between the basic units (modeled as group objects, see Sec. III) during a phase.

It is also noted in Fig. 1 that stochastic modeling tools (such as those described in later sections) and resource allocation procedures (such as the Multi-Mission Area C2 System-MMACS-program, see Section VI) are employed to carry out performance analysis and effectiveness evaluations and to allocate system resources. Tests and experiments are performed to calibrate the parameters of the SCG model through the use of large scale simulation and/or field tests, as illustrated in Section V.

Such models and programs serve as effective tools for planning, resource allocation, training and real-time decision aiding for the commander and the staff. One is able to investigate key issues relating to the effectiveness of specified weapon systems against other weapon systems, the efficacy of a command policy vs other command policies, the effectiveness of offensive and defensive plans, and the identification of the underlying bottleneck resources and/or policies that limit the effective performance of a C2 system.

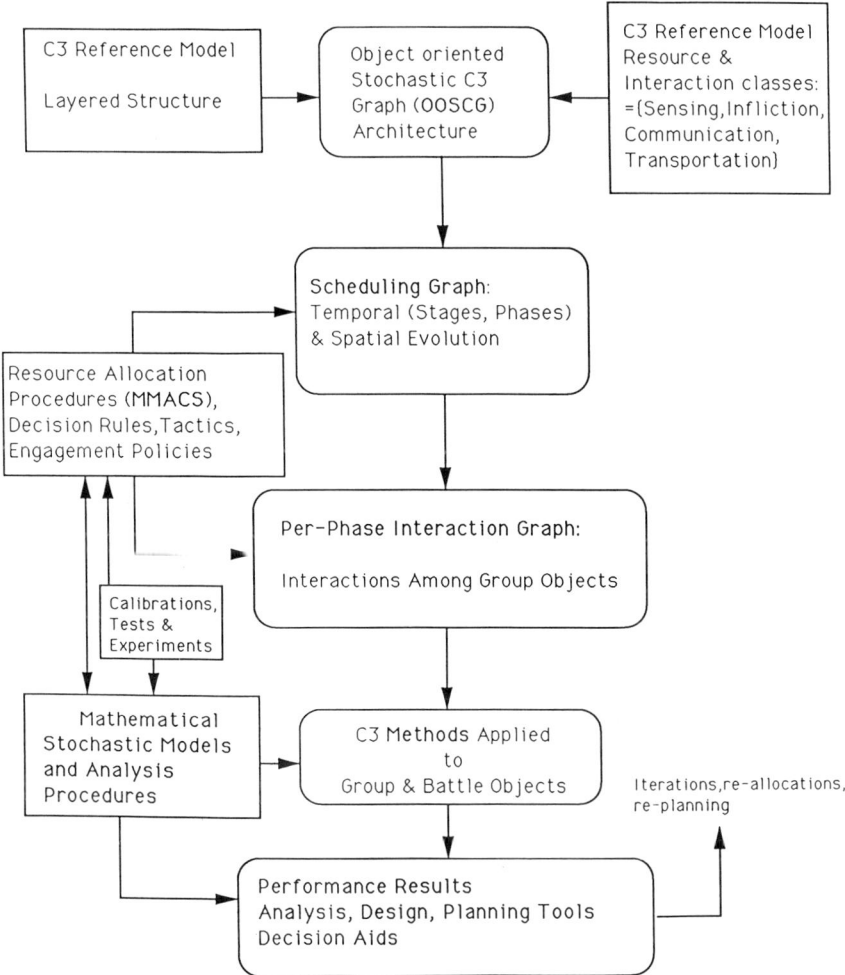

Fig. 1 The OOSCG model development process

III. Object Oriented SCG Implementation

The SCG model has been configured and implemented as an object-oriented program structure. Key elements of this object oriented SCG (OOSCG) architecture and program are described in the following (see Figs. 1-2).

"Unit" is used to designate the lowest level resource element used in the quantification of the four basic actions, and their consequences; i.e., firing (and resulting attritions, a unit is thus the lowest entity for which an attrition event is quantified), sensing, maneuvering/transporting and communicating. The units are combined

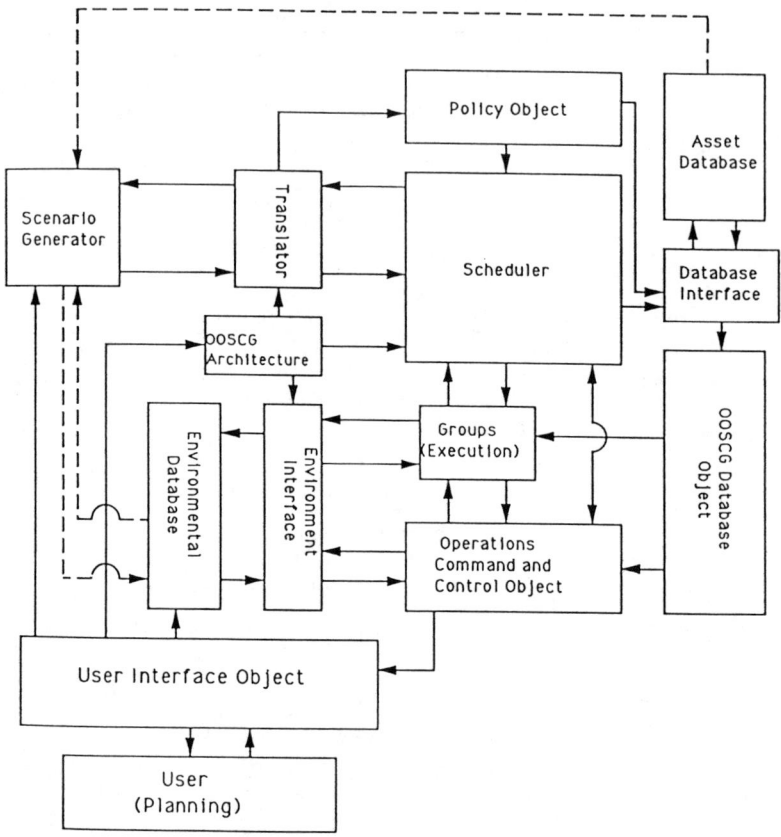

Fig. 2 Flow of command and information in an OOSCG program

into "groups," in accordance with selected commonality criteria. Groups can thus be set-up in accordance with shared features involving aspects such as command, geographical location, organizational association, functionality, and resource attributes. For example, a group of units can consist of units which are similar in their infliction and sensing characteristics that are maneuvering as a single task force along the same route and are under the same local command (such as a group of M1 tanks moving together in the time-space domain). A group is the basic "object" of the OOSCG model. The group object thus represents a military formation consisting of unit resources; the latter can be individual resource elements (such as soldiers, tanks, infantry fighting vehicles, ships, aircraft) and their various formations into organizational units such as companies, brigades, and divisions. Hence, dependent on the view of interest, a unit attrition can represent an attrition of a single resource (say a M1 tank) or a whole collection of resources and organizational formations.

In addition to group objects, an environment object is defined. It represents the underlying static (combat independent) state of the environment (topography, weather and trafficability conditions), and dynamic (combat dependent) environmental effects (smoke screens, mines, etc.)

An object is an entity which fuses data and the related procedures (methods, functions) that can be invoked to operate with the data. The data elements represent the "attributes" of the object elements. An object can interact with another object through the use of messages. A message indicates the type of interaction (in our C2 interaction space S_I) and may include a set of parameters featuring this interaction. The message arriving at an object invokes an object method. The outcome of the invoked method can induce changes in the attributes of the object.

The methods associated with a group object and its interactions consist of the basic four SCG actions of sensing and identification, infliction and attrition, transportation and maneuver, and communications and dissemination. For each interaction and the involved methods, the activation, termination, and allocation of the resources, as well as the doctrine and operational procedures employed, are selected in accordance with the command policy dictated by the local command associated with the underlying group object.

Identification and sensing methods involve mathematical models and algorithms that incorporate the types and parameters of the sensors (of the sensing group) and targets (which may be part of a sensed adversary group object) for the interacting groups. The related object attributes characterize the features of the sensors for the sensing group and the characteristics of the targets in the target group that are relevant to the process of sensing and identifying this target group. Environmental parameters are also invoked. In this fashion, depending on the types of sensors and targets involved and the associated environmental conditions, messages (carrying the proper parameters issued by the involved group and environment objects) are invoked to activate the sensing method of the object. Basic parameters can include probability of detection (P_d), probability of acquisition (P_a), time to acquisition (T_a) for a single sensor vs a single target, under the specified environmental conditions. These are further used by the method to mathematically compute the statistical performance measures of detection and acquisition for a group-on-group engagement.

Infliction and attrition methods are included to provide for the probabilistic computation of losses related to infliction interactions between groups. The attrition procedure embedded in the group object involves a mathematical model that probabilistically describes the nature and outcome of infliction actions (which can involve sensing, firing, maneuvering and communication events), and then computes the distribution of casualties (as well as means, standard deviations, 95-percentile and other statistical measures of interest), for each resource class of interest. The attrition model and performance evaluation procedure takes into account the attributes of each of the interacting groups as they relate to features such as force level, weapon capabilities, locations, temporal and spatial limits, and environmental effects. Thus, when group A is firing at group B, the A-B infliction interaction is initiated. The latter calculates the parameters needed by the attrition algorithm to statistically compute the outcome of the firing process. Such parameters include one-on-one probability of kill given shot $P_{K|S}$ for the engaged resources, weapon firing characteristics (including firing rates), sensor delays (time to acquisition) and identification features (P_d). The need for the latter can induce the activation of the groups' own identification procedure. Also involved are command policies and doctrines in fire allocation and weapon-to-target allocation (to specify what fraction of the fire is directed at which target groups). An infliction message directed from A to B (or to a number of target groups) carries these parameters and invokes at B the attrition procedure that subsequently evaluates the results of the infliction action, computing for example the distribution of the number of resource losses at B. Following this computation, the attributes of the involved groups are updated. For example, incorporating the distribution of the number of losses incurred at the end of the SCG engagement phase, new distribution of force levels are computed; at the same time, ammunition and fuel supplies are updated.

Note that we employ statistical descriptions for the outcomes of the underlying interactions. The uncertainties of the battle, leading to the distinct (however small) possibility of a successive occurrence of seemingly isolated lower probability events can contribute significantly in affecting the final outcome of the battle. Our applications of these models to army and naval scenarios have well demonstrated this point. Combat planning, resource allocation, and selection of command policies for sensing, infliction, maneuvering and communications cannot be done effectively when based solely on mean value outcomes. Our analyses have shown outcome distributions to be many times multimodal, involving distinct higher level probabilities of the occurrence of a number of states; the commander and the command staff, at any level, must be able to limit the probability of occurrence of all (significant) such states of events that are undesirable, rather then just ensure that an average outcome state is achieved.

Transportation methods relate to the transfer of resources of a group. They involve actions related to change of speed, route selection, reinforcing, and splitting of forces and supply processes. Note that under the SCG model, a group moves from one node to another as its space-time boundaries are crossed (e.g., when one time phase is over and another is to begin, and/or when the key space of operations is changed). Under the OOSCG implementation, the group is always described by its object; however, as it moves in time and space, its corresponding attributes are changed. The time-space movements of the groups are supervised by a scheduling object (and are described by a scheduling graph, see Fig. 1). Such movements can be controlled to occur in a static or dynamic fashion. Under a static movement policy, the movement of a group is prescribed, deterministically or probabilistically, at the start of the modeled combat process (so that the next time-space cell that the surviving group will move into is predetermined, or one out of a number of potential new cells may be occupied by the group in accordance with a specified probability distribution). In turn, under a dynamic maneuvering policy, the group is able to select the next location in its path in accordance with the observed (heretofore) outcomes of the conflict, following a defined maneuvering policy. The mathematical model presented in the next section incorporates such a dynamic maneuvering policy with the associated sensing and attrition procedures.

Communications methods are used to convey the effects of disseminating data using the given communications systems and networks, under combat stress conditions. The communications methods provide for the calculation of the communications measures of performance which involve indices of precision, timeliness (message queueing and transmission delays), throughput, coordination, fusion, and filtering.

The environment object represents the environment in which the groups operate. A geographical grid is used. It involves attributes associated with the topology, terrain, trafficability, and weather, and it provides information about the distribution of obscurants, line-of-sight conditions between specified group locations, and the presence of NBC (nuclear, biological and chemical) contaminations and mine fields. The environment object interacts with the group objects in providing them with parameter information essential to the calculations carried out by their native methods. For example, an attrition method requires environmental information relating to the location of the firing group and the target groups to select the corresponding attrition parameters and to allocate fire among the selected target groups.

A version of such an OOSCG program has recently been implemented by IRI and calibrated against the large finer grade combat simulation model, The General Dynamics Tactical Force-on-force Effectiveness Model (TACFEM), considering an Army based scenario[15]. The calibration process enabled the OOSCG model to acquire interaction parameters that are relevant to the underlying scenario, such as basic (weapon and environment dependent) kill and detection one-on-one parameters. Using such captured parameters, the OOSCG model can then employ its own stochastic based procedures in evaluating the outcome of the battle at the termination of each engagement phase. These results can then be compared with results obtained under repetitive runs of the large simulation model to validate the analytical OOSCG model. Once calibrated and validated, the OOSCG model can be used independently, in a highly time effective fashion, to investigate and compare various resource allocations, determine sensitivities to subsystem parameters, establish specific effectiveness of key resources and subsystems, derive performance utility in new resource introductions, and exhibit the performance features of alternate command policies dictating doctrines and strategies for identification, infliction, maneuvering and communications.

IV. Stochastic C3 Model for Maneuverable Battle Systems

A. Introduction

Stochastic models serve as effective tools for describing dynamic multiforce C3 systems. Incorporated in the description are the key subsystem and command decision elements governing the operation and characterization of a C3 system. The dynamical thread of the stochastic processes involved is structured upon the basic observation-decision-action cycle paradigm governing the evolution of a C3 process. The resources, actions, interactions and command decision policies are those associated with the elements of identification/sensing, infliction/firing, transportation/maneuvering and communications/coordination.

A basic Markovian stochastic C3 model employing canonical C3 system components has been presented by us in Ref. 4. Multiforce stochastic system models have been investigated in Ref. 6. Other related continuous-time and discrete-time, Markov and semi-Markov based models which we have been studying are noted in the references. The model presented in Sect. V illustrates the structure of our multiphase and multistage models.

Each model is characterized in terms of well defined sets of parameters that can be measured and estimated from an existing simulation program, experiment, test-bed, or through the use of system based assumptions and doctrines. Our models then provide, in a time-effective fashion, detailed performance curves in terms of the capability functions of the selected C3 subsystem. These models can thus serve as the corresponding C3 methods embedded within the description of each group object of the OOSCG model, as well as for the modeling and analysis of the outcomes of a conflict in carrying out a stated mission or while evolving over a specified time-space segment (and thus covering a set of nodes involved in the SCG description of the battle).

In this section, we consider a stochastic C3 model for a system that explicitly incorporates dynamic force movements, at decision times or upon the occurrence of force attritions. Force maneuvering policies, command rules of engagement and battle evolution can depend on the complete (known or estimated) current state information (force-size, location, force type, speeds, terrain type) and various system parameters (such as those involving communications and identification capabilities, kill probability functions and capacities, and movement capabilities and constraints) of the friendly and adversary forces. We present here a discrete-time dynamic semi-Markov pure-attrition C3 (DSMPAC) stochastic model for characterizing such a battle, and carry out its performance analysis. An illustrative example for the use of this model is given in Ref. 8. The performance equations have been used there to demonstrate the sensitivity of the system performance to variations in the capabilities of the underlying resource subsystems, as reflected by variations in key C3 system parameter values.

B. C3 System Model

The number of friendly (F) and adversary (G) force units at time t is denoted as X_t^F and X_t^G, respectively. Their relative position (with respect to each other) at time t is described by the variable R_t. Note that these variables can be taken to be vectors to represent each force as composed of multiple components, as described in Ref. 6. For simplicity of presentation of the dynamic model, we assume here the above variables to be scalars.

The stochastic point process $A = \{A_n, n=1,2,3,...\}$ represents the command decision times regarding maneuvering and movement of forces. Thus, the commander carries out the n-th movement decision (and execution, if needed) at time A_n. Note that A_n is a random variable which depends upon the present and past values of the identified state process $Z = \{Z_t = (\underline{X}_t, R_t), t \geq 0\}$, where $\underline{X}_t = \{X_t^F, X_t^G\}$; as such it is called a Markov time or a stopping time. Thus, based on the evolution and progress of the battle, a commander can decide when to move the forces. For presentation simplicity, we assume here that decision times are identical with attrition times of either side, so that A_n also represents the time of the n-th attrition (of either friendly or adversary forces). Attrition times are assumed to be observable by both sides. The granularity of the basic unit that can undergo attrition can be selected in accordance with the view desired by the modeler, as noted for the definition of a group and its basic units for the OOSCG model.

Consider the state variables embedded at times of decision/attrition. We set

$$\underline{Y}_n = \underline{X}_{A_n+} \qquad R_n = R_{A_n-} \qquad (1)$$

to denote the corresponding states of the system, representing the number of force units and their relative location, just following and just before (or during, assuming that any movement will take place following an attrition), respectively, the n-th attrition/decision time. Note that $\underline{Y}_n = \{Y_n^F, Y_n^G\}$.

The dynamic behavior (stochastic evolution) of the battle is defined by the statistical characterization of the underlying C3 state process Z, or alternatively (as followed in this section), of its associated discrete-time embedded C3 process (Y, R, A), where

$$(Y, R, A) = \{(\underline{Y}_n, R_n), A_n, n \geq 1\}$$

We assume here that (Y, R, A) is a Markov-Renewal process. Thus, we have

$$P\{\underline{Y}_{n+1} = \underline{j}, R_{n+1} = r_{n+1}, A_{n+1} \leq t \mid \underline{Y}_n = \underline{i}, R_n = r_n, A_n = s, \underline{Y}_k, R_k, A_k,$$
$$k = 0, 1,..., n-1\}$$
$$= P\{\underline{Y}_{n+1} = \underline{j}, R_{n+1} = r_{n+1}, A_{n+1} \leq t \mid \underline{Y}_n = \underline{i}, R_n = r_n, A_n = s\}$$
$$\triangleq K(\underline{j}, r_{n+1}, t, n+1 \mid \underline{i}, r_n, s) \qquad (2)$$

where $K(\cdot|\cdot)$ denotes the Markov-Renewal kernel characterizing the statistics of the underlying C3 stochastic process.

In accordance with Eq. (2), the C3 process evolves stochastically as follows. Following the n-th attrition, occurring at time A_n, the force levels are $\underline{Y}_n$. At A_{n-} the forces are still located at a relative distance R_n. Then, at A_{n+}, the commanders move their forces so that their new relative location is R_{n+1}. Subsequently the battle resumes (or continues) until the next attrition occurs, at time A_{n+1}, leading to a new force state level $\underline{Y}_{n+1}$; the force commanders then exercise their maneuvering policies leading their forces to relative position R_{n+2}. As stated by Eq. (2), given the states of the system at the n-th attrition, as well as the complete past evolution of the process, the states of the system (force levels and position) at the (n + 1) attrition depend only on the present state (and possibly time), $(\underline{Y}_n = \underline{i}, R_n = r_n, A_n = s, n)$. Similarly, for the time to next attrition, $A_{n+1} - A_n$. The determination of the new location $R_{n+1} = r_{n+1}$ is made in accordance with the maneuvering function

$$r_{n+1} = r(\underline{Y}_n = \underline{i}, R_n = r_n, A_n = s, n) = r(\underline{i}, r_n, s, n) \quad (3)$$

so that force movements can depend on the current force levels, relative location, elapsed battle time, and previous number of attrition/decision events.

Hence, to further characterize the statistical dynamics of this C3 process, it remains to describe the stochastic behavior of the system during an interattrition/decision period (A_n, A_{n+1}). Such a description is presented in the next section.

C. C3 Process: Parameters and Stochastic Analysis

To illustrate the evolution of a C3 process within the interattrition/decision period (A_n, A_{n+1}), we present the following model. Following the n-th attrition event, force movements take place. A movement time parameter Δ_m represents the average time involved in movements. A sophisticated random movement time description is also readily incorporated.

Forces then use their identification (sensing) resources and communications (coordinating) systems and networks to evaluate the force level and location of the adversary forces (as well as of their own forces). For example, assume the following identification/communications model, whereby forces need to detect adversary forces and communicate this information across their communications networks that are subjected to noise and jamming interferences, while they know to sufficient accuracy their own force level and the relative location of their forces with respect to detected adversary forces. We let N_D denote the number of detected adversary targets by the force's sensing assets, during a detection unit period of duration Δ_D. The time constant Δ_D depends on the mechanics and dynamics of the underlying identification resources, communications network delay-throughput and error-rate behavior, force maneuvers, and environmental conditions. When the number of adversary targets (units) is equal to $X = j$, and the relative distance is $R = r$, the probability of detecting k targets is represented by

$$P_D(k|j,r) = P\{N_D = k | X = j, R = r\} \qquad (4)$$

Note that, in general, we could have $k > j$, as might be the case if the adversary is deceptively employing dummy/decoy targets. (However, unless stated otherwise, we henceforth set $P_D(k|j) = 0$ for $k > j$.)

For example, if we assume that detection events of different targets are statistically independent (which would tend to be the case if the targets are randomly uniformly distributed in the engagement area, relative to the distribution of the sensors), and no special consideration for dummy targets is involved, we conclude that N_D is governed by a binomial distribution

$$P_D(k|j,r) = \binom{j}{k} P_D^k [1 - P_D]^{j-k}, \quad k=0,1,2,\ldots,j \qquad (5)$$

where $\binom{j}{k}$ is the Binomial coefficient, and

$P(D) = P_D = $ Detection probability
= Probability that an adversary unit is detected in a detection time slot of duration Δ_D time units.

As another example, consider a situation whereby the j units associated with a target group are tightly co-located, a detection of a single unit belonging to the group would lead to the detection of the whole group. In this case the time to detection of the group could be taken to last for a random number of slots, governed, say, by a geometric distribution, whereby the probability of a group detection at the end of a slot is set equal to $1 - (1 - P_D)^j$.

In general, we permit the detection probability to depend on the complete current state $\{\underline{Y}_n, R_n, A_n, n\}$, so that $P_D = P_D(\underline{Y}_n, R_n, A_n, n)$ for detection operations taking place during the period (A_n, A_{n+1}). Note that P_D directly depends on the characteristics of the targets and sensors, environmental conditions, and frequently on the quality of the associated communications systems and the employed signal processing, detection, and networking algorithms. Degraded operation of the sensors and communications systems, target obscurants, and adverse visibility and weather conditions lead to lower P_D levels.

When detection of G (adversary) forces is carried-out by F (friendly) forces, we use the corresponding parameters P_D^F, N_D^G. Similarly, when detection of F forces by G forces is considered, the corresponding variables are P_D^G, N_D^F.

Thus, for example, for detection by the F force of G force units, we have

$$P_D^F(k|j,r) = P\{N_D^G = k | X^G = j, R = r\}$$
$$= \binom{j}{k} [P_D^F]^k [1 - P_D^F]^{j-k}, \quad k = 0,1,2,\dots,j \quad (6)$$

Note that in Eqs. (5)-(6) we further condition on the full or partial set of the underlying state and time ($\underline{Y}_n$, R_n, A_n, n), in accordance with the dependencies of P_D.

We define an illustrative attrition model as follows. Following detection of at least a single adversary unit, and after an additional start-shooting delay Δ_{SS} which accounts for the time it takes for the forces to prepare their weapon systems and supply for shooting, an engagement battle starts. It is assumed that then every Δ_{IS} units of time a single shot (of any desired magnitude, and thus possibly representing the cumulative effect of batches of shots or missile launches) is taken by each side, until an attrition occurs.

For a single shot aimed at a single target, considering a given firing weapon/sensor system and a target type, the probability of target hit (which is also set here to be equal to the probability of kill) is denoted as $P_h = P_h(\underline{Y}_n, R_n, A_n, n)$, when the underlying system state is ($\underline{Y}_n$, R_n, A_n, n), for shooting (countering) operations taking place during the period (A_n, A_{n+1}). We also write $P_h = P_h(r)$ to indicate the explicit dependence on the relative distance $R_n = r$. We let $P_h^F(r)$ and $P_h^G(r)$ denote the corresponding hit probabilities when the shots are taken by the friendly and adversary forces, respectively. Note that hit probabilities are typically dependent on the quality of the underlying sensor data precision and thus on the communications/radar environment and the associated jamming interference conditions.

The probability that the F force detects k adversary targets, out of the existing j targets, given that at least a single target is detected, is given from Eq. (6) by

$$P_{DC}^F(k|j) = \binom{j}{k}(P_D^F)^k[1 - P_D^F]^{j-k}[1 - (1 - P_D^F)^j]^{-1},$$
$$1 \leq k \leq j \quad (7)$$

Similarly, we express the conditional distribution of the number of detected targets by the G forces, $p_{DC}^G(k|i)$. We further condition in Eq. (7) on additional current state ($\underline{Y}_n$, R_n, A_n, n) information, in accordance with PD dependencies.

One needs then to specify the fire allocation discipline, so that the weapon-target allocation and the fire power assigned to each such weapon-target association (in time and in resources) is defined. For example, assume a target selection discipline (TSD) which associates with each force unit (each unit considered to have a single weapon system) at most a single target (also being identified here as a single force unit), on a full-time dedicated basis. Assume a dual shooting scenario, so that each side is shooting at its adversary. Then, given that k_F and k_G targets have been detected by the G and F forces, respectively, and that there are presently i friendly forces and j adversary forces ($X_n^F = i$, $X_n^G = j$), the friendly forces engage $\min(i,k_G) \triangleq i \cap k_G$ targets by employing $i \cap k_G$ force units, while at the same time the adversary forces engage $\min(j,k_F) \triangleq j \cap k_F$ targets by employing $j \cap k_F$ force units. The corresponding probabilities that a single infliction action results with at least a single kill (during a firer time slot Δ_{IS}), are given by the following conditional kill probabilities:

$$p_K^F(i,j,r,k_G) = 1 - [1 - P_h^F]^{i \cap k_G}, \quad 1 \leq k_G \leq j \quad (8)$$

$$p_K^G(i,j,r,k_F) = 1 - [1 - P_h^G]^{j \cap k_F}, \quad 1 \leq k_F \leq i \quad (9)$$

Note that these probabilities can also depend on the full state ($\underline{Y}_n$, R_n, A_n, n) through the dependence of P_h on the latter information; this dependence is however not shown here explicitly to simplify notations, but is incorporated in the performance analysis. This holds as well for other notations used below.

Given states (i,j), detected target numbers k_G and k_F, and given that an attrition has occurred, the conditional probabilities that the attrition(s) is(are) of an F target(s), G target(s) or both G and F targets are denoted as $\delta_{ij}^F(r,k_F,k_G)$, $\delta_{ij}^G(r,k_F,k_G)$ and $\delta_{ij}^{FG}(r,k_F,k_G)$. These probabilities are computed to be given by the following equations:

$$\delta_{ij}^F(r,k_F,k_G) = p_K^G(i,j,r,k_F)[1 - p_K^F(i,j,r,k_G)]$$
$$/\{1 - [1 - p_K^F(i,j,r,k_G)][1 - p_K^G(i,j,r,k_F)]\} \quad (10)$$

$$\delta_{ij}^G(r,k_F,k_G) = p_K^F(i,j,r,k_G)[1 - p_K^G(i,j,r,k_F)]$$
$$/\{1 - [1 - p_K^F(i,j,r,k_G)][1 - p_K^G(i,j,r,k_F)]\} \quad (11)$$

$$\delta_{i,j}^{FG}(r,k_F,k_G) = p_K^G(i,j,r,k_F)p_K^F(i,j,r,k_G)$$
$$/\{1 - [1 - p_K^F(i,j,r,k_G)][1 - p_K^G(i,j,r,k_F)]\} \quad (12)$$

Upon the occurrence of an attrition, a random number of force units may be killed. The distribution of the attrition batches is characterized by the probabilities $u^F(n|i,j,r,k_F)$ and $u^G(n|i,j,r,k_G)$ denoting the probability that the attrition is of n force units, given at least a single force unit is hit, for F and G forces, respectively. The corresponding attrition batch distributions, given (i,j,r), k_F, k_G, are then given by

$$u^F(n|i,j,r,k_F) = \left\{\binom{j\cap k_F}{n}(P_h^G)^n[1 - P_h^G]^{j\cap k_F - n}\right\}$$
$$\times \{[1 - (1 - P_h^G)^{j\cap k_F}]\}^{-1}, \quad j\cap k_F \geq n \geq 1 \quad (13)$$

$$u^G(n|i,j,r,k_G) = \left\{\binom{i\cap k_G}{n}(P_h^F)^n[1 - P_h^F]^{i\cap k_G - n}\right\}$$
$$\times \{[1-(1- P_h^F)^{i\cap k_G}]\}^{-1}, \quad i\cap k_G \geq n \geq 1 \quad (14)$$

D. Effectiveness Analysis

We consider the following key performance measures [as measures of effectiveness (MOEs)] for a C3 system:

1. Probability of win, P_w, by the friendly (F) forces.
2. Mean time to termination T_T or average battle duration, representing the average time to termination of the underlying confrontation.
3. Average loss to win L_W representing the average number of losses to be incurred by the F forces, given that the F forces win.

A number of cost and system state functions can be similarly calculated (See Sect. V for the calculation of state probability distributions at the termination of each battle phase and stage.) One can also examine effectiveness indices that integrate these three measures to yield additional measures of effectiveness, as required by the mission. For this purpose, one notes that it is desirable to develop the battle plans and scenarios and allocate properly the available C3 resources, so that the highest probability of win is attained, at a minimal cost of unit losses (i.e., high survivability), while the battle duration is kept within acceptable limits.

To calculate the above effectiveness measures, we need to prescribe the initial and final states for the underlying C3 process (Y,A). For example, in considering a battle where the fighting effectiveness and win/loss factors are determined by the underlying force level (Y_n^F, Y_n^G), we set the initial states for the force levels to be $[Y^F(0), Y^G(0)]$, and the final force level states to be $[Y^F(F), Y^G(F)]$. We can then assume that when the size of the force reaches its prescribed final state, it cannot operate effectively any more, and it is said to have lost the confrontation; its adversary is then said to have won the confrontation, provided its force size is then above its own final level (and provided it has not previously been reduced to a size which is below its own final value). In the multi-force (vector based) case, win-termination and loss-termination sets (rather than scalars) are defined (in accordance with the aim of the underlying mission) to characterize win and loss conditions. When modeling a single engagement, such boundary conditions can be set to designate the limits for the termination of this engagement, at which time the commander can decide on the next step.

We set $P_W(i,j,r)$ to denote the probability of win, given the initial states of force levels and distance are (i,j,r). Similarly, $T_T(i,j,r)$ and $L_W(i,j,r)$ are defined.

The following recurrence equations are then derived for the calculation of the win probability $P_W(i,j,r)$.

$$P_W(i,j,r) = \sum_{k_F=1}^{i} P_D^G(k_F|i,\bar{r}) \sum_{k_G=1}^{j} P_D^F(k_G|j,\bar{r})$$

$$\cdot \{\delta_{ij}^F(\bar{r}, k_F, k_G) \sum_{n=1}^{j \cap k_F} u^F(n|i,j,\bar{r},k_F) \cdot P_W(i-n,j,\bar{r})$$

$$+ \delta_{ij}^G(\bar{r}, k_F, k_G) \sum_{n=1}^{i \cap k_G} u^G(n|i,j,\bar{r},k_G) \cdot P_W(i,j-n,\bar{r})$$

$$+ \delta_{ij}^{FG}(\bar{r}, k_F, k_G) \sum_{n_F=1}^{j \cap k_F} \sum_{n_G=1}^{i \cap k_G} u^F(n_F|i,j,\bar{r},k_F) u^G(n_G|i,j,\bar{r},k_G)$$

$$\cdot P_W(i-n_F, j-n_G, \bar{r}) \} \tag{15}$$

for (i,j) values above the loss/win boundary, i.e., for $i > Y^F(F)$, $j > Y^G(F)$, where

$$\bar{r} = r(i,j,r) \tag{16}$$

describes the underlying maneuvering function, expressing the new relative position of the forces after movement, when the states prior to this movement are (i,j,r).

The boundary conditions are presented as

$$P_W(i,j,r) = 0, \quad \text{for } i \leq Y^F(F) \tag{17}$$

$$P_\omega(i,j,r) = 1, \quad \text{for } i > Y^F(F), \ j \leq Y^G(F) \quad (18)$$

Similarly, recursive equations are obtained for $T_T(i,j)$ and $L_\omega(i,j)$.

These equations have been solved to yield effective performance results characterizing the behavior of various C3 systems.[5,8] Such equations, as well as other analytical models based on similar approaches, have also been embedded into the SCG and OOSCG models described above. Note that the models allow the examination of the effectiveness of a multitude of dynamic maneuvering policies, as well as assessing their dependence on the system detection, communications, transportation, and weapon resource capabilities (through the one-on-one P_h and P_d parameters, environmental conditions, and the associated sensing and fire allocation command strategies).

V. Stochastic Multistage and Multiphase Battle C2 Model

A. Introduction

In this section, we present a stochastic C2 model for a multistage and multiphase C2 battle system. Under our multiphase battle model, the battle is described to evolve as the serial succession of battle stages. During a stage, the forces engage in a certain type of combat. Following the termination of a stage, a new battle stage is initiated, so that the currently available forces now engage in a different type of combat. The state of the system at the termination of a battle stage provides a key input for the determination of the related state of the system at the start of the next stage.

Each battle stage is, in turn, divided into a serial succession of combat phases. During a phase, a combat takes place in accordance with the underlying system state conditions. The state of the system at the termination of a combat phase provides a key input for the determination of the related state of the system at the start of the next combat phase, within the same battle stage. The combat phases involve similar combat types conducted under varying system and force state conditions.

Performance equations are derived to describe the state distribution at the termination of each combat phase and battle stage, as well as to compute mission win probabilities, asset losses and battle durations.

In this section, we apply our multistage/multiphase stochastic C2 models to describe and analyze the performance of naval outer-air/inner-air battle stages. A battle group is attacked by enemy bombers, within a specified sector. The sector is further divided into multiple concentric zones (levels). Zone 6 is furthest away from the battle group, while zone 1 is closest to the battle group.

The battle is divided into two stages. During the first stage (Stage 1), representing the outer-air combat, the penetrating enemy bombers are engaged by friendly fighter planes on patrol, aided by radar and other electronics sensing assets supplying information concerning the size and location of enemy bombers. Phase 6 of the outer-air battle stage takes place during the engagement occurring within Zone 6. The surviving enemy bombers would then continue towards Zone 5, whereby they will be engaged by the friendly fighters delegated to this zone. The ensuing engagement then constitutes Phase 5 of the outer-air combat. At the termination of the outer-air phase-2 engagement, the surviving (if any) enemy bombers will initiate stage-2 which is the inner-air combat. During the latter, bombers continue to penetrate for a short time while launching air-to-surface missiles. The latter are engaged by counter fire and surface-to-air missile response. Any enemy missiles surviving the inner-air combat can cause damage to battle group assets.

We develop a stochastic mathematical model to describe the states and dynamics of these battles. Parameters are selected to describe the probabilistic performance of the units, their firing and sensing probabilistic capabilities and their maneuvering policies, fighter distribution strategies and target allocation courses of action. Analytic performance equations are then written, providing for the computation of the battle's measures of effectiveness. We calculate the probability distributions of the states (such as numbers of surviving enemy bombers and friendly extent of damage) at the termination of each phase and stage. These results allow the calculations of the probability distribution of battle outcome or mission success under any underlying set of system parameters, friendly force allocations, maneuvering strategies and courses of action.

In this section, we present a brief outline of the model, of its calibration through the use of a Navy large simulation program [Naval Ocean Systems Center (NOSC) Battle Group Tactical Trainer (BGTT) Research Evaluation and System Analysis (RESA) facility] and comparisons between the performance results exhibited by our calibrated models and this large simulation program. More detailed presentations are given in Refs. 9 and 11. The same model structure has been used in Ref. 3 in the modeling and analysis of a two-stage multiphase Army scenario. Under the latter, during stage 1, friendly forces advance over enemy terrain toward their final destination; they are occasionally engaged by enemy forces; however, they do not actively initiate firing actions during stage 1. Different phases during stage 1 correspond to traversal of different terrain and enemy infliction conditions. When the friendly forces arrive at a certain distance from the eventual target (or at a specified, or dynamically determined space-time boundary), stage 2 of the battle is initiated. During this stage, the friendly forces actively fire at the target, inducing enemy resistance. The model is shown to allow the computation of the distribution of the force level states at any phase and stage. It further illustrates the computation of the optimum boundary at which stage 2 should be initiated to yield maximum target damage at acceptable battle duration and casualty levels. Note that a too early transition to stage 2 will result with savings of stage-1 casualties but may position the friendly forces too far a distance from the enemy target. Under a dynamic decision policy, the transition into stage 2 is considered at the end of each phase, and is decided based on the observed outcomes of the heretofore engagements.

The model presented here can further be extended for multibattle, multistage and multiphase descriptions, as illustrated in Refs. 9 and 10. Under such a model, two battle processes are modeled, each in terms of a multistage multiphase model, as described above. At the end of each stage, for each battle, the two battle systems can interact through the interchange of status information, development of combined strategy and resource allocation command decisions, re-allocation of (sensor, weapon, transportation or communications) resources (reinforcing, combining and splitting). The next stage then proceeds under the new allocations.

The multiphase and multistage models presented here serve well as the procedures/methods embedded within the group object of the OOSCG model. They also provide an analytical methodology for the evaluation of the outcome of the conflict over a series or series/parallel topologies of time-space cells (and SCG nodes) for a single mission/conflict or for interrelated missions and conflicts.

B. Scenario and System Topology

The system topology and the underlying scenario are described as follows. The space is divided into sectors (Fig. 3). At the center are positioned the friendly carriers and ships. It is assumed here that enemy bombers attack through sector 1. During the outer-air combat (stage 1) enemy bombers penetrate the air space. The carrier group (including support ships, reconnaissance aircraft and prepositioned fighter planes) receives early warning of the approaching bombers. During stage 1, the friendly fighters attack the enemy bombers. Stage 1 ends when the surviving bombers have penetrated far enough to launch their air-to-surface missiles against the carrier group members. During the inner-air combat (stage 2), the ships in the carrier group launch surface-to-air anti-missiles against the incoming missiles. Any air-to-surface missiles that survive stage 2 either miss completely, strike the support ships or strike the carrier. A missile that hits a target can (stochastically) cause one out of a number of specified damage types (such as damage to the hull, to aircraft, to supplies, or to radar and communications equipment).

The space around the ships and carriers is divided into six concentric spheres. The outmost level, level 6, is at the longest range at which the underlying battle is said to start. The levels extend to specified ranges. Stage 1 spans phases 6-2, with the corresponding engagements occurring in layers 6-2, respectively (where layer i extends between level i and level i - 1). During phase 6 of the battle, the bombers move from level 6 to level 5, and fighters which are positioned within layer 6 are able to attack these bombers. During phase 5, the surviving bombers penetrate through layer 5. They are attacked by fighters which are now positioned within layer 5 (or neighboring layers, if ranges permit). Because of the underling distances and speeds involved, fighters are constrained in terms of their extent of movement between phases. For example, the layers have been selected so that during a single phase a fighter can move from a position within a layer to only a neighboring layer. As a result, the initial positioning of the fighters (as impacted by other constraints, such as fueling and launch rates) is of key importance.

Stage 1 proceeds similarly through phases 6-2, across corresponding layers 6-2. At the termination of phase 2, stage 2 is initiated. The space cell covering the stage-2 period is layer 1, which is further divided into sublayers 1.2, 1.1 and 1.0. Layer 1.0 designates the surface of the carriers and ships.

The division of each stage into phases allows the model equations to incorporate the movement characteristics of fighters, bombers, and missiles and to account for variations in the C2 system parameters (such as detection and kill probabilities) as a function of state level and time and space positions. It also provides a time-space framework for segmented description of the various rules of engagement, fire allocation, and resource allocation command policies.

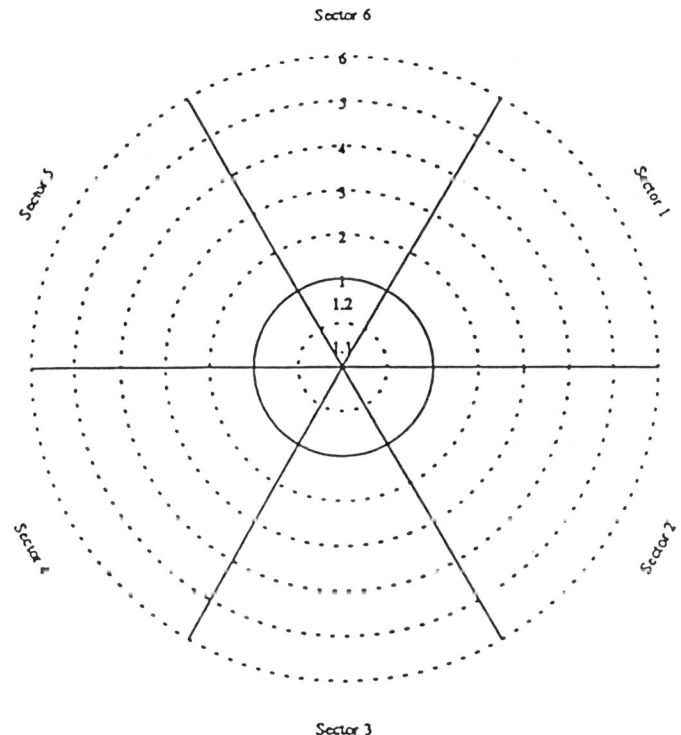

Fig. 3 System topology for the outer-air/inner-air battle model

C. Modeling and Performance Analysis

1. Approach

The performance of the C2 system is described through the solution of analytic recurrence equations. Initial values for the model parameters and the initial placement of fighter aircraft are specified. Rules for launching and moving fighters during the course of the battle are defined. We then write equations which express the probability distribution of the number of bombers which survive each phase of stage 1 in terms of the number that have survived the previous phase. We numerically solve these recursive equations for the distribution of the number of bombers that survive stage 1. We then proceed in a similar fashion to determine the distribution of the number of air-to-surface missiles that survive stage 2, and the distribution of the destruction which they cause.

Performance measures computed thus include the following:

a) The distribution of the number of surviving bombers at the termination of each phase of stage 1.

b) The distribution of the number of surviving missiles, at the termination of stage 2.

c) The distribution of the damage caused to ships and carriers.

2. Stage-1 Variables

The stage-1 variables include the following. We use the index ℓ to denote phase ℓ of the combat and index i to designate the layer i space cell. Adversary bombers are assumed to move together as a group. In phase ℓ they are at layer $i = \ell$.

a_F	= fraction of fighters at layer ℓ which pursue bombers to level $\ell - 1$
$b(\ell)$	= number of bombers in layer ℓ at the start of phase ℓ
$b_D(\ell)$	= corresponding number of detected bombers
b_F	= fraction of fighters at layer $\ell - 1$ which participate in the phase ℓ battle
$b_K(\ell)$	= number of bombers killed during phase ℓ
$b_R(\ell)$	= number of bombers surviving phase $\ell = b(\ell) - b_K(\ell)$; $b(\ell) = b_R(\ell+1)$, for $\ell=2,3,4,5$
$f_\ell(i)$	= numberof fighters in layer i at start of phase ℓ
$f_P(\ell)$	= number of fighters at layer ℓ which pursue bombers and join level $\ell - 1$ engagement during phase $\ell - 1$
$f_{s\ell}(i)$	= number of fighters at layer i which shoot at bombers in level ℓ during phase ℓ, $i = \ell - 1, \ell$
N_{mf}	= number of missiles per fighter

$N_{Ns\ell}(i_1,i_2)$ = number of new fighters moving at the start of phase ℓ from layer i_1 to layer i_2

$P_{DB\ell}$ = probability of detection of a bomber, during phase ℓ battle

$P_{hf\ell}$ = probability of a single fighter missile hitting a bomber, for phase ℓ battle

3. Stage-1 Command Policies

The following friendly command policies require specification.

a) <u>Fighter distribution policy</u>. Specifies the allocation of fighter planes across levels initially, and at the start of each phase with the following constraints applied: 1) at most N_{NP} new planes can be sent from the carriers during a single phase and 2) only fighter planes from neighboring layers can be allocated to a cell during a single phase.

b) <u>Fighter missile and target selection policies</u> 1) The missile allocation policy determines the number of missiles shot during phase ℓ, denoted as $f_{ms}(\ell)$, in terms of the number of detected bombers, $b_D(\ell)$, and the number of available missiles, $f_m(\ell)$. 2) The target selection discipline (TSD) is the strategy for allocating the selected fighter missiles to the bomber targets.

4. Stage-1 Initial Conditions, Parameters, and Maneuvering Functions

The following initial conditions are specified:

$b_0 = b(6)$ = initial number of bombers

f_0 = total initial number of fighters available for participation in the battle

$\{f_6(i), i = 1.0, 1.1, 1.2, 2, 3, 4, 5, 6\}$ = Initial distribution of the fighter planes

The stage-1 maneuvering function describes the numbers of fighter planes moved at each phase between levels: $\{f_{N\ell}(i), i=1.1, \ldots, \ell, \ell=2,3,\ldots,5\}$, where $f_{N\ell}(i)$ designates the number of fighters moved from level i-1 to level i at the start of phase ℓ. System parameters include $P_{hf\ell}$, $P_{DB\ell}$, a_F and b_F, as defined above.

5. Performance Equations

5.a. <u>Detection</u>: The number of bombers detected during a phase depends on many system conditions and parameters, including the number of bombers and fighters involved, distances, types and locations of sensors used, and correlations with previous recorded information. For example, assuming statistical independence between detections of individual bombers and taking the probability of detection of a single bomber during phase ℓ to be $P_{DB\ell}$, we express the probability that j bombers are detected in phase ℓ, given that k bombers have survived to the start of phase ℓ, as

$$P_{D\ell}(j|k) = \binom{k}{j} P_{DB\ell}^{j}(1 - P_{DB\ell})^{k-j}, \quad j=0,1,\ldots k$$

Other relations can also be invoked.

5.b. <u>Missile distribution and allocation:</u> The following is a simplified model as it relates to the missile distribution and allocation. Each fighter participating in the battle during a phase is assigned to a detected bomber. The fighter then pursues the bomber and fires at a given rate at the bomber until the bomber is killed, the phase ends or the fighter exhausts its supply of missiles. A fighter which kills its target is reassigned to another bomber.

For illustrative purposes, assume that the layers are selected so that a fighter can fire at a bomber a maximum of four shots during a phase. Assume further that a fighter can have a number of missiles which is less than or equal to 8. Thus, a fighter can participate in at most two phases of the battle. Fighters with full load of missiles are called "new," whereas fighters which have already fired four missiles are called "old." We track the numbers of old and new fighters at each level, during each phase, in accordance with the following equations.

The number of new fighters in layer i during phase ℓ are given by

$$f_\ell(i) = f_{\ell+1}(i) + f_{N\ell}(i)I(i > 1.0) - f_{N\ell}(i+1) - f_{s,\ell+1}(i)I(i = \ell),$$
$$\ell = 6,5,\ldots 2; \quad i = 1.0, 1.1, 1.2, 2, \ldots, \ell$$

where I(A) is the indicator function of event A, I(A) = 1 if A holds and I(A) = 0, otherwise. In the expression above, the first term accounts for the fighters present at the start of the previous phase in layer i; the second term accounts for fighters coming up from below, whereas the third term represents fighters moving up to the next level. The fourth term describes the number of fighters which fired up at bombers in the previous phase; it is expressed as

$$f_{s,\ell+1}(\ell) = \lfloor a_\ell f_{\ell+1}(\ell) \rfloor$$

where $\lfloor x \rfloor$ designates the largest integer lower than x.

The number of old fighters participating in the battle during phase ℓ is given as

$$g_\ell = \lfloor a_\ell f_{\ell+1}(\ell+1) \rfloor + f_{s,\ell+1}(\ell)$$

The number of missiles fired at bombers during the phase ℓ battle is then given by

$$N_{s\ell} = 4[f_\ell(\ell) + f_{s,\ell}(\ell-1)] + g_\ell(N_{mf}-4)$$

5.c. One-step transition probabilities: We compute the distribution of the number of bombers that survive each phase of the battle. The initial conditions stated above specify the number of bombers present at the start of phase 6. In the following, we present a recursive equation which expresses the distribution of the number of bombers at the start of a phase in terms of the distribution of the number of bombers present at the start of the previous phase.

$$P_{B,\ell-1}(k) = \sum_{i=k}^{b_0} P(\text{phase starts with i bombers})$$

$$\sum_{j=i-k}^{i} P(\text{detect j out of i bombers}) P(\text{kill i - k out of j bombers with } N_{s\ell} \text{ shots})$$

$$= \sum_{i=k}^{b_0} P_{B\ell}(i) \left[\sum_{j=i-k+1}^{i} \binom{i}{j} P_{DB\ell}^{\,j}(1 - P_{DB\ell})^{i-j} \binom{N_{s\ell}}{i-k} P_{hf\ell}^{\,i-k}(1 - P_{hf\ell})^{N_{s\ell}-i+k} \right.$$

$$\left. + \binom{i}{i-k} P_{DB\ell}^{\,i-k}(1 - P_{DB\ell})^k \sum_{m=i-k}^{N_{s\ell}} \binom{N_{s\ell}}{m} P_{hf\ell}^{\,m}(1 - P_{hf\ell})^{N_{s\ell}-m} \right]$$

for $k=0,\ldots,b_0$.

Using this recursive set of equations, we calculate the distribution of the number of surviving bombers at the termination of each phase. The corresponding distribution for the number of surviving bombers at the end of phase 2 is used to initiate stage 2 of the battle. The latter number is used to calculate the distribution of the number of air-to-surface missiles which are launched by the surviving bombers against the ships and carrier. A corresponding set of recursive equations (see Ref. 11 for details) is then written to provide for the probabilistic performance outcomes of the missile against missile inner-air combat, leading to the calculation of the number of surviving air-to-surface missiles at the end of phase 1.1. This number is then used to calculate the distribution of the damage incurred to the ships and carrier by any surviving missiles which may hit their target (in accordance with properly specified probabilistic parameters).

D. Calibration and Comparison with a Large Simulation Model

Extensive tests have been carried out by us at the U.S. Navy NRaD, San Diego, using the Navy's large-scale simulation and training program (BGTT-RESA) which describes the detailed operation of such outer-air/inner-air battles. This program was used to calibrate our mathematical model by capturing and computing the required system parameters, through the repeated running of sample scenarios under certain system conditions. Following the calibration phase, we used our program to analyze various other system scenarios. We then carried out comparisons between the outcomes of combats, at the termination of each phase, as exhibited by repeated Monte-Carlo simulations using the large-scale simulation program and as predicted by our analytical model. The latter results have been shown to provide precise system behavior and performance descriptions while demanding significantly shorter run times. The analytical model has then been employed to provide timely performance characterizations of battles subjected to various other patterns of force sizes, asset parameters, and system operational conditions.

For a particular scenario, 20 runs of the BGTT program were carried out. These runs were used to estimate the parameters required by our analytical model, such as the per-phase detection and kill probabilities. Under this scenario, 35 bombers were opposed by seven fighters. When the bombers enter level 6, three fighters are positioned in level 4, two fighters are positioned in level 3, and two fighters are positioned at level 2. Some of the bombers are detected during phase 6, but since no fighters are within range, no engagement takes place. In later phases, virtually all of the bombers are detected. Under this scenario, $b_F = 0$ so that no fighters fire missiles at bombers resident in a level above them, and $a_F = 1$, so that all new fighters pursue bombers into the next phase until their supply of missiles is exhausted.

Using the results collected from these runs, the model parameters were obtained. For example, the following values were obtained for this scenario. The probability of a single missile hitting its target was $P_{hf\ell} = 0.4845$, for each phase of stage-1. The probability that an adversary missile is detected during stage-2 was estimated to be 0.9947, whereas 42.4 surface-to-air missiles were determined to be launched during stage-2. The probability that a surface-to-air missile hits and kills an air-to-surface missile was estimated to be 0.3632.

A number of cases were examined as to the fighter movements and fighter firing schedules. All cases were used in calibrating the model through the proper derivation of the parameter estimates. Using these parameters, we have then used the analytical model to provide for the performance of the battle under various system conditions. The large simulation model was then run under these new system conditions, and the results obtained were compared with those exhibited by the much faster analytical program.

In Fig. 4, our model results are exhibited in showing the distributions of the number of bombers which survive stage-1, when the

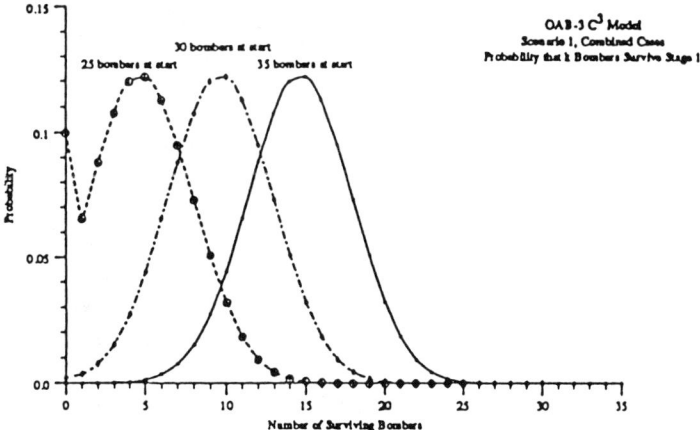

Fig. 4 Probability distribution of the number of surviving bombers with the initial number of bombers as a parameter

bomber force starts with 25, 30, or 35 bombers. Figure 5 shows the distributions of the number of bombers surviving each phase of stage 1, in accordance with the analytical model.

In Figs. 6 and 7, we compare the calculated distribution using our model with the corresponding experimental distributions exhibited by the BGTT-RESA runs, for the numbers of bombers surviving phases 2 and 5 of stage 1. These results (and many other not shown here, as well as χ^2 tests conducted) indicate that the analytical multistage (combat) multiphase (engagement) model used provides a faithful tracking of the course of the battle and its probabilistic outcomes.

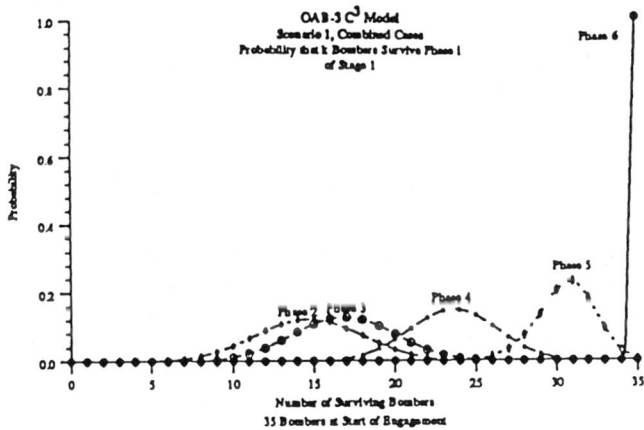

Fig. 5 Probability distribution of the number of surviving bombers for each phase with 35 initial bombers

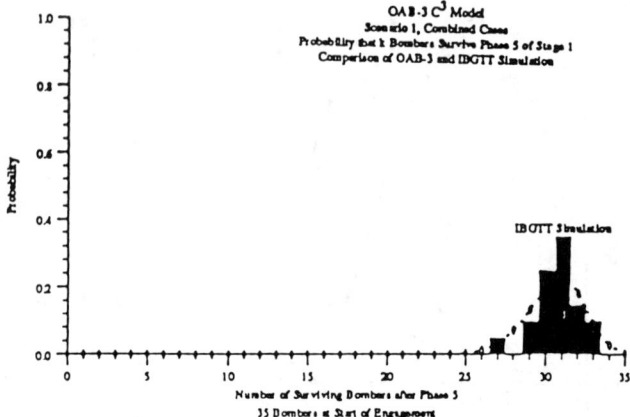

Fig. 6 Probability distributions of the number of surviving bombers after phase 5 with 35 initial bombers, as obtained under the analytical program and under the BGTT simulations

It is noted that the state distributions obtained using the stochastic model provide information of key importance to the commander, and staff, in comparison with results provided by mean value analysis which exhibit just the average outcome of each phase of combat and stage of battle. For example, note from Fig. 4 that if an attacker fails to use a sufficiently large number of attack weapon systems (25 bombers, in this case) not only that its average number of surviving units (bombers) is going to be much lower, but also there is a distinct high probability that it will lose all of its attacking units. The latter probability is reduced considerably by increasing the number of attacking units (in this example, in increasing the number of bombers from 25 to just 30). This illustrates the multi-modal nature of state distributions characteristic of many multistage, multiphase battle system outcomes due to the wide uncertainties embedded in the modeling, operation, and realization of any combat system and its C2 functions.

VI. Resource Allocation

Once a SCG model is constructed for an underlying battle system and its associated C2 scenarios and missions (or sub-missions), the performance of the involved combat systems and the corresponding statistical outcomes of the battles can be calculated. The C^2 system architect and commanders (and their staff) must then examine and trade the allocation and sharing of their available resources for C2 among the various system missions to be carried out. Many such models have been investigated for various combat systems. An example of such a generic

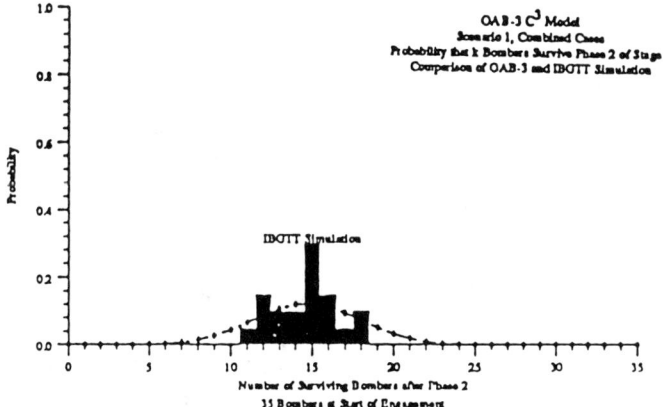

Fig. 7 Probability distributions of the number of surviving bombers after phase 2 with 35 initial bombers, as obtained under the analytical program and under the BGTT simulations

model is provided by the multimission area C2 (MMAC) resource allocation model presented in Ref. 14. The latter model provides for the evaluation of the throughput performance of various categories of resource allocation policies. The resources are distributed over distinct geographical, functional and organizational areas. Resources are divided into classes and types. Resource classes involve identification, infliction, transportation and communications resources. Each one of these four classes is further divided into resource types, to account for assets of different characteristics, capabilities, capacities and possible association and attachments.

Each policy dictates a method for sharing distributed resources among randomly occurring missions; each mission lasts for a random duration and requires predefined set of resources of various classes and types within the underlying areas. As a result of the limitation in system resources, only certain missions (or sub-missions) can be accommodated (to their required resource levels). A model, such as MMAC, allows us to evaluate the probabilistic extent to which missions are supported when using various resource allocation policies. An object-oriented program implementation of the MMAC tool is presented in Ref. 16.

The improvement in C2 combat system performance attained through the proper sharing of key system resources among multiple missions can then be assessed. A model for the analysis, planning, design, and performance evaluation of C3 systems, which integrates combat analysis and resource allocation analytical tools is presented in Ref. 17. It has been implemented as an integrated combat analysis, resource allocation and planning (ICAP) object-oriented program. Considering company level ARMY scenarios, under which fire support resources are shared among multiple battles, we demonstrate in Ref. 17 how this program can be used to identify the most effective integrated combat and resource allocation policies and schemes.

References

[1] Joint Directors of Laboratories,Technical C3 Panel, Basic Research Group, "The C2RM" *Coordination Draft #6*, May 11, 1992.

[2] Mayk, I., and Rubin, I., "The C3 Reference Model (C3RM), Recent Developments," *Proceedings of the Symposium on Command and Control Research*, Naval Postgraduate School, Monterey, CA, June 1988.

[3] Mayk, I., and Rubin, I., "Paradigms for Understanding C3, Anyone?" *Science of Command and Control: Coping with Uncertainty*, edited by S.E. Johnson and A.H. Levis, National Defense Univ./AFCEA International Press, Washington, D.C., 1988, pp. 48-61.

[4] Rubin, I., and Mayk, I., "Markovian Modeling of Canonical C3 Systems Components," *Proceedings of the 8th MIT/ONR Workshop on C3 Systems*, MIT, Cambridge, MA, June 1985.

[5] Rubin, I., and Mayk, I., "Dynamic Stochastic C3 Models and Their Performance Evaluation," *9th Workshop on C3 Systems*, Naval Postgraduate School, Monterey, CA, June 1986.

[6] Rubin, I., and Mayk, I., "Modeling and Analysis of Markovian Multi-Force C3 Processes," *9th Workshop on C3 Systems*, Naval Postgraduate School, Monterey, CA, June 1986.

[7] Rubin, I., Baker, J., and Mayk, I., "Stochastic C3 Modeling and Analysis for Multi-Phase Battle Systems," *Proceedings of the Symposium on Command and Control Research*, National Defense Univ., Fort McNair, Washington, D.C., June 1987.

[8] Rubin, I., and Mayk, I., "Stochastic C3 Modeling and Analysis for Maneuverable Battle Systems," *Proceedings of IEEE Military Communications Conference (MILCOM'87)*, Washington, D.C., Oct. 1987.

[9] Rubin, I., and Baker, J. E., "Stochastic Modeling and Analysis for Maneuverable Multi-Battle and Multi-Phase C^3 Systems", *IRI Tech. Rept.* IRI-CEC-FR-8702, Oct. 1987.

[10] Rubin, I., Baker, J. E., and Mayk, I., "Stochastic Modeling and Analysis for Multi-Battle and Multi-Stage C3 Systems," *Proceedings of the Symposium on Command and Control Research*, Naval Postgraduate School, Monterey, CA, June 1988.

[11] Rubin, I., Baker, J. E., and Mayk, I., "A Stochastic C3 Model for the Naval Multi-Phase Outer Air and Inner Air Battles," *Proceedings of the Symposium on Command and Control Research*, National Defense Univ., Fort McNair, Washington, D.C., June 1989.

[12] Rubin, I., "A Stochastic C3 Graph (SCG) Model and Architecture," *Tech. Rept.*, IRI Corp., May 1989.

[13] Rubin, I., and Shambayati, S., "An Object Oriented Stochastic C3 Graph (OOSCG) Architecture and Program," *Tech. Rept.*, IRI Corp., April 1990.

[14] Rubin, I., Mayk, I., and Ratkovic, A., "A Stochastic Model for Planning and Performance Evaluation of Multi-Mission Area C2 (MMAC2) Systems," *Proceedings of the Symposium on Command and Control Research*, Naval Postgraduate School, Monterey, CA, June, 1990.

[15] Rubin, I., Shambayati, S., Mayk, I., Aouate, A., and Atkinson, A., "A Stochastic Object Oriented C3 Model and its Calibration for Planning, Control and Performance Evaluation of Combat Systems," *Proceedings of the Symposium on*

Command and Control Research, National Defense Univ., Fort McNair, Washington, D.C., June 1991.

[16] Rubin, I., Ratkovic, A., and Mayk, I., "A Resource Allocation Tool for Multi-Mission Multi-Area C2 Systems," *Proceedings of the Symposium on Command and Control Research*, Naval Postgraduate School, Monterey, CA, June 1992.

[17] Rubin, I., Shambayati, S., and Mayk, I., "Integrated Combat Analysis and Resource Allocation Program and Tool," *Proceedings of the Symposium on Command and Control Research*, National Defense Univ., Fort McNair, Washington, D.C., June 1993.

Author Index

Burton, G. 21
Choisser, R. 39
Dockery, J. T 63
Goodman, I. R. 97
Ingber, L. 117
Jin, V. Y. 151
Jones, C. R. 1, 167
Kleinman, D. L. 21
Levis, A. H. 151, 181, 193
Mayk, I. 221, 239
Monguillet, J.-M. 193
Rubin, I. 239
Shaw, J. 39
Woodcock, A. E. R. 63

PROGRESS IN ASTRONAUTICS AND AERONAUTICS SERIES VOLUMES

*1. **Solid Propellant Rocket Research** (1960)
Martin Summerfield
Princeton University

*2. **Liquid Rockets and Propellants** (1960)
Loren E. Bollinger
Ohio State University
Martin Goldsmith
The Rand Corp.
Alexis W. Lemmon Jr.
Battelle Memorial Institute

*3. **Energy Conversion for Space Power** (1961)
Nathan W. Snyder
Institute for Defense Analyses

*4. **Space Power Systems** (1961)
Nathan W. Snyder
Institute for Defense Analyses

*5. **Electrostatic Propulsion** (1961)
David B. Langmuir
Space Technology Laboratories, Inc.
Ernst Stuhlinger
NASA George C. Marshall Space Flight Center
J.M. Sellen Jr.
Space Technology Laboratories, Inc.

*6. **Detonation and Two-Phase Flow** (1962)
S.S. Penner
California Institute of Technology
F.A. Williams
Harvard University

*Out of print.

*7. **Hypersonic Flow Research** (1962)
Frederick R. Riddell
AVCO Corp.

*8. **Guidance and Control** (1962)
Robert E. Roberson,
Consultant
James S. Farrior
Lockheed Missiles and Space Co.

*9. **Electric Propulsion Development** (1963)
Ernst Stuhlinger
NASA George C. Marshall Space Flight Center

*10. **Technology of Lunar Exploration** (1963)
Clifford I. Cummings
Harold R. Lawrence
Jet Propulsion Laboratory

*11. **Power Systems for Space Flight** (1963)
Morris A. Zipkin
Russell N. Edwards
General Electric Co.

*12. **Ionization in High-Temperature Gases** (1963)
Kurt E. Shuler, Editor
National Bureau of Standards
John B. Fenn,
Associate Editor
Princeton University

*13. **Guidance and Control—II** (1964)
Robert C. Langford
General Precision Inc.
Charles J. Mundo
Institute of Naval Studies

*14. **Celestial Mechanics and Astrodynamics** (1964)
Victor G. Szebehely
Yale University Observatory

*15. **Heterogeneous Combustion** (1964)
Hans G. Wolfhard
Institute for Defense Analyses
Irvin Glassman
Princeton University
Leon Green Jr.
Air Force Systems Command

*16. **Space Power Systems Engineering** (1966)
George C. Szego
Institute for Defense Analyses
J. Edward Taylor
TRW Inc.

*17. **Methods in Astrodynamics and Celestial Mechanics** (1966)
Raynor L. Duncombe
U.S. Naval Observatory
Victor G. Szebehely
Yale University Observatory

*18. **Thermophysics and Temperature Control of Spacecraft and Entry Vehicles** (1966)
Gerhard B. Heller
NASA George C. Marshall Space Flight Center

*19. Communication Satellite Systems Technology (1966)
Richard B. Marsten
Radio Corporation of America

*20. Thermophysics of Spacecraft and Planetary Bodies: Radiation Properties of Solids and the Electromagnetic Radiation Environment in Space (1967)
Gerhard B. Heller
NASA George C. Marshall Space Flight Center

*21. Thermal Design Principles of Spacecraft and Entry Bodies (1969)
Jerry T. Bevans
TRW Systems

*22. Stratospheric Circulation (1969)
Willis L. Webb
Atmospheric Sciences Laboratory, White Sands, and University of Texas at El Paso

*23. Thermophysics: Applications to Thermal Design of Spacecraft (1970)
Jerry T. Bevans
TRW Systems

24. Heat Transfer and Spacecraft Thermal Control (1971)
John W. Lucas
Jet Propulsion Laboratory

25. Communication Satellites for the 70's: Technology (1971)
Nathaniel E. Feldman
The Rand Corp.
Charles M. Kelly
The Aerospace Corp.

26. Communication Satellites for the 70's: Systems (1971)
Nathaniel E. Feldman
The Rand Corp.
Charles M. Kelly
The Aerospace Corp.

27. Thermospheric Circulation (1972)
Willis L. Webb
Atmospheric Sciences Laboratory, White Sands, and University of Texas at El Paso

28. Thermal Characteristics of the Moon (1972)
John W. Lucas
Jet Propulsion Laboratory

*29. Fundamentals of Spacecraft Thermal Design (1972)
John W. Lucas
Jet Propulsion Laboratory

30. Solar Activity Observations and Predictions (1972)
Patrick S. McIntosh
Murray Dryer
Environmental Research Laboratories, National Oceanic and Atmospheric Administration

31. Thermal Control and Radiation (1973)
Chang-Lin Tien
University of California at Berkeley

32. Communications Satellite Systems (1974)
P.L. Bargellini
COMSAT Laboratories

33. Communications Satellite Technology (1974)
P.L. Bargellini
COMSAT Laboratories

*34. Instrumentation for Airbreathing Propulsion (1974)
Allen E. Fuhs
Naval Postgraduate School
Marshall Kingery
Arnold Engineering Development Center

35. Thermophysics and Spacecraft Thermal Control (1974)
Robert G. Hering
University of Iowa

36. Thermal Pollution Analysis (1975)
Joseph A. Schetz
Virginia Polytechnic Institute
ISBN 0-915928-00-0

37. Aeroacoustics: Jet and Combustion Noise; Duct Acoustics (1975)
Henry T. Nagamatsu, Editor
General Electric Research and Development Center
Jack V. O'Keefe, Associate Editor
The Boeing Co.
Ira R. Schwartz, Associate Editor
NASA Ames Research Center
ISBN 0-915928-01-9

38. Aeroacoustics: Fan, STOL, and Boundary Layer Noise; Sonic Boom; Aeroacoustics Instrumentation (1975)
Henry T. Nagamatsu, Editor
General Electric Research and Development Center
Jack V. O'Keefe, Associate Editor
The Boeing Co.
Ira R. Schwartz, Associate Editor
NASA Ames Research Center
ISBN 0-915928-02-7

SERIES LISTING

39. **Heat Transfer with Thermal Control Applications** (1975)
M. Michael Yovanovich
University of Waterloo
ISBN 0-915928-03-5

*40. **Aerodynamics of Base Combustion** (1976)
S.N.B. Murthy, Editor
J.R. Osborn,
Associate Editor
Purdue University
A.W. Barrows
J.R. Ward,
Associate Editors
Ballistics Research Laboratories
ISBN 0-915928-04-3

41. **Communications Satellite Developments: Systems** (1976)
Gilbert E. LaVean
Defense Communications Agency
William G. Schmidt
CML Satellite Corp.
ISBN 0-915928-05-1

42. **Communications Satellite Developments: Technology** (1976)
William G. Schmidt
CML Satellite Corp.
Gilbert E. LaVean
Defense Communications Agency
ISBN 0-915928-06-X

*43. **Aeroacoustics: Jet Noise, Combustion and Core Engine Noise** (1976)
Ira R. Schwartz, Editor
NASA Ames Research Center
Henry T. Nagamatsu,
Associate Editor
General Electric Research and Development Center
Warren C. Strahle,
Associate Editor
Georgia Institute of Technology
ISBN 0-915928-07-8

*44. **Aeroacoustics: Fan Noise and Control; Duct Acoustics; Rotor Noise** (1976)
Ira R. Schwartz, Editor
NASA Ames Research Center
Henry T. Nagamatsu,
Associate Editor
General Electric Research and Development Center
Warren C. Strahle,
Associate Editor
Georgia Institute of Technology
ISBN 0-915928-08-6

*45. **Aeroacoustics: STOL Noise; Airframe and Airfoil Noise** (1976)
Ira R. Schwartz, Editor
NASA Ames Research Center
Henry T. Nagamatsu,
Associate Editor
General Electric Research and Development Center
Warren C. Strahle,
Associate Editor
Georgia Institute of Technology
ISBN 0-915928-09-4

*46. **Aeroacoustics: Acoustic Wave Propagation; Aircraft Noise Prediction; Aeroacoustic Instrumentation** (1976)
Ira R. Schwartz, Editor
NASA Ames Research Center
Henry T. Nagamatsu,
Associate Editor
General Electric Research and Development Center
Warren C. Strahle,
Associate Editor
Georgia Institute of Technology
ISBN 0-913928-10-8

47. **Spacecraft Charging by Magnetospheric Plasmas** (1976)
Alan Rosen
TRW Inc.
ISBN 0-915928-11-6

48. **Scientific Investigations on the Skylab Satellite** (1976)
Marion I. Kent
Ernst Stuhlinger
NASA George C. Marshall Space Flight Center
Shi-Tsan Wu
University of Alabama
ISBN 0-915928-12-4

49. **Radiative Transfer and Thermal Control** (1976)
Allie M. Smith
ARO Inc.
ISBN 0-915928-13-2

50. **Exploration of the Outer Solar System** (1976)
Eugene W. Greenstadt
TRW Inc.
Murray Dryer
National Oceanic and Atmospheric Administration
Devrie S. Intriligator
University of Southern California
ISBN 0-915928-14-0

51. **Rarefied Gas Dynamics, Parts I and II** (two volumes) (1977)
J. Leith Potter
ARO Inc.
ISBN 0-915928-15-9

52. **Materials Sciences in Space with Application to Space Processing** (1977)
Leo Steg
General Electric Co.
ISBN 0-915928-16-7

53. **Experimental Diagnostics in Gas Phase Combustion Systems** (1977)
Ben T. Zinn, Editor
Georgia Institute of Technology
Craig T. Bowman, Associate Editor
Stanford University
Daniel L. Hartley, Associate Editor
Sandia Laboratories
Edward W. Price, Associate Editor
Georgia Institute of Technology
James G. Skifstad, Associate Editor
Purdue University
ISBN 0-015928-18-3

54. **Satellite Communications: Future Systems** (1977)
David Jarett
TRW Inc.
ISBN 0-915928-18-3

55. **Satellite Communications: Advanced Technologies** (1977)
David Jarett
TRW Inc.
ISBN 0-915928-19-1

56. **Thermophysics of Spacecraft and Outer Planet Entry Probes** (1977)
Allie M. Smith
ARO Inc.
ISBN 0-915928-20-5

57. **Space-Based Manufacturing from Nonterrestrial Materials** (1977)
Gerard K. O'Neill, Editor
Brian O'Leary, Assistant Editor
Princeton University
ISBN 0-915928-21-3

58. **Turbulent Combustion** (1978)
Lawrence A. Kennedy
State University of New York at Buffalo
ISBN 0-915928-22-1

59. **Aerodynamic Heating and Thermal Protection Systems** (1978)
Leroy S. Fletcher
University of Virginia
ISBN 0-915928-23-X

60. **Heat Transfer and Thermal Control Systems** (1978)
Leroy S. Fletcher
University of Virginia
ISBN 0-915928-24-8

61. **Radiation Energy Conversion in Space** (1978)
Kenneth W. Billman
NASA Ames Research Center
ISBN 0-915928-26-4

62. **Alternative Hydrocarbon Fuels: Combustion and Chemical Kinetics** (1978)
Craig T. Bowman
Stanford University
Jorgen Birkeland
Department of Energy
ISBN 0-915928-25-6

63. **Experimental Diagnostics in Combustion of Solids** (1978)
Thomas L. Boggs
Naval Weapons Center
Ben T. Zinn
Georgia Institute of Technology
ISBN 0-915928-28-0

64. **Outer Planet Entry Heating and Thermal Protection** (1979)
Raymond Viskanta
Purdue University
ISBN 0-915928-29-9

65. **Thermophysics and Thermal Control** (1979)
Raymond Viskanta
Purdue University
ISBN 0-915928-30-2

66. **Interior Ballistics of Guns** (1979)
Herman Krier
University of Illinois at Urbana-Champaign
Martin Summerfield
New York University
ISBN 0-915928-32-9

*67. **Remote Sensing of Earth from Space: Role of "Smart Sensors"** (1979)
Roger A. Breckenridge
NASA Langley Research Center
ISBN 0-915928-33-7

68. **Injection and Mixing in Turbulent Flow** (1980)
Joseph A. Schetz
Virginia Polytechnic Institute and State University
ISBN 0-915928-35-3

69. **Entry Heating and Thermal Protection** (1980)
Walter B. Olstad
NASA Headquarters
ISBN 0-915928-38-8

70. **Heat Transfer, Thermal Control, and Heat Pipes** (1980)
Walter B. Olstad
NASA Headquarters
ISBN 0-915928-39-6

*71. **Space Systems and Their Interactions with Earth's Space Environment** (1980)
Henry B. Garrett
Charles P. Pike
Hanscom Air Force Base
ISBN 0-915928-41-8

72. **Viscous Flow Drag Reduction** (1980)
Gary R. Hough
Vought Advanced Technology Center
ISBN 0-915928-44-2

SERIES LISTING

73. Combustion
Experiments in a Zero-
Gravity Laboratory (1981)
Thomas H. Cochran
NASA Lewis
Research Center
ISBN 0-915928-48-5

74. Rarefied Gas
Dynamics, Parts I and II
(two volumes) (1981)
Sam S. Fisher
University of Virginia
ISBN 0-915928-51-5

75. Gasdynamics
of Detonations and
Explosions (1981)
J.R Bowen
University of Wisconsin
at Madison
N. Manson
Université de Poitiers
A.K. Oppenheim
University of California
at Berkeley
R.I. Soloukhin
Institute of Heat and Mass
Transfer, BSSR Academy
of Sciences
ISBN 0-915928-46-9

76. Combustion in
Reactive Systems (1981)
J.R. Bowen
University of Wisconsin
at Madison
N. Manson
Université de Poitiers
A.K. Oppenheim
University of California
at Berkeley
R.I. Soloukhin
Institute of Heat and Mass
Transfer, BSSR Academy
of Sciences
ISBN 0-915928-47-7

77. Aerothermodynamics
and Planetary Entry (1981)
A.L. Crosbie
University of Missouri-
Rolla
ISBN 0-915928-52-3

78. Heat Transfer and
Thermal Control (1981)
A.L. Crosbie
University of Missouri-
Rolla
ISBN 0-915928-53-1

79. Electric Propulsion
and Its Applications
to Space Missions (1981)
Robert C. Finke
NASA Lewis
Research Center
ISBN 0-915928-55-8

80. Aero-Optical
Phenomena (1982)
Keith G. Gilbert
Leonard J. Otten
Air Force Weapons
Laboratory
ISBN 0-915928-60-4

81. Transonic
Aerodynamics (1982)
David Nixon
Nielsen Engineering &
Research, Inc.
ISBN 0-915928-65-5

82. Thermophysics of
Atmospheric Entry (1982)
T.E. Horton
University of Mississippi
ISBN 0-915928-66-3

83. Spacecraft Radiative
Transfer and Temperature
Control (1982)
T.E. Horton
University of Mississippi
ISBN 0-915928-67-1

84. Liquid-Metal Flows
and
Magnetohydrodynamics
(1983)
H. Branover
Ben-Gurion University
of the Negev
P.S. Lykoudis
Purdue University
A. Yakhot
Ben-Gurion University
of the Negev
ISBN 0-915928-70-1

85. Entry Vehicle
Heating and Thermal
Protection Systems: Space
Shuttle, Solar Starprobe,
Jupiter Galileo Probe
(1983)
Paul E. Bauer
McDonnell Douglas
Astronautics Co.
Howard E. Collicott
The Boeing Co.
ISBN 0-915928-74-4

86. Spacecraft Thermal
Control, Design, and
Operation (1983)
Howard E. Collicott
The Boeing Co.
Paul E. Bauer
McDonnell Douglas
Astronautics Co.
ISBN 0-915928-75-2

87. Shock Waves,
Explosions, and
Detonations (1983)
J.R. Bowen
University of Washington
N. Manson
Université de Poitiers
A.K. Oppenheim
University of California
at Berkeley
R.I. Soloukhin
Institute of Heat and Mass
Transfer, BSSR Academy
of Sciences
ISBN 0-915928-76-0

88. Flames, Lasers, and
Reactive Systems (1983)
J.R. Bowen
University of Washington
N. Manson
Université de Poitiers
A.K. Oppenheim
University of California
at Berkeley
R.I. Soloukhin
Institute of Heat and Mass
Transfer, BSSR Academy
of Sciences
ISBN 0-915928-77-9

89. **Orbit-Raising and Maneuvering Propulsion: Research Status and Needs** (1984)
Leonard H. Caveny
Air Force Office of Scientific Research
ISBN 0-915928-82-5

90. **Fundamentals of Solid-Propellant Combustion** (1984)
Kenneth K. Kuo
Pennsylvania State University
Martin Summerfield
Princeton Combustion Research Laboratories, Inc.
ISBN 0-915928-84-1

91. **Spacecraft Contamination: Sources and Prevention** (1984)
J.A. Roux
University of Mississippi
T.D. McCay
NASA Marshall Space Flight Center
ISBN 0-915928-85-X

92. **Combustion Diagnostics by Nonintrusive Methods** (1984)
T.D. McCay
NASA Marshall Space Flight Center
J.A. Roux
University of Mississippi
ISBN 0-915928-86-8

93. **The INTELSAT Global Satellite System** (1984)
Joel Alper
COMSAT Corp.
Joseph Pelton
INTELSAT
ISBN 0-915928-90-6

94. **Dynamics of Shock Waves, Explosions, and Detonations** (1984)
J.R. Bowen
University of Washington
N. Manson
Université de Poitiers
A.K. Oppenheim
University of California at Berkely
R.I. Soloukhin
Institute of Heat and Mass Transfer, BSSR Academy of Sciences
ISBN 0-915928-91-4

95. **Dynamics of Flames and Reactive Systems** (1984)
J.R. Bowen
University of Washington
N. Manson
Université de Poitiers
A.K. Oppenheim
University of California at Bereley
R.I. Soloukhin
Institute of Heat and Mass Transfer, BSSR Academy of Sciences
ISBN 0-915928-92-2

96. **Thermal Design of Aeroassisted Orbital Transfer Vehicles** (1985)
H.F. Nelson
University of Missouri-Rolla
ISBN 0-915928-94-9

97. **Monitoring Earth's Ocean, Land, and Atmosphere from Space — Sensors, Systems, and Applications** (1985)
Abraham Schnapf
Aerospace Systems Engineering
ISBN 0-915928-98-1

98. **Thrust and Drag: Its Prediction and Verification** (1985)
Eugene E. Covert
Massachusetts Institute of Technology
C.R. James
Vought Corp.
William F. Kimzey
Sverdrup Technology AEDC Group
George K. Richey
U.S. Air Force
Eugene C. Rooney
U.S. Navy Department of Defense
ISBN 0-930403-00-2

99. **Space Stations and Space Platforms — Concepts, Design, Infrastructure, and Uses** (1985)
Ivan Bekey
Daniel Herman
NASA Headquarters
ISBN 0-930403-01-0

100. **Single- and Multi-Phase Flows in an Electromagnetic Field: Energy, Metallurgical, and Solar Applications** (1985)
Herman Branover
Ben-Gurion University of the Negev
Paul S. Lykoudis
Purdue University
Michael Mond
Ben-Gurion University of the Negev
ISBN 0-930403-04-5

101. **MHD Energy Conversion: Physiotechnical Problems** (1986)
V.A. Kirillin
A.E. Sheyndlin
Soviet Academy of Sciences
ISBN 0-930403-05-3

102. **Numerical Methods for Engine-Airframe Integration** (1986)
S.N.B. Murthy
Purdue University
Gerald C. Paynter
Boeing Airplane Co.
ISBN 0-930403-09-6

103. **Thermophysical Aspects of Re-Entry Flows** (1986)
James N. Moss
NASA Langley Research Center
Carl D. Scott
NASA Johnson Space Center
ISBN 0-930403-10-X

104. **Tactical Missile Aerodynamics** (1986)
M.J. Hemsch
PRC Kentron, Inc.
J.N. Nielsen
NASA Ames Research Center
ISBN 0-930403-13-4

105. **Dynamics of Reactive Systems Part I: Flames and Configurations; Part II: Modeling and Heterogeneous Combustion** (1986)
J.R. Bowen
University of Washington
J.-C. Leyer
Université de Poitiers
R.I. Soloukhin
Institute of Heat and Mass Transfer, BSSR Academy of Sciences
ISBN 0-930403-14-2

106. **Dynamics of Explosions** (1986)
J.R. Bowen
University of Washington
J.-C. Leyer
Université de Poitiers
R.I. Soloukhin
Institute of Heat and Mass Transfer, BSSR Academy of Sciences
ISBN 0-930403-15-0

107. **Spacecraft Dielectric Material Properties and Spacecraft Charging** (1986)
A.R. Frederickson
U.S. Air Force Rome Air Development Center
D.B. Cotts
SRI International
J.A. Wall
U.S. Air Force Rome Air Development Center
F.L. Bouquet
Jet Propulsion Laboratory, California Institute of Technology
ISBN 0-930403-17-7

108. **Opportunities for Academic Research in a Low-Gravity Environment** (1986)
George A. Hazelrigg
National Science Foundation
Joseph M. Reynolds
Louisiana State University
ISBN 0-930403-18-5

109. **Gun Propulsion Technology** (1988)
Ludwig Stiefel
U.S. Army Armament Research, Development and Engineering Center
ISBN 0-930403-20-7

110. **Commercial Opportunities in Space** (1988)
F. Shahrokhi
K.E. Harwell
University of Tennessee Space Institute
C.C. Chao
National Cheng Kung University
ISBN 0-930403-39-8

111. **Liquid-Metal Flows: Magnetohydrodynamics and Applications** (1988)
Herman Branover,
Michael Mond, and
Yeshajahu Unger
Ben-Gurion University of the Negev
ISBN 0-930403-43-6

112. **Current Trends in Turbulence Research** (1988)
Herman Branover,
Michael Mond, and
Yeshajahu Unger
Ben-Gurion University of the Negev
ISBN 0-930403-44-4

113. **Dynamics of Reactive Systems Part I: Flames; Part II: Heterogeneous Combustion and Applications** (1988)
A.L. Kuhl
R & D Associates
J.R. Bowen
University of Washington
J.-C. Leyer
Université de Poitiers
A. Borisov
USSR Academy of Sciences
ISBN 0-930403-46-0

114. **Dynamics of Explosions** (1988)
A.L. Kuhl
R & D Associates
J.R. Bowen
University of Washington
J.-C. Leyer
Université de Poitiers
A. Borisov
USSR Academy of Sciences
ISBN 0-930403-47-9

115. **Machine Intelligence and Autonomy for Aerospace** (1988)
E. Heer
Heer Associates, Inc.
H. Lum
NASA Ames Research Center
ISBN 0-930403-48-7

116. **Rarefied Gas Dynamics: Space-Related Studies** (1989)
E.P. Muntz
University of Southern California
D.P. Weaver
U.S. Air Force Astronautics Laboratory (AFSC)
D.H. Campbell
University of Dayton Research Institute
ISBN 0-930403-53-3

117. **Rarefied Gas Dynamics: Physical Phenomena** (1989)
E.P. Muntz
University of Southern California
D.P. Weaver
U.S. Air Force Astronautics Laboratory (AFSC)
D. Campbell
University of Dayton Research Institute
ISBN 0-930403-54-1

118. **Rarefied Gas Dynamics: Theoretical and Computational Techniques** (1989)
E.P. Muntz
University of Southern California
D.P. Weaver
U.S. Air Force Astronautics Laboratory (AFSC)
D.H. Campbell
University of Dayton Research Institute
ISBN 0-930403-55-X

119. **Test and Evaluation of the Tactical Missile** (1989)
Emil J. Eichblatt Jr.
Pacific Missile Test Center
ISBN 0-930403-56-8

120. **Unsteady Transonic Aerodynamics** (1989)
David Nixon
Nielsen Engineering & Research, Inc.
ISBN 0-930403-52-5

121. **Orbital Debris from Upper-Stage Breakup** (1989)
Joseph P. Loftus Jr.
NASA Johnson Space Center
ISBN 0-930403-58-4

122. **Thermal-Hydraulics for Space Power, Propulsion and Thermal Management System Design** (1989)
William J. Krotiuk
General Electric Co.
ISBN 0-930403-64-9

123. **Viscous Drag Reduction in Boundary Layers** (1990)
Dennis M. Bushnell
Jerry N. Hefner
NASA Langley Research Center
ISBN 0-930403-66-5

124. **Tactical and Strategic Missile Guidance** (1990)
Paul Zarchan
Charles Stark Draper Laboratory, Inc.
ISBN 0-930403-68-1

125. **Applied Computational Aerodynamics** (1990)
P.A. Henne
Douglas Aircraft Company
ISBN 0-930403-69-X

126. **Space Commercialization: Launch Vehicles and Programs** (1990)
F. Shahrokhi
University of Tennessee Space Institute
J.S. Greenberg
Princeton Synergetics Inc.
T. Al-Saud
Ministry of Defense and Aviation Kingdom of Saudi Arabia
ISBN 0-930403-75-4

127. **Space Commercialization: Platforms and Processing** (1990)
F. Shahrokhi
University of Tennessee Space Institute
G. Hazelrigg
National Science Foundation
R. Bayuzick
Vanderbilt University
ISBN 0-930403-76-2

128. **Space Commercialization: Satellite Technology** (1990)
F. Shahrokhi
University of Tennessee Space Institute
N. Jasentuliyana
United Nations
N. Tarabzouni
King Abulaziz City for Science and Technology
ISBN 0-930403-77-0

129. **Mechanics and Control of Large Flexible Structures** (1990)
John L. Junkins
Texas A&M University
ISBN 0-930403-73-8

130. **Low-Gravity Fluid Dynamics and Transport Phenomena** (1990)
Jean N. Koster
Robert L. Sani
University of Colorado at Boulder
ISBN 0-930403-74-6

131. **Dynamics of Deflagrations and Reactive Systems: Flames** (1991)
A. L. Kuhl
Lawrence Livermore National Laboratory
J.-C. Leyer
Université de Poitiers
A. A. Borisov
USSR Academy of Sciences
W. A. Sirignano
University of California
ISBN 0-930403-95-9

132. **Dynamics of Deflagrations and Reactive Systems: Heterogeneous Combustion** (1991)
A. L. Kuhl
Lawrence Livermore National Laboratory
J.-C. Leyer
Université de Poitiers
A. A. Borisov
USSR Academy of Sciences
W. A. Sirignano
University of California
ISBN 0-930403-96-7

133. **Dynamics of Detonations and Explosions: Detonations** (1991)
A. L. Kuhl
Lawrence Livermore National Laboratory
J.-C. Leyer
Université de Poitiers
A. A. Borisov
USSR Academy of Sciences
W. A. Sirignano
University of California
ISBN 0-930403-97-5

134. **Dynamics of Detonations and Explosions: Explosion Phenomena** (1991)
A. L. Kuhl
Lawrence Livermore National Laboratory
J.-C. Leyer
Université de Poitiers
A. A. Borisov
USSR Academy of Sciences
W. A. Sirignano
University of California
ISBN 0-930403-98-3

135. **Numerical Approaches to Combustion Modeling** (1991)
Elaine S. Oran
Jay P. Boris
Naval Research Laboratory
ISBN 1-56347-004-7

136. **Aerospace Software Engineering** (1991)
Christine Anderson
U.S. Air Force Wright Laboratory
Merlin Dorfman
Lockheed Missiles & Space Company, Inc.
ISBN 1-56346-005-5

137. **High-Speed Flight Propulsion Systems** (1991)
S. N. B. Murthy
Purdue University
E. T. Curran
Wright Laboratory
ISBN 1-56347-011-X

138. **Propagation of Intensive Laser Radiation in Clouds** (1992)
O. A. Volkovitsky
Yu. S. Sedunov
L. P. Semenov
Institute of Experimental Meteorology
ISBN 1-56347-020-9

139. **Gun Muzzle Blast and Flash** (1992)
Günter Klingenberg
Fraunhofer-Institut für Kurzzeitdynamik, Ernst-Mach-Institut (EMI)
Joseph M. Heimerl
U.S. Army Ballistic Research Laboratory (BRL)
ISBN 1-56347-012-8

140. **Thermal Structures and Materials for High-Speed Flight** (1992)
Earl A. Thornton
University of Virginia
ISBN 1-56347-017-9

141. **Tactical Missile Aerodynamics: General Topics** (1992)
Michael J. Hemsch
Lockheed Engineering & Sciences Company
ISBN 1-56347-015-2

142. **Tactical Missile Aerodynamics: Prediction Methodology** (1992)
Michael R. Mendenhall
Nielsen Engineering & Research, Inc.
ISBN 1-56347-016-0

143. **Nonsteady Burning and Combustion Stability of Solid Propellants** (1992)
Luigi De Luca
Politecnico di Milano
Edward W. Price
Georgia Institute of Technology
Martin Summerfield
Princeton Combustion Research Laboratories, Inc.
ISBN 1-56347-014-4

144. **Space Economics** (1992)
Joel S. Greenberg
Princeton Synergetics, Inc.
Henry R. Hertzfeld
HRH Associates
ISBN 1-56347-042-X

145. **Mars: Past, Present, and Future** (1992)
E. Brian Pritchard
NASA Langley Research Center
ISBN 1-56347-043-8

146. **Computational Nonlinear Mechanics in Aerospace Engineering** (1992)
Satya N. Atluri
Georgia Institute of Technology
ISBN 1-56347-044-6

147. **Modern Engineering for Design of Liquid-Propellant Rocket Engines** (1992)
Dieter K. Huzel
David H. Huang
ISBN 1-56347-013-6

148. Metallurgical Technologies, Energy Conversion, and Magnetohydrodynamic Flows (1993)
Herman Branover
Yeshajahu Unger
Ben-Gurion University of the Negev
ISBN 1-56347-019-5

149. Advances in Turbulence Studies (1993)
Herman Branover
Yeshajahu Unger
Ben-Gurion University of the Negev
ISBN 1-56347-018-7

150. Structural Optimization: Status and Promise (1993)
Manohar P. Kamat
Georgia Institute of Technology
ISBN 1-56347-056-X

151. Dynamics of Gaseous Combustion (1993)
A. L. Kuhl
Lawrence Livermore National Laboratory
J.-C. Leyer
Université de Poitiers
A. A. Borisov
Russian Academy of Sciences
W. A. Sirignano
University of California
ISBN 1-56347-060-8

152. Dynamics of Heterogeneous Combustion and Reacting Systems (1993)
A. L. Kuhl
Lawrence Livermore National Laboratory
J.-C. Leyer
Université de Poitiers
A. A. Borisov
Russian Academy of Sciences
W. A. Sirignano
University of California
ISBN 1-56347-058-6

153. Dynamic Aspects of Detonations (1993)
A. L. Kuhl
Lawrence Livermore National Laboratory
J.-C. Leyer
Université de Poitiers
A. A. Borisov
Russian Academy of Sciences
W. A. Sirignano
University of California
ISBN 1-56347-057-8

154. Dynamic Aspects of Explosion Phenomena (1993)
A. L. Kuhl
Lawrence Livermore National Laboratory
J.-C. Leyer
Université de Poitiers
A. A. Borisov
Russian Academy of Sciences
W. A. Sirignano
University of California
ISBN 1-56347-059-4

155. Tactical Missile Warheads (1993)
Joseph Carleone
Aerojet General Corporation
ISBN 1-56347-067-5

156. Toward a Science of Command, Control, and Communications (1993)
Carl R. Jones
Naval Postgraduate School
ISBN 1-56347-068-3

(Other Volumes are planned.)